T0192802

Heat Pumps in Chemical Process Industry

Heat Pumps in Chemical Process Industry

Anton A. Kiss
Carlos A. Infante Ferreira

CRC Press
Taylor & Francis Group
Boca Raton London New York

CRC Press is an imprint of the
Taylor & Francis Group, an **informa** business

CRC Press
Taylor & Francis Group
6000 Broken Sound Parkway NW, Suite 300
Boca Raton, FL 33487-2742

First issued in paperback 2020

ISBN 13: 978-0-367-57449-9 (pbk)
ISBN 13: 978-1-4987-1895-0 (hbk)

Library of Congress Cataloging-in-Publication Data

Names: Kiss, Anton Alexandru, author. | Infante Ferreira, Carlos A., author.
Title: Heat pumps in th chemical process industry / Anton A. Kiss and Carlos A. Infante Ferreira.
Description: Boca Raton : Taylor & Francis, a CRC title, part of the Taylor & Francis imprint, a member of the Taylor & Francis Group, the academic division of T&F Informa, plc, [2017] | Includes bibliographical references and index.
Identifiers: LCCN 2016008989 | ISBN 9781498718950 (alk. paper)
Subjects: LCSH: Chemical plants--Equipment and supplies. | Chemical engineering--Equipment and supplies. | Heat pumps.
Classification: LCC TP363 .K476 2017 | DDC 621.402/5--dc23
LC record available at https://lccn.loc.gov/2016008989

Visit the Taylor & Francis Web site at
http://www.taylorandfrancis.com

and the CRC Press Web site at
http://www.crcpress.com

Contents

Preface

Modern society faces a variety of challenges to meet the energy requirements of a growing population. Considering that the chemical process industry is among the most energy-demanding sectors, chemical engineers have embarked on a quest for shaping a sustainable future. Due to the limitation of fossil fuels, the need for energy independence, as well as the environmental problem of the greenhouse gas effect, there is a large, increasing interest in the research and development of chemical processes that require less capital investment and reduced operating costs and lead to high ecoefficiency. The use of heat pumps is a hot topic due to many advantages, such as low energy requirements as well as an increasing number of industrial applications. Although the research and development carried out in academia and industry in this field are expanding quickly, there is still no book currently available focusing on the use of heat pumps in the chemical industry.

Therefore, we feel that there is a significant gap that can be addressed with this book and that it will be of immense interest to readership across the world. This book is the first to provide an overview of heat pump technology applied in the chemical process industry, covering both theoretical and practical aspects: working principle, applied thermodynamics, theoretical background, numerical examples and case studies, as well as practical applications in the chemical process industry. The worked-out examples will instruct students, engineers and process designers about how to design various heat pumps used in the industry. The abundant valuable resources included here – relevant equations, diagrams, figures and references that reflect the current and upcoming heat pump technologies – will be of great help to all readers from the chemical and petrochemical industry, biorefineries and other related areas.

The key benefits of this work for the reader include the following:

- Learn more about heat pumps, what they are, when and how to use them properly

- Obtain more information about the theoretical and practical background of heat pumps

- Understand how to identify the need and how to select, design and apply heat pumps

- Discover the existing and potential applications of heat pumps in a process

- Find the specifics of heat pump applications in the chemical process industry

The following provides a brief description of each chapter:

1. **Introduction to Heat Pumps:** This chapter provides an overview of heat pumps and the possible sources of heat and cold usable by heat pumps in the chemical process industry.

2. **Thermodynamics of Heat Pump Cycles:** This chapter defines the main concepts used in thermodynamics (e.g. fundamentals, enthalpy, entropy, equations of state, chemical and phase equilibrium) and their practical applications in heat pump cycles.

3. **Entropy Production and Exergy Analysis:** This chapter describes the use of exergy analysis and the minimization of the entropy production in heat pump cycles.

4. **Pinch Analysis and Process Integration:** Chapter 4 describes the techniques for designing a process to minimize energy usage and maximize heat recovery by calculating the thermodynamically attainable energy targets for a given process and identifying how to achieve them. In this context, the proper use and placement of heat pumps are explained.

5. **Selection of Heat Pumps:** This chapter explains how to select the appropriate heat pump technology for various processes in the chemical industry and what the required steps are during development, demonstration and deployment of heat pumps in industry.

6. **Mechanically Driven Heat Pumps:** An overview of the mechanically driven heat pumps used in the chemical process industry is provided, including sustainability and working fluid–related issues: vapor recompression, vapor compression, compression-resorption, transcritical and Stirling heat pumps.

7. **Thermally Driven Heat Pumps:** This chapter provides an overview of the thermally driven heat pumps that are mostly used in the chemical process industry, including environmental impact. Heat pump technologies included are liquid-vapor absorption, solid-vapor adsorption and ejector-based heat pumps.

8. **Solid-State Heat Pumps:** Chapter 8 provides an overview of more exotic solid-state heat pumps that are developed for the chemical process industry: magnetic, thermoelectric and thermoacoustic heat pumps.

9. **Industrial Applications of Heat Pumps:** Various applications of heat pump technologies in the chemical process industry are discussed.

10. **Case Studies:** This chapter includes case studies of heat pump technologies applied in the chemical process industry, such as distillation, evaporation and refrigeration.

The main targets of this book are senior graduate students, research scientists from universities and various research institutes, professionals in the fields of (petro)chemical engineering and biorefinery, as well as many university libraries. The book can also serve as a reference book for scientists, researchers, engineering procurement and contracting organizations, operators of chemical and biorefinery production facilities, research and development engineering departments and other industry practitioners involved in heat pumps and energy-efficient technologies. Undergraduate engineering students and engineers worldwide in industrial companies, consulting companies, government offices and nonprofit organizations can also use this material in their daily operations.

We are fully aware of the unavoidable presence of some minor mistakes in this book and would like to express in advance gratitude for any observation and suggestion toward further improving this material.

We thank our colleagues and students from the Netherlands, AkzoNobel, the University of Twente and Delft University of Technology for their support. We are thankful to CRC Press for the professional support and kind assistance and to our collaborators in the research programs that we are carrying out. Finally, we are indebted to our families and close friends for their long-standing encouragement and understanding, namely, for the late evenings and long working weekends.

Anton A. Kiss
Carlos A. Infante Ferreira

Authors

Prof. Dr. Ir. Anton Alexandru Kiss works as a part-time professor of separation technology at the University of Twente and holds a scientific position of RD&I specialist and senior project leader at AkzoNobel – Research, Development and Innovation (the Netherlands). Here he acts as the principal scientist in separation technologies, process intensification, reactive-separation processes, heat pumps and integrated sustainable processes with 'green' attributes, while mastering, as well, process modeling and simulation of industrial processes. He is also a member of established professional institutes and associations: American Institute of Chemical Engineers (AIChE), Institution of Chemical Engineers (IChemE), European Society of Mathematical Chemistry (ESMC), Society of Chemical Industry (SCI), European Federation of Chemical Engineering (EFCE), Working Party on Computer Aided Process Engineering (CAPE-WP), and Process Systems Engineering – The Netherlands (PSE-NL).

Tony Kiss holds a PhD in chemical engineering from the University of Amsterdam, and he was also employed as a postdoctoral research fellow at Delft University of Technology and the University of Amsterdam while briefly working as a consultant engineer for Unilever Food Research Center.

During the past decade, he has carried out numerous research and industrial projects and supervised many graduation projects while also publishing several textbooks and book chapters (published by Wiley, Elsevier, Springer), patents and over 100 scientific articles in top peer-reviewed journals (mainly Elsevier, Wiley and Springer journals). In addition, he acts as a reviewer for the 25 top journals in chemical engineering, energy and applied chemistry and is a member of the advisory editorial board for the *Journal of Chemical Technology and Biotechnology*.

For the pioneering work and remarkable achievements in his area of scientific research, he was rewarded the Hoogewerff Jongerenprijs 2013, a prestigious award recognizing the most promising young scientist in the Netherlands. The same year, he also received the AkzoNobel Innovation Excellence Award 2013 for the most successful industrial innovation. More information is available at his personal webspace: www.tonykiss.com.

Dr. Ir. Carlos A. Infante Ferreira works as an associate professor of mechanical engineering at the Delft University of Technology, the Netherlands. He holds an MSc degree in mechanical engineering from the Technical University of Lisbon (Portugal) and an MSc and a PhD degree from Delft University of Technology (the Netherlands). He also has 7 years of industrial experience at Apparatenfabriek Helpman (Groningen, the Netherlands), a manufacturing company of equipment for the commercial and industrial refrigeration sectors, where he lead the Research and Development Department, being responsible for the introduction of new product lines. During the past 20 years, he worked at the Mechanical Engineering Department of the Delft University of Technology on research topics focusing on thermodynamics and transport phenomena related to refrigeration, heat pump and heat transformer cycles.

In his current role as associate professor at Delft University of Technology, he conducted many projects related to heat pumps, and the MSc and PhD studies that he has supervised are related to compression resorption and liquid sorption heat pumps, efficient use of energy by using innovative thermodynamic cycles or equipment, taking entropy minimization as a criterion for optimization. He is extensively involved in the BSc-ME and MSc-ME education of the Mechanical Engineering Department, teaching topics such as thermodynamics, energy engineering, refrigeration and fluid machinery. Carlos Infante Ferreira is also president of Commission B1 (Thermodynamics) of the International Institute of Refrigeration, and regional editor of the *International Journal of Refrigeration*. He has authored or coauthored numerous academic journal papers related to heat pump cycles.

CHAPTER 1

Introduction to Heat Pumps

1.1 INTRODUCTION

A *heat pump* (HP) is a device that transfers energy from a *heat source* to a *heat sink* (destination) and upgrades the energy to a higher temperature level. Heat pumps are designed to move heat in the opposite direction of normal heat flow by taking heat from a colder space and releasing it to a warmer one. Although the overall process seems to allow the heat to flow from the cold source to the warm destination, the normal heat flow from high to low temperature is respected in each step of the process. Note that a heat pump uses a certain amount of work (external power) to accomplish the transfer of energy from the heat source to the heat sink.

The most common examples are refrigerators, air conditioners (ACs) and reversible-cycle heat pumps for providing thermal comfort. Notably, the term *heat pump* is more generic; it applies to many heating, ventilating and air conditioning (HVAC) devices used for space heating or cooling. When used for heating, a heat pump employs the same refrigeration-type cycle used by an air conditioner, but in the opposite direction – thus drawing heat from the ground or external air and releasing the heat into the air-conditioned space rather than the surrounding environment.

Heat pumps have several key advantages: very high efficiency as compared to gas-heated systems; the possibility to use environmental renewable energy from the air, water or ground; large energy savings of 50%–70% translated in reduced final and primary energy demand and significant reductions in greenhouse gas (GHG) emissions (e.g. CO_2). Nonetheless, heat pumps also have some drawbacks: An initial investment is needed (so payback times might be an issue), an electrical connection is required to be present typically and some refrigerants used in a heat pump are toxic or flammable (health, safety, environment [HSE] issues). It is worth noting that the investment costs of a heat pump depend strongly on the application type and location, being influenced by several key factors: required temperature (higher temperature requires expensive components), required heat capacity (higher capacity needs more expensive installation), number of installations (build a number of small installations or one large installation) or available space to connect a heat pump to an existing installation (especially in an existing chemical plant).

The next sections give a historical perspective on the developments of heat pumps and describe the working principle and fundamentals, performance limitations and major types of heat pumps and their use in HVAC applications as well as in the chemical process industry (CPI).

1.2 HISTORICAL PERSPECTIVE

While heating has been no secret to humankind ever since the discovery and use of fire, the problem of artificial cooling was more complex. With the exception of evaporative cooling, there was no possibility for artificial cooling until about 150 years ago. Natural ice was transported on a global scale, but due to shortage problems, the heat pump development priority was on the refrigeration side. The problem of artificial cooling was not solved before about 1850, when the first refrigeration machines were invented. Of course, the same machines can be also used as heat pumps for heating. Yet, the huge demand for cooling was mainly responsible for the rapid development of heat pumps and their spread around the globe. Nowadays, over 130 million heat pumps for cooling and heating are in operation worldwide, so the importance of heat pump technologies is undeniable.

The scientific approach to heat pump technologies started with Carnot [1824], who was the first to establish a precise relationship between heat and work. Carnot's ideas were reformulated later by Clausius [1850], but the basic statement is that *mechanical energy may be transformed completely into heat energy*, but that *heat energy may be only partially transformed into mechanical energy*.

Other contributions came from von Mayer, who established the principle of equivalence between work and heat [1842]; Joule, who gave the experimental proof of the principle [1843] and von Helmholtz, who expressed the principle of conservation of energy in general terms [1847] – hence firmly establishing the first law of thermodynamics. Considered one of the key founders of thermodynamics, Clausius restated Carnot's principle (known as the *Carnot cycle*), thus proving the sound basis of the theory of heat. Also, Clausius was the first to state the basic ideas of the second law of thermodynamics [1850] and later explicitly introduced the concept of entropy [1865].

Independently of Clausius, Lord Kelvin derived a more general formula for the second principle [1851] and introduced the thermodynamic scale of temperature [1852]. Also, Kelvin [1852] remarked that a *reverse heat engine* could be used not only for cooling but also for heating and pointed out that such a heating device would need less primary energy due to the extraction of heat from the environment. Later, Boltzmann [1866] linked the concepts of entropy and probability in statistical physics, thus clarifying the Carnot principle, knowing that entropy represents the degree of disorder. Gibbs introduced enthalpy into theoretical thermodynamics [1873–1878], while Mollier brought it into applied thermodynamics [1902], using it as one co-ordinate (along with entropy or pressure) of his thermodynamic diagrams. These diagrams provided a graphic visualization and an easy method of calculation for the vapor compression (VC) cycle.

The concept of *exergy* – defined as the work that would be delivered by a reversible process that would bring the flow in equilibrium with the environmental conditions, this process consisting of an isentropic expansion to environment pressure and an isothermal expansion to the entropy state of the environment – was derived from the ideas of Zeugner [1859] and Lorenz [1896] and was taken up again by Bosnjakovic [1935] and after 1950 by Grassmann and Nesselmann. On the basis of a thermodynamic comparison, Linde [1870] pointed out that the compression system is more efficient than systems based on the absorption system and other principles. In addition, Swarts is widely considered as establishing the foundations of organofluorine chemistry by his work on aliphatic fluorocarbons [1890–1893]. Altenkirch [1910] carried out a study of binary mixtures for absorption refrigeration machines, and the two-stage machine had very good output. Other key contributions to the development of heat pumps include the first VC machine by Perkins [1834], first commercially successful ammonia absorption cooling system and introduction of ammonia as a refrigerant by Carré [1851], first commercial ice-making plant by Twining [1855], first pilot heat pump for heating only by von Rittinger [1855–1857], first thermostatically controlled refrigeration system by van der Weyde [1870], diffusion-absorption cycle by Geppert [1899], rolling piston rotary compressor by Rolaff [1920], plate heat exchanger by Seligman [1923], thermostatic expansion valve by Diffinger [1923], and capillary tube refrigerant control by Carpenter [1927], as well as the introduction of new refrigerants such as dimethyl ether [Tellier, 1863], carbon dioxide [Lowe, 1866], sulphur dioxide [Pictet, 1874], methyl chloride [Vincent, 1878]; chlorofluorocarbon (CFC) refrigerants [Midgley, Henne and McNary, 1928] and hydrofluorocarbon refrigerants [Henne, 1936].

During the industrialization period [1876–1918], demo units were replaced with more reliable and optimized machines to take advantage of scientific advancements and manufacturing ability. The refrigeration systems started to become industrial products on a large scale, and Linde played the most important role in this change. Around 1900, most fundamental inventions were made, and by 1918, there were many compressor manufacturers in the United States and Europe. During the period 1919–1950, the heat pumps for space heating and domestic hot water heating developed to become reliable, efficient and sometimes even economically viable heating devices. The period 1950–1970 was characterized by a falling oil price, which dramatically slowed the activities related to heat pumps for heating, so there was stagnation in development and market penetration.

The oil crisis started in 1973 and had a damaging effect on economies, and developed nations had to rethink their dependence on fossil fuels. Worldwide, new energy strategies included solar, wind, biomass and geothermal energy, as well as the use of ambient heat by heat pumps. The International Energy Agency (IEA) identified heat-pumping technologies as one of the key strategies for building energy conservation. But, the rapid growth of the heat pump business brought too many competitors with insufficient know-how, and this was one of the reasons for the collapse of the heat pump boom by the end of the 1980s, stimulated also by the oil price decline after 1982. However, after the 1990s, more efficient and reliable heat pumps became available, and growing environmental concerns are supporting the idea of saving primary energy and money by heat pumps. Nowadays, heat pump developments are backed by significant research and development efforts, quality control and strong industrial interest. A detailed history of heat pumps was well presented in the report of Zogg (2008).

1.3 WORKING PRINCIPLE

The operation principle of a heat pump is based on the physical property that the boiling point of a fluid depends on the pressure. By lowering the pressure, a medium can be evaporated at low temperature, while an increase of pressure leads to condensation at high temperature. Basically, this enables the transport of heat from a low-temperature waste source towards a high-temperature heat destination. A heat pump transfers heat from one location (the 'source') to another (the 'sink' or 'heat sink') using mechanical work, while a heat transformer (HT) and an absorption heat pump use heat as input instead of work (see Figure 1.1) (Kiss, 2013). Most heat pump technologies move heat from a low-temperature heat source to a high-temperature heat sink. Heat pumps can also operate in reverse, providing heat, so they can also be considered as a heat engine that is operating in reverse. One common type of heat pump works by exploiting the physical properties of an evaporating and condensing fluid known as a refrigerant (Reay and MacMichael, 1988; Silberstein, 2002). The following types of heat pumps are most common: mechanically driven heat pump, heat-driven HT and heat-driven heat pump.

The heat-driven heat pumps require a high-temperature source to drive the system, and they are less efficient than their mechanically driven counterparts. The mechanically driven heat pumps make use of an evaporating/condensing fluid. The heat pump compresses the refrigerant to make it hotter on the side to be heated and releases the pressure at the side where the heat is absorbed. The working fluid (gas phase) is pressurized and circulated through the system by a compressor. On the discharge side of the compressor, the heated and pressurized vapor is cooled in a condenser until it turns into a liquid at high pressure and moderate temperature. The condensed refrigerant then passes through a pressure-lowering (metering) device, such as an expansion valve, capillary tube or turbine. The low-pressure liquid refrigerant then enters an evaporator, in which the fluid absorbs heat and boils to become vapor. The refrigerant then returns to the compressor, and the cycle is repeated.

The refrigerant must reach a suitably high temperature when compressed to release heat through the 'hot' heat exchanger (HHX) or condenser. Similarly, the fluid must reach a suitably low temperature when allowed to expand or else the heat cannot flow from the ambient cold region into the fluid in the cold heat exchanger (CHX) or evaporator. In particular, the pressure difference must be sufficiently large for the fluid to condense at the hot side and still evaporate in the lower-pressure region at the cold side. The larger the temperature difference, the larger the required pressure difference and thus more energy is needed to compress the fluid. As with all heat pumps, the coefficient of performance (COP) (defined as the amount of thermal energy moved per unit of input work required) decreases with increasing temperature difference. Insulation is used to reduce the work and energy required to achieve a high enough temperature in the space to be heated.

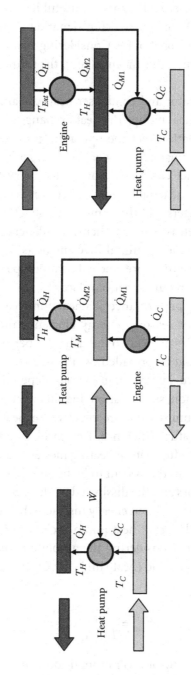

FIGURE 1.1 Mechanically driven heat pump (left). Heat-driven heat transformer (centre) and heat-driven heat pump (right).

1.4 EFFICIENCY AND PERFORMANCE

As the word *efficiency* has a specific thermodynamic definition, it is best to be avoided when comparing the performance of heat pumps. A better term is the *coefficient of performance* (COP), which is used to describe the ratio of useful heat delivered per work input. In general, heat pumps have a COP of 4–6, which is remarkable but still behind combined heat and power (CHP) with a COP of about 9. Considering that current heat pumps have an efficiency of about 50% of Carnot, this means that in the future the energy efficiency of heat pumps could almost double.

Also, the design of the evaporator and condenser heat exchangers plays an important role in the overall efficiency of a heat pump. The heat exchange areas and the corresponding temperature differential directly affect the operating pressures and hence the work required to provide the same heating or cooling effect. A larger heat exchange area means that a lower temperature difference can be used and thus the system becomes more efficient. Also, the COP increases as the temperature lift between heat source and destination decreases. Note that heat exchangers and the compressor generating work are expensive, and the industry generally competes on price rather than efficiency. Therefore, heat pumps have a price drawback when it comes to initial investment as compared to conventional heating solutions, so the drive towards more efficient heat pumps is often led by legislative measures on minimum efficiency requirement standards.

In cooling mode, the operating performance of a heat pump can be described as energy efficiency ratio (EER or COP_R) or seasonal energy efficiency ratio (SEER or $SCOP_R$), a larger ratio indicating better performance. To describe the performance of a heat pump in heating mode, the COP_H should be provided, while for the cooling mode the EER or SEER should be included. The actual performance varies, depending on many factors (e.g. temperature differences, installation, site elevation, maintenance).

It is worth noting that heat pumps are more effective for heating than for cooling an interior space, at the same temperature difference. The reason is that the input energy of the compressor is also converted to useful heat (in heating mode) and is discharged along with the moved heat via the condenser to the system to be heated. However, in case of cooling, the condenser is normally outdoors and the dissipated work (waste heat) of the compressor is transported to outdoors using more input energy instead of being put to a useful aim.

Note that the COP of an ideal Carnot heat pump cycle (COP_{rev}) represents the maximum theoretical COP achievable. Considering the absolute temperatures T_C of the cold side (heat source) and T_H of the hot side (heat sink), the COP of a reversible Carnot heat pump cycle is given by

$$COP_{rev} = \frac{T_H}{T_H - T_C} \tag{1.1}$$

The COP for a heat pump in a heating or cooling application with steady-state operation is

$$COP_H = \frac{\Delta \dot{Q}_H}{\Delta \dot{W}} \leq \frac{T_H}{T_H - T_C} \tag{1.2}$$

$$COP_R = \frac{\Delta \dot{Q}_C}{\Delta \dot{W}} \leq \frac{T_C}{T_H - T_C} \qquad (1.3)$$

where $\Delta \dot{Q}_C$ is the amount of heat extracted from a cold reservoir at temperature T_C, and $\Delta \dot{Q}_H$ is the amount of heat delivered to a hot reservoir at temperature T_H, while $\Delta \dot{W}$ is the input power of the compressor. Note that all temperatures are given as absolute temperatures (measured in Kelvin).

For example, a $COP_H = 4$ means that 1 kW of electric energy is needed to release 4 kW of heat at the condenser. At the evaporator side of the heat pump, 3 kW of heat is extracted, the additional heat being generated by the compressor. However, in case of a refrigeration system with $COP_R = 4$, an amount of 1 kW of electric energy is needed to extract 4 kW of heat.

In heat pump processes, the temperatures of the heat sink and the heat source would remain constant only in the case of infinite flow rates. For benchmarking of heat pumps, it is convenient to take the average outlet temperature of heat sink $T_{H,out}$ and the average inlet temperature of heat source $T_{C,in}$ as references. This leads to the approximate exergetic efficiency, known as the Lorenz efficiency:

$$COP_{Lorentz} = \frac{T_{H,out}}{T_{H,out} - T_{C,in}} \qquad (1.4)$$

For the comparison of complete systems, including the production of electric energy, the primary energy ratio (PER) allows a comparison of electric-driven heat pumps with absorption heat pumps (AHPs) and boilers, while for fuel-driven systems the primary energy input is equal to the fuel energy input.

$$PER = Useful\_heat\_output/Primary\_energy\_input \qquad (1.5)$$

1.5 HEAT SOURCES AND SINKS

All heat sources for a heat pump must be colder than the space to be heated. Most commonly, heat pumps draw heat from the air, ground, water sources and waste heat process streams.

Air is a key source or sink for heat in smaller installations, as used by an air source heat pump. An air source heat pump extracts heat from outside air and transfers heat to inside air (air-air heat pump) or to a heating circuit and a tank of domestic hot water (air-water heat pump). Air-air heat pumps are the most common type of heat pumps and the least expensive, being similar to air conditioners operating in reverse. Air-water heat pumps are similar to air-air heat pumps, but they transfer the extracted heat into a water-heating circuit (e.g. floor heating) or into a domestic hot water tank (usable in showers and hot water taps). Air source heat pumps are relatively easy and inexpensive to install and therefore have been the most widely used heat pump type. However, they suffer limitations due to their use of the outside air as a heat source. The higher temperature differential during periods of extreme cold leads to declining efficiency. In mild weather, the COP may be around 4.0, while at temperatures below 0°C an air source heat pump may still achieve a COP of 2.5. The average COP over seasonal variation is typically 2.5–2.8. The heating output of low-temperature-optimized heat pumps (and hence their energy efficiency) still declines dramatically as the temperature drops, but the threshold at which the decline starts is lower than conventional pumps. Exhaust air heat pumps extract heat from the exhaust air of a building (requires mechanical ventilation) and include the exhaust air-air heat pump (transfers heat to intake air) and exhaust air-water heat pump (transfers heat to a heating circuit and a tank of domestic hot water).

The *ground* is a key source for geothermal heat pumps that use shallow underground heat exchangers as a heat source or sink, and water as the heat transport medium. This is possible because below ground level, the temperature is relatively constant across the seasons, and the earth can provide or absorb a large amount of heat. Ground source heat pumps work in the same way as air source heat pumps but exchange heat with the ground via water pumped through pipes in the ground. Ground source heat pumps are simpler and therefore more reliable than air source heat pumps as they do not need fan or defrosting systems and can be housed inside.

Typically, ground source heat pumps have high efficiencies, as they draw heat from the relatively constant-temperature ground or groundwater. This means that the temperature differential is lower, leading to higher efficiency. The COP is 4.0 at the beginning of the heating season, with a lower seasonal COP of around 3.0. The trade-off for this improved performance is that a ground source heat pump is more expensive to install due to the need for the drilling of boreholes for vertical placement of heat exchanger piping or the digging of trenches for horizontal placement of the piping that carries the heat exchange fluid (water with some antifreeze). Although a ground heat exchanger requires a higher initial capital cost, the annual running costs are lower.

Ground source heat pump types include the ground-air heat pump (transfers heat from ground to inside air), soil-air heat pump (transfers heat from soil to air), rock-air heat pump (using rocks as a source of heat), ground-water heat pump (transfers heat from the ground to heating or domestic water), soil-water heat pump (transfers heat from soil to water) and rock-water heat pump (transfers heat from rocks to water).

Water is a key source or sink for heat in larger installations handling more heat or in tight physical spaces. The heat is sourced or rejected in water flow that – compared to airflows – can carry much larger amounts of heat through a given pipe or duct cross section. The water may be heated at a remote location by boilers, solar energy or other means. Alternatively, when needed, the water may be cooled by using a cooling tower or discharged into a large body of water (sea, lake or stream). Heat pumps using water include the water-water heat pump (using a body of water as a source of heat) and water-air heat pump (using a body of water as a source of heat).

Water source heat pumps (WSHPs) are installed with single-pass (body of water or a water stream as source) or recirculation cooling (closed-loop heat transfer medium to a central cooling tower or chiller) and heating (closed-loop heat transfer medium from central boilers generating heat from combustion).

Dual heat pumps use twin sources (e.g. when outdoor air is above 4°C–8°C, they use air and when air is colder they use the ground source). These twin-source systems can also store summer heat by running ground source water through the air exchanger or through the building heat exchanger, even when the heat pump itself is not running. This has clear advantages: It functions at a low running cost for air cooling, and it cranks up the ground source temperature, which improves the energy efficiency of the heat pump by about 4% for each degree in temperature rise of the ground source.

An air-water–brine-water heat pump is a dual heat pump that uses only renewable energy sources, combining air and geothermal heat in one compact device. One commercial type of heat pump has two evaporators (an outside air evaporator and a brine evaporator) connected to the heat pump cycle, thus allowing the use of the most economical heating source for the current external conditions. The unit automatically selects the most efficient operating mode: air, geothermal heat or both of them.

Process streams are the most used heat sources and sinks in the CPI. Typical heat sources include the top vapor streams of distillation columns (which are upgraded to higher temperatures and used to drive the reboiler of the same column); low-grade waste heat (LGWH) streams (which are used to provide heat or cold to other process streams or to generate low-pressure steam); vapor streams (e.g. from solvents in liquid flows; evaporated water from products in drying processes) or various streams from the food industry. More details and examples are provided in the sections that follow, as well as the rest of the book.

1.6 HEATING, VENTILATING AND AIR CONDITIONING APPLICATIONS

In HVAC applications, the term *heat pump* usually refers to vapor-compression refrigeration devices optimized for high efficiency in both directions of heat transfer. Note that HVAC is the technology of indoor and vehicular environmental comfort, which aims to provide thermal comfort and adequate indoor air quality. HVAC is important in the design of medium-to-large industrial and office buildings, where safe and healthy building conditions are regulated with respect to air temperature and humidity, using fresh air from outdoors. Figure 1.2 illustrates the major HVAC application of heat pumps.

Reversible heat pumps work in either thermal direction to provide heating or cooling to the internal space. They employ a reversing valve to reverse the flow of refrigerant from the compressor through the condenser and evaporation coils. The reversing valve switches the direction of refrigerant through the cycle; therefore, the heat pump may deliver either heating or cooling. In warmer climates, the default setting of the reversing valve is cooling, while in cooler climates the default setting is heating. Because the two heat exchangers (condenser and evaporator) must switch functions, they are optimized to perform adequately in both modes. Therefore, the efficiency of a reversible heat pump is typically slightly less than that of two separately optimized machines.

Note that in heating mode, the outdoor coil is an evaporator, while the indoor is a condenser. The refrigerant flowing from the evaporator (outdoor coil) carries the heat from outside air (or soil) indoors, after the temperature of the fluid has been increased by compressing it. The indoor coil then transfers the heat – including energy from the

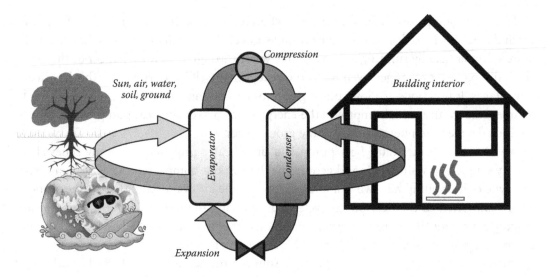

FIGURE 1.2 The HVAC application of heat pumps.

compression – to the indoor air (which is then moved around the inside of the building) or to water (which is then used for domestic hot water consumption or to heat the building via radiators or floor heating). The refrigerant is then allowed to expand, cool and absorb heat to reheat to the outdoor temperature in the outside evaporator, and the cycle repeats. In cooling mode, the cycle is similar, but the outdoor coil is now the condenser and the indoor coil is the evaporator. This is the familiar mode in which air conditioners operate.

More information about the use of heat pumps in HVAC applications is available elsewhere, in specialized books (Langley, 2001; Silberstein, 2002; Brumbaugh, 2004; Cantor, 2011; Petit and Collins, 2011; Whitman et al., 2012); hence, this topic is out of the scope of this book.

1.7 HEAT PUMP TYPES AND INDUSTRIAL APPLICATIONS

Heat pumps are used in the chemical processes industry for upgrading low-grade waste heat (e.g. vapor recompression, VRC), providing cooling (e.g. absorption chillers, absorption refrigeration) or in combined heat and power (CHP) applications. The major applications of heat pumps include the following:

- *Distillation* is the most used separation technology, and it is responsible for over 40% of the energy usage in chemical plants. For that reason, many of the heat pump applications in the CPI are tailored to distillation. Heat pumps can be used to upgrade the low-quality energy in the condenser to drive the reboiler of the column and thus reduce the consumption of valuable utilities (Kiss et al., 2012; Kiss, 2013).

- *Drying processes*. Typically, dry air is heated with steam, gas or hot water and then circulated over a wet product. As air picks up moisture from the wet product, its humidity increases. Instead of exhausting this humid air or dehumidify it, a heat pump can be used to extract the heat from the humid air. The air is cooled and dehumidified, while the extracted heat can be upgraded to a higher temperature level and used to heat the dryer. Hence, the heat pump heats the dryer as well as dehumidifies and recycles the air, at high efficiencies.

- *Washing processes* involve typically spraying of hot water (mixed with a solvent) over a product. An air discharge fan is used to prevent the installation from vapor flowing out of the washing machine through the opening. The air discharge maintains a sub-atmospheric pressure inside the washing machine and blows humid hot air (which contains plenty of energy) to the ambient. A heat pump can use the heat from the discharge air to heat the washing water.

- *Heating of process streams* with waste heat from a refrigeration system is typically applied in the food industry. While there are products that need to be cooled or frozen before transport or consumption, hot water is needed for process and cleaning purposes. Waste heat from a refrigeration system (available at 25°C–30°C) can be upgraded with a heat pump and used to heat water, up to 80°C. Note that the heat pump further increases the pressure of the refrigerant from the refrigeration system to achieve high condensation temperatures.

- *Pasteurization* heats a product (above 70°C) and afterwards cools it. The product temperature thus varies from cold to hot and then back to cold after pasteurization. While heat integration is already employed in most pasteurization processes, a heat pump could be the ideal solution to extract heat from the product that needs to be cooled and supply this heat at a higher temperature to the product that needs to reach pasteurization temperature.

1.7.1 Compression Heat Pumps

Figure 1.3 shows the main types of vapor compression (VC) and recompression technologies (Kiss et al., 2012) applied to a distillation column: VC and mechanical vapor recompression (MVR) or thermal vapor recompression (TVR).

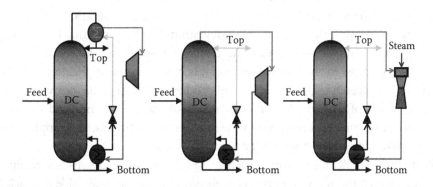

FIGURE 1.3 Vapor compression (left), mechanical vapor recompression (centre) and thermal vapor recompression (right). DC, distillation column.

Vapor compression is a classical heat pump design already proven at an industrial scale. It uses a specific fluid as a heat transfer medium that runs between the heat source and sink through a pipeline. A compressor is installed in between to provide the required work input, while a flash valve closes the cycle. Because all these elements are external to the distillation process, the distillation column does not require major modifications, except for the adjustments in heat exchangers for the change in utilities. VC is particularly beneficial when dealing with corrosive or fouling compounds. However, the design is dependent on the ability of the heat transfer fluid to meet stringent operational, environmental and safety requirements. For most applications, there is no adequate alternative – and even when an acceptable heat transfer fluid is found, the energy savings are not always translated into overall economic savings. On one side, the compressors are expensive and hard-to-maintain equipment, while in contrast the equipment and inefficiencies in obtaining mechanical energy (work) have a heavy impact on the final energy bill (Kiss et al., 2012).

Mechanical vapor recompression is a state-of-the-art industrial system for binary distillation that is widely applied in separation of close boiling components. In such a system, the top vapor is used as heat transfer medium, being fed directly to the compressor. Accordingly, the heat pump also performs the function of the condenser, thus saving one heat exchanger as compared to the classic alternative. Moreover, it avoids the need for cooling the heat transfer fluid below the boiling point of the top product – an issue of importance in the VC scheme for heat transfer purposes. Notably, MVR features slightly higher efficiency and lower investment costs than VC. However, MVR does not tackle directly the main drawback of VC: the economics involved in the compressor usage. Similarly, the distillate still has to meet at least the operational requirements for the heat transfer medium, not to mention the criteria for safe and economic compressor operations.

Thermal vapor recompression is a particular variant of the MVR in which the compressor is replaced by a steam ejector as the work input mechanism. In view of the advantages of the steam ejector, TVR has been widely implemented in industry. The steam ejector uses the Venturi effect to increase the pressure to the condensation level by steam injection into a special variable-diameter pipeline. This makes TVR a robust design with reduced capital and maintenance expenditures, as there are no rotating pieces involved. However, the steam ejector has a relatively low efficiency in increasing the pressure. Moreover, the design of the steam ejector is crucial in achieving economical operation. There are wide changes in steam consumption even at small deviations from the optimal operating point. Notably, the steam input is mixed with the distillate to generate the required pressure. As steam is being added to the vapor distillate, it is clear that the applications of TVR are mostly for systems producing water as the top product. In theory, the motive fluid for the ejector can be (part of) the distillate flow, which could be boiled and used to pressurize the vapor to the pressure level required in the reboiler. Nevertheless, such applications are rarely encountered due to the potential heat transfer losses.

An alternative to VRC heat pumps (such as MVR and TVR) is self-heat recuperation technology. However, the addition of two compressors leads to unacceptable payback times. Chemical heat pumps were also proposed, but they do involve the addition of endothermic and exothermic chemical reactors, rendering them unfeasible economically (Kiss et al., 2012).

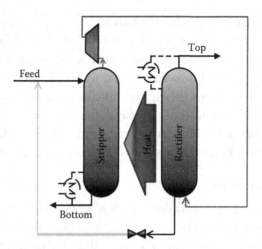

FIGURE 1.4   Heat-integrated distillation column (HIDiC).

The *heat-integrated distillation column* (*HIDiC*) is the most radical approach to heat pump design, making use of internal heat integration and MVR. Instead of using a single-point heat source and sink, the whole rectifying section of a distillation column becomes the heat source, while the stripping part of the distillation column acts as a heat sink, as shown in Figure 1.4 (Kiss et al., 2012). The problem of different sizes for rectifying and stripping sections can be relatively easily tackled by using one of the many alternative HIDiC configurations (Kiss and Olujic, 2014). This internal heat integration widely enhances the reachable COP because the required temperature difference for heat transfer is kept low, with gliding temperatures across both parts. The work input is provided by a compressor installed at the top outlet of the stripper section, while the heat pump cycle is closed by the valve flashing the liquid bottom outlet of the rectifier section. The HIDiC success relies actually on good hardware performance for both heat and mass transfer tasks at the same time. Currently, there is only a precommercial 15-kiloton/year implementation of HIDiC in Japan (Kiss and Olujic, 2014).

1.7.2 Transcritical Heat Pumps

The transcritical heat pump (TCHP) is based on a transcritical cycle, which is a thermodynamic cycle where the working fluid goes through both subcritical and supercritical states – this being typically the case when CO_2 is the refrigerant. Note that CO_2 is transcritical above 31°C. Heat intake at the evaporator takes place below the critical pressure (71 bar), while the release of heat within the gas cooler takes place above the critical pressure. For this reason, heat is released over a temperature range, and this is clearly in contrast with classic compressor systems, where heat is released at a fixed temperature. Consequently, the term *gas cooler* is used instead of condenser, as the refrigerant is not condensed, but only cooled. Note that to calculate the COP of a TCHP, one has to use the mean temperature in the temperature range where heat is released.

The main advantage of a CO_2 TCHP is that its efficiency can be higher than conventional heat pumps, for a high-temperature lift at the side of the gas cooler (at least 30–40 K).

However, the temperature of the fluid that needs to be heated has to be sufficiently low to make an efficient heat pump. Also, when CO_2 is not cooled far enough due to the fluid flowing in, the process will not be efficient; in this situation, an ammonia VC heat pump is a better solution. Another potential drawback is that the TCHP works with high refrigerant pressures of more than 100 bar, so the equipment cost is rather high. For the TCHP, the evaporator operates at the low-temperature outlet of the waste stream (minus driving force). The gas cooler inlet is taken at the high temperature outlet of the waste stream (plus driving force). An internal recovery heat exchanger is used to recuperate as much heat as possible.

1.7.3 Absorption Heat Pumps

An *absorption heat pump* (*AHP*) considers thermochemical conversion to enhance operational efficiency. In this case, absorption pairs are used as heat transfer fluids, as for example, ammonia and water or lithium bromide and water. The AHP is widely applied in refrigeration, although there are also stand-alone (pilot plant) implementations of AHP suitable for distillation or multistage evaporation processes. As shown in Figure 1.5 (left), the heart of the AHP is the steam-driven desorber that separates the absorption pair (Kiss et al., 2012). One of the components condenses first in the reboiler of the distillation column, and then it is flashed to cool the condenser. Afterwards, it is mixed with the other component from the regenerator to deliver heat in the reboiler through exothermic absorption (Ziegler, 2002). The resulting liquid is then pressurized and heated to displace the equilibrium, and the whole cycle is repeated. An AHP is preferred as it avoids the inconvenience and expense of using a compressor as the driver for the heat pump. However, the requirement of five heat exchangers gives the AHP a pricy installation cost and therefore long payback times.

1.7.4 Heat Transformers

Heat transformers are devices that transform a large heat resource available at a rather low temperature in a smaller amount of heat delivered at a higher-temperature level. In contrast to traditional heat pumps, HTs use no electrical power or work. For example, absorption heat transformers (AHTs) work on the principle of an absorption inverse cycle, but the net effect is that of transferring only a part of the amount of heat available to a higher-temperature level. As such, AHTs are not suitable for small-scale

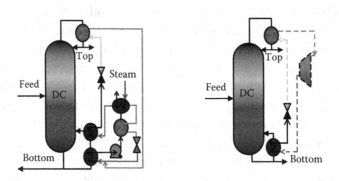

FIGURE 1.5   Absorption heat pump (left) and compression-resorption heat pump (right).

applications, but they might be attractive when dealing with large heat flow rates (>200 kW) available at low temperatures. The commercial attractiveness of HTs lies in building a compact unit that can effectively upgrade large amounts of heat without needing a large solar field, for example. Examples of HTs can be found in synthetic rubber plants or applications related to renewable energies and as useful add-ons to industrial waste heat recovery.

1.7.5 Hybrid Heat Pumps

A *compression-resorption heat pump* (*CRHP*) – also called a hybrid heat pump – is a recent approach used to take advantage of thermochemical sorption processes. A CRHP can achieve high temperature levels and lifts, with relatively high COP. The CRHP uses the VC scheme, in which the working fluid is replaced by an absorption pair. When the vapor zeotropic mixture approaches the reboiler, the condensation and absorption processes run at the same time, thus enhancing the heat transfer. This gives the CRHP enhanced overall efficiency and reduced energy requirements, a critical issue for economic operations in wider temperature ranges. After the reboiler, the rich mixture is then flashed to take heat from the condenser, evaporating both components from the working pair. The evaporation in the desorber is incomplete, so solution recirculation between the desorber and the resorber or wet compression is possible. When all the solution is recirculated, the cycle is called the Osenbrück cycle. The *hybrid wet compression* (HWC) term is used for the cycles where all the solution is sent to the compressor, avoiding the use of a solution pump. Nonetheless, solving the issue of wet compression is of crucial importance for the wider implementation of CRHPs. At the moment, the issue is still under research, with successful pilot plant implementations of stand-alone CRHPs already in place (e.g. Delft University of Technology).

The advantages of the CRHP are related to the large temperature lift possible with a high COP and the use of environmentally friendly refrigerants (e.g. water, ammonia and CO_2), which can significantly contribute to the improvement of the performance of heating processes. Specifically for industrial heating processes, the CRHP allows energy performance gains of more than 20% as compared to VC. The use of a mixture allows lower pressure levels and condensation and evaporation at gliding temperatures, which can result in higher efficiency. Wet compression has the effect of suppressing vapor superheating, and it can also improve heat pump efficiency – if the technical problems surrounding it are solved. Ammonia-water mixtures can be used as efficient working fluids in CRHPs, showing a number of advantages: Higher COP occurs because of the use of nonisothermal phase transition of the mixture in the heat exchangers at constant pressure; the mixture allows the achievement of high-temperature operation at relatively low operating pressures; the cycle can be designed to show a temperature glide in the resorber that corresponds to the temperature glide of the industrial flow that has to be heated and for specific operating conditions, the cycle performance is significantly better than for the VC cycle. However, the technical complexity causes the CRHP to be expensive to install. Moreover, there are only a limited number of manufacturers offering hybrid heat pumps for industrial use (e.g. Hybrid Energy, www.hybridenergy.no/en).

TABLE 1.1 Comparison of Solid Sorption Cycles

Solid Sorption Cycle	Salt-Ammonia	Carbon-Ammonia	Metal-Hydrogen	Silica Gel–Water	Zeolite-Water
Drive	Thermally	Thermally	Thermally Hybrid/ composite	Thermally	Thermally
Operating window	Limited, salts dependent	Limited Hybrid systems	Limited, metal dependent	Limited 0–150°C	Limited 0–250°C
Temperature lift	>100°C	~60°C	>50°C	~40°C	>50°C
Development stage	Early stage	PoC for cooling	Early stage	Commercially available for cooling	SorTech commercial product domestic heating 10 kW
Performance	$COP_H \le 1.7$ $COP_{HT} \le 0.4$	Suitable only for cooling	$COP_H \le 1.7$ $COP_{HT} \le 0.4$	$COP_H \le$ N/A $COP_{HT} \le 0.4$	$COP_H \le 1.6$ $COP_{HT} \le 0.2$
Reliability	No moving parts but stability is an issue	No moving parts but stability is an issue	No moving parts but stability is an issue	No moving parts Maintain vacuum conditions	No moving parts Maintain vacuum conditions

1.7.6 Adsorption Heat Pumps

Although based on the same principles as the AHP, the adsorption heat pumps use a solid instead of a fluid as adsorption medium. Solid sorption has a working principle based on the reversible sorption reaction between a gas or vapor and a solid (porous) material. Table 1.1 conveniently illustrates the comparison between the solid sorption cycles (Kiss, 2013). Note that the adsorption principle is used increasingly in small heat pump systems (70–500 kW), which are mainly used for cooling.

The sorption principle is already widely applied in separation and purification processes, such as pressure swing adsorption and temperature swing adsorption. The key characteristics of solid sorption are physical adsorption (onto surface), chemical absorption (into bulk material), temperature lifts achieved by 'thermal' compression, batch processes and no moving parts. Some of the known working pairs are salt-ammonia, salts-water, carbon-ammonia, carbon-methanol, metal-hydrogen, silica gel–water, zeolite-water; for example:

$$MgCl_2 \cdot 2NH_3(s) + 4NH_3(g) \leftrightarrow MgCl_2 \cdot 6NH_3(s) \tag{1.6}$$

1.7.7 Solid-State Heat Pumps

The use of *magnetic heat pumps* or magnetic refrigeration is a cooling technology based on the magnetocaloric effect that is a magnetothermodynamic phenomenon in which a temperature change of a suitable material (e.g. iron) is caused by exposing the material to a changing magnetic field. This technique can be used to attain extremely low temperatures, along with the typical ranges used in refrigerators. Some commercial ventures to implement this technology are under way, claiming to reduce energy usage by 40% compared to current domestic refrigerators.

The magnetic refrigeration cycle is performed with changes in the applied magnetic field strength instead of pressure (as is the case in the Brayton refrigeration cycle). A fluid is circulated through a bed of magnetic material, which is externally periodically magnetized and demagnetized while the fluid flows in opposite directions. The cycle sequentially follows the next four steps:

- *Heat delivery to application*: Isothermal magnetization occurs from state A (no field applied) to state B (field applied); the entropy changes from s_A to s_B so that the heat delivered to the application is $T_H(s_A - s_B)$.

- *Isofield cooling process*: The field remains applied while heat is released to the regenerator bed as the fluid flows through it. The removed heat is obtained by integrating *Tds* between the start entropy s_B and the end entropy s_C.

- *Heat uptake from source*: Isothermal demagnetization occurs from state C (field applied) to state D (no field applied); heat flow is given by $T_C(s_D - s_C)$.

- *Isofield heating process*: While no field is applied, the fluid (now flowing in the opposite direction) heats up by removing heat from the regenerator bed. The consumed heat is obtained by integrating *Tds* between the start entropy s_D and the end entropy s_A.

Thermoelectric (TE) heat pumps use the thermoelectric effect, and they have improved over time to the point at which they are useful for certain refrigeration tasks. Thermoelectric (Peltier) heat pumps are generally only around 10%–15% as efficient as the ideal heat pump (Carnot cycle), as compared to 40%–60% achieved by conventional compression cycle systems (reverse Rankine systems using compression/expansion). However, this area of technology is currently the subject of active research in materials science. A major reason for its popularity is the long lifetime, as there are no moving parts, and if the thermoelectric module uses renewable energy, then it would be better for the environment.

A *thermoacoustic heat pump (TAHP)* relates to the physical phenomenon that a temperature difference can create and amplify a sound wave and vice versa that a sound wave is able to create a temperature difference. A sound wave is associated with changes in pressure, temperature and density of the medium through which the sound wave propagates. In addition, the medium itself is moved around an equilibrium position. An acoustic wave is brought into interaction with a porous structure with a much higher heat capacity compared to the propagation. This porous structure acts as a kind of heat storage-regenerator. Within thermoacoustics, a distinction is made between a thermoacoustic engine or prime mover (TA engine) and a TAHP. The first relates to a device creating an acoustic wave by a temperature difference, while in the second an acoustic wave is used to create a temperature difference.

The TAHP is the front runner in using a different mechanism for the work input. Although it is a relatively new technology, the proof-of-principle stage has been successfully

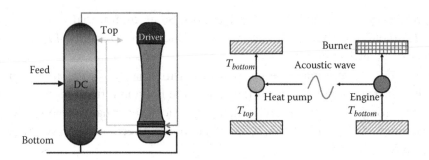

FIGURE 1.6 Thermoacoustic heat pump applied to distillation (left) and schematics (right).

completed – with scaling up currently undergoing intense efforts. Basically, the TAHP is a thermoacoustic device that uses high-amplitude sound waves to pump heat from one place to another. Figure 1.6 illustrates the TAHP application to a distillation column as well as the working principle (Kiss et al., 2012). The thermoacoustic device consists of heat exchangers, a resonator and a regenerator (on traveling wave devices) or stack (on standing wave devices). Depending on the type of engine, a driver or loudspeaker might be used as well to generate sound waves. To limit the space used, an electric driver (linear motor) generates the acoustic power cased inside a resonator, the temperature lifts being determined by the size of resonator, as well as the properties and pressure of the acoustic medium. The resonator, housing the TA engine and the TA heat pump, determines the operating frequency, acts as a pressure vessel and transports the acoustic power between the components. Note that there is no concern about the noise levels because these are similar to the current industrial standards (e.g. below 85 dB). Moreover, as the medium used is a gas (air, noble gas [e.g. helium]), the system complies with safety and environmental concerns, while having virtually no limitations in its applications.

1.8 FEASIBILITY CHECK

The use of heat pumps should be considered within the larger context of an industrial site and taking into account the results of the pinch analysis (described in a further chapter). However, in many cases a quick check is possible by simply answering the questions that follow. If the answer is yes to all questions, then it might be feasible to apply a heat pump for a specific process.

- *Is the heat demand of the process optimized already?* If not, then first try to reduce the heat and utilities demand (by pinch analysis) and only afterwards consider heat pumps.

- *Is there any heat source (e.g. waste heat stream) available with sufficient heat capacity?* If not, then a heat pump cannot be directly applied as it requires a heat source with enough capacity.

- *Is the waste heat available at the same time as the heat demand?* If not, then consider the use of a heat buffer for storage of thermal energy (e.g. in case of batch processes).

- *Is the temperature of the heat source lower than the required heating temperature?* If not, then most likely a direct heat exchange is possible, so probably a heat pump is not needed.

- *Is the temperature difference between the heat source and the heat sink lower than 80 K?* If not, then a heat pump will have a low COP, and thus long payback times are to be expected.

- *Is the heat demand constant and for long periods?* If not, then the heat pump will have a reduced load, so less savings will be possible; thus, long payback times are to be expected.

1.9 CONCLUDING REMARKS

Heat pumps can be used in many industrial processes that have both waste heat sources (wastewater, hot humid air, condenser heat) and heat sinks (process water, heating systems, dryers or reboilers). Heat pumps have several advantages: very high efficiency; the possibility of using renewable energy from air, water or ground; large energy savings of 50%–70% and significant reduction of GHG emissions (e.g. CO_2). Nonetheless, heat pumps also have some drawbacks: An initial investment is needed (so payback times might be an issue), an electrical connection is typically required to be present and some refrigerants used in a heat pump are toxic or flammable (HSE issues).

Heat pump technologies were proposed and developed to upgrade low-level energy and reuse it in the process, thus reducing the consumption of valuable utilities. With the exception of solid sorption systems, the heat pumps presented here have no intrinsic (temperature) limitations; therefore, the application range is quite large within the CPI. Temperature lifts of up to 30–100 K are feasible, leading to significant savings in the primary energy requirements of up to 80% (Kiss et al., 2012).

LIST OF SYMBOLS

s	Entropy	kJ/kgK
T	Temperature	K
$\dot{Q}$	Heat flow	kW
$\dot{W}$	Power	kW

Subscripts

C	Cold stream
Ext	External hot stream
in	Inlet conditions
H	Hot stream; heat pump cycle
HT	Heat transformer cycle
M	Intermediate condition
out	Outlet conditions
R	Refrigeration cycle
rev	Reversible

Abbreviations

AC	Air conditioner
AHP	Absorption heat pump
AHT	Absorption heat transformer
CHP	Combined heat-power
CHX	Cold heat exchanger
COP	Coefficient of performance
CPI	Chemical process industry
CRHP	Compression-resorption heat pump
DC	Distillation column
EER	Energy efficiency ratio
GHG	Greenhouse gas
HHX	Hot heat exchanger
HIDiC	Heat-integrated distillation column
HP	Heat pump
HSE	Health, safety, environment
HT	Heat transformer
HVAC	Heating, ventilation and air conditioning
HWC	Hybrid wet compression
IEA	International Energy Agency
LGWH	Low-grade waste heat
MVR	Mechanical vapor recompression
PER	Primary energy ratio
PoC	Proof of concept
SCOP	Seasonal coefficient of performance
SEER	Seasonal energy efficiency ratio
TA	Thermoacoustic
TE	Thermoelectric
TAHP	Thermoacoustic heat pump
TCHP	Transcritical heat pump
TVR	Thermal vapor recompression
VC	Vapor compression
VRC	Vapor recompression
WSHP	Water source heat pump

REFERENCES

Brumbaugh J. E., *HVAC fundamentals. Volume 3: Air conditioning, heat pumps and distribution systems*, 4th edition, Audel, New York, 2004.

Cantor J., *Heat pumps for the home*, Crowood Press, Ramsbury, UK, 2011.

Kiss A. A., *Advanced distillation technologies – Design, control and applications*, Wiley, Chichester, UK, 2013.

Kiss A. A., Flores Landaeta S. J., Infante Ferreira C. A., Towards energy efficient distillation technologies – Making the right choice, *Energy*, 47 (2012), 531–542.

Kiss A. A., Olujić Ž., A review on process intensification in internally heat-integrated distillation columns, *Chemical Engineering and Processing*, 86 (2014), 125–144.

Langley B. C., *Heat pump technology*, 3rd edition, Prentice Hall, Englewood Cliffs, NJ, 2001.

Petit R. F. Sr., Collins T. L., *Heat pumps: Operation, installation, service*, Esco Press, Mt. Prospect, IL, 2011.

Reay D. A., MacMichael D. B. A., *Heat pumps*, 2nd edition, Pergamon Press, Oxford, UK, 1988.

Silberstein E., *Heat pumps*, Delmar Cengage Learning, Independence, KY, 2002.

Whitman B., Johnson B., Tomczyk J., Silberstein E., *Refrigeration and air conditioning technology*, 7th edition, Cengage Learning, Independence, KY, 2012.

Ziegler F., State of the art in sorption heat pumping and cooling technologies, *International Journal of Refrigeration*, 25 (2002), 450–459.

Zogg M., *History of heat pumps – Swiss contributions and international milestones*, Swiss Federal Office of Energy SFOE, Ittigen, 2008.

CHAPTER 2

Thermodynamics of Heat Pump Cycles

2.1 INTRODUCTION

Chemical engineers use computer simulations to perform a variety of tasks, ranging from calculations of mass and energy balances to rating, sizing, optimization, dynamics modeling and performance evaluation of process alternatives that could reduce investment and operating costs. The application of heat pumps in the chemical process industry is clearly intimately linked to the chemical process itself. Therefore, the process design and simulation of a heat pump should take place within the context of the chemical process (e.g. heat pump–assisted distillation). Nowadays, due to significantly improved computing power, an engineer can relatively quickly set up the basic simulation specifications and process conditions and define a complex flowsheet. However, the solid base of any process modeling and simulation is represented by the physical properties models. Missing or inadequate physical properties can undermine the accuracy of a model or plant design. Finding good values for inadequate or missing physical property parameters is the key to success, but this depends strongly on choosing the right estimation methods – an issue already recognized by the axiom "garbage in, garbage out," which means that the simulation results have the same quality as the input data or parameters (Carlson, 1996).

This chapter aims to provide an overview of the fundamental concepts of thermodynamics, covering both physical properties and phase equilibrium. While *chemical engineering thermodynamics* deals mostly with computation of physical properties and of equilibrium states of fluids involved in chemical processes, *engineering thermodynamics* handles applications of the thermodynamic principles in the field of energy production or use, processes and machineries, deep cooling and liquefaction, compression of gases, and so on. The key thermodynamic notions described hereafter are used in further chapters (e.g. for minimization of entropy production, exergy analysis, pinch analysis, and process integration). A more detailed treatment of thermodynamics can be found in dedicated textbooks by Kyle (1999), Smith et al. (2005), Sandler (2006), Moran et al. (2011), and Elliott and Lira (2012).

The descriptions of the major methods for estimating fundamental properties are available in the works of Reid et al. (1987) and Poling et al. (2001), while Gmehling et al.'s works (1997, 2012) focus on thermodynamics applied in process simulation.

The chapter starts by reviewing the laws of thermodynamics to ensure a coherent presentation of the fundamental equations. Attention is given to the network of thermodynamic relations that makes it possible to navigate between different properties, many of them not accessible by measurements. The general conditions of phase equilibrium are also formulated, and the concepts of fugacity and activity are highlighted. The pressure-volume-temperature (*PVT*) behavior of process fluids is examined, generic models are introduced (e.g. equations of state [EOSs], principle of corresponding states), and the generalized computational methods for generating charts and tables for various properties are described. Distinct sections are devoted to the EOSs and liquid activity (LACT) models.

Selecting a suitable property model and supplying the adequate parameters are key issues. Modern thermo methods can tackle complex mixtures, including nonpolar and polar species, supercritical and subcritical components, water, large and small molecules, electrolytes, and polymers. Attention is given to model selection and regression of parameters from experimental data. Key topics include vapor-liquid equilibrium (VLE), vapor-liquid-liquid equilibrium (VLLE) and gas-liquid equilibrium (GLE), which cover the largest part of chemical industry applications, including heat pump–assisted processes.

2.2 FUNDAMENTALS OF THERMODYNAMICS

Thermodynamics deals with heat and temperature and their relation to energy and work, and it is the science that studies the changes in the state or condition of a system when changes in its internal energy are important. *Internal energy* designates all forms of energy associated with motions, interactions and bonding of molecules forming a material body. This is different from forms of *external energy* (e.g. kinetic and potential energy) that are of primary interest in mechanics. The central problem of thermodynamics is the determination of the equilibrium state that is eventually attained after the removal of internal constraints in a closed, composite system.

The sign conventions used in this chapter are as follows:

- *Energy* (heat, work): positive (+) means *into* the system, negative (−) means *out* of the system.

- *Work* ($dW = -PdV$): compression is positive (volume decreases, $dV < 0$), while expansion is negative (volume increases, $dV > 0$).

- Capital letters designate *molar values* of properties, such as volume V, internal energy U, enthalpy H, entropy S and free energy functions A, G. Consequently, extensive values for a system containing n moles are noted by (nV), (nU), and so on.

2.2.1 Laws of Thermodynamics

Zeroth law of thermodynamics: If two systems are each in thermal equilibrium with a third, they are also in thermal equilibrium with each other.

Note that two systems are considered to be in the state of thermal equilibrium if they do not change over time and are linked by a wall permeable only to heat. This law helps to define the notion of temperature. Another statement of the zeroth law is that all diathermal walls are equivalent. Note that a diathermal wall between two thermodynamic systems allows heat but not matter to pass across it. This law is important for the mathematical formulation of thermodynamics, which needs the assertion that the relation of thermal equilibrium is an equivalence relation. This information is in fact needed for the mathematical definition of temperature that will agree with the physical existence of valid thermometers.

First law of thermodynamics: The increase in internal energy of a closed system is equal to the difference of the heat supplied to the system and the work done by it.

Basically, the first law postulates the *conservation of energy*, which states that energy can be transformed (changed from one form to another) but cannot be created or destroyed. The system is the part of the space where the process occurs. Everything not included in the system is considered *surroundings*. Considering a closed system where work W and heat Q are the only forms of energy exchanged between system and surroundings, the *first law of thermodynamics* is given by

$$\Delta \ (energy \ of \ the \ surroundings) + \Delta \ (energy \ of \ the \ system) = 0 \qquad (2.1)$$

The first term is given by the net amount of heat and work exchanged $-(Q + W)$, while the second term takes into account the change in energy of the system itself. The most significant contribution is the *internal energy U* that includes all forms of energy associated with the activity of atoms and molecules forming a system, except those related to macroscopic movement (*kinetic energy*) or change in position (*potential energy*). Internal energy is a state function, so its absolute value cannot be known, but its variation (ΔU) between two states is always the same regardless of the path followed. In differential form, the first law postulates that in a closed system the *internal energy U* can be varied only by exchange of heat or of work:

$$dU = \delta Q + \delta W \qquad (2.2)$$

Note that the differential dU is 'exact', meaning that ΔU depends only on the initial and final states (pressure, volume, temperature, composition), while δQ and δW are not exact differentials because Q and W depend on the path of transformation. An important measurable physical property is the *constant-volume heat capacity C_V*, defined as follows:

$$\left(\frac{\partial U}{\partial T}\right)_V = \left(\frac{\partial Q}{\partial T}\right)_V = C_V \qquad (2.3)$$

As most industrial operations take place at fairly constant pressure, it is useful to associate U and PV in a new state function *enthalpy (H)*, which leads to *the first law* expressed in differential form:

$$H = U + PV \qquad (2.4)$$

$$dH = \delta Q + VdP \tag{2.5}$$

Thus an important measurable property can be defined, the *constant-pressure heat capacity* (C_P):

$$\left(\frac{\partial H}{\partial T} \right)_P = \left(\frac{\partial Q}{\partial T} \right)_P = C_P \tag{2.6}$$

Note that C_V and C_P are functions of temperature, and for an ideal gas, $C_P^0 = C_V^0 + R$.

The energy balance in the chemical industry is typically formulated as an enthalpy balance. For open (flowing) systems, when thermal effects predominate, the first law becomes

$$\Delta H = Q + W_s \tag{2.7}$$

The mechanical energy W_s is exchanged as *shaft work* (see Figure 2.1). The volumetric work is included directly in the calculation of enthalpy. If only temperature changes are involved, so no phase change occurs, the variation of the (sensible) enthalpy is

$$\Delta H = H(T_2) - H(T_1) = \int_{T_1}^{T_2} C_P(T) dT \tag{2.8}$$

This formula is acceptable for condensed phases (solid or liquids) without phase transition, but not for real gases, for which the influence of pressure has to be accounted for. While this formula is used in hand calculations, process simulators calculate enthalpy based on EOSs or using the corresponding states principle.

An ideal gas is taken usually as the reference for the thermodynamic properties of a real fluid. Because the enthalpy of an ideal gas does not depend on pressure, its variation between two states depends only on temperature. Thus, it is convenient to define an arbitrary *reference state*, which may be different from a standard state, a concept used in equilibrium reactions. The enthalpy of an ideal gas H^0 can be set at zero when $T = T_0$. A

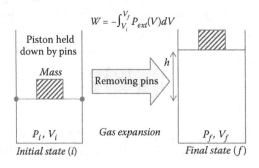

FIGURE 2.1 Mechanical work during gas expansion.

convenient reference temperature is the absolute zero Kelvin. In this case, the enthalpy of an ideal gas can be calculated as follows:

$$H^0 = \int_0^T C_P^0(T)dT = aT + bT^2/2 + cT^3/3 + \dots \tag{2.9}$$

Note that the difference in enthalpy between a real fluid and an ideal gas (noted as *departure function $H - H^0$*), at the same temperature T and a reference pressure P_0 can be calculated only from *PVT* information. Thus, the enthalpy of a real fluid, including contributions of both sensible heat and phase transition, can be calculated using the following approach, which can be extended to other thermodynamic properties:

$$H = H^0 + (H - H^0) \tag{2.10}$$

Some common situations related to the first law of thermodynamics are graphically illustrated in Figures 2.1 through 2.3.

Second law of thermodynamics: Heat cannot spontaneously flow from a colder location to a hotter location.

Processes evolve toward states of minimum energy and maximum disorder, and these two tendencies are in competition. The second law deals with the spontaneous evolution of a thermodynamic process and the efficiency of conversion between different forms of energy (particularly between work and heat), being intimately linked with the notion of *entropy S*. Work W can be transformed spontaneously and integrally into heat (such that $Q = W/J$, where J is a conversion factor), but heat conversion into work is never spontaneous and by no means complete.

In other words, the second law of thermodynamics postulates that *it is impossible for a device operating in a cyclic manner to completely convert heat into work*. If the heat flows spontaneously from higher to lower temperature, the opposite process requires a *heat engine*. The Caratheodory principle (another statement of the second law) states that in the neighborhood of any equilibrium state of a system, there are states that cannot be reached by reversible adiabatic processes. The Caratheodory's statement is equivalent to the existence of the entropy function.

Carnot demonstrated that the maximum work from a heat engine is given by a cycle formed by two reversible adiabates and two isotherms (see Figure 2.4), with the efficiency given by

$$\eta = \frac{Q_H - Q_C}{Q_H} = \frac{T_H - T_C}{T_H} = 1 - \frac{T_C}{T_H} \tag{2.11}$$

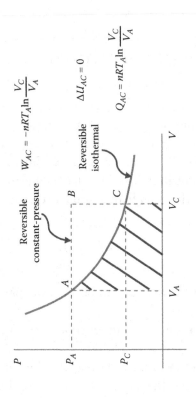

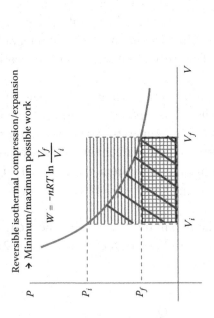

FIGURE 2.2 Reversible isothermal expansion/compression of an ideal gas (left); work and heat along reversible isothermal expansion for an ideal gas (right).

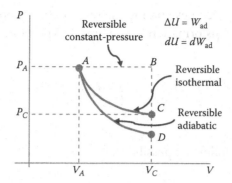

FIGURE 2.3   Adiabatic process in which no heat is transferred (bottom right).

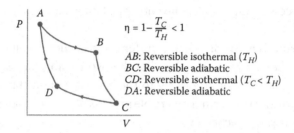

FIGURE 2.4   Carnot cycle: One can never use all the thermal energy given to the engine by converting it into mechanical work.

where T_H is the absolute temperature at which the heat Q_H is added, and T_C is the temperature at which the heat Q_C is removed. It follows that the following relation holds:

$$\frac{Q_H}{T_H} - \frac{Q_C}{T_C} = 0 \qquad (2.12)$$

Equation 2.12 suggests that the ratio between the heat and temperature is a system property (a new state function: *entropy*) that characterizes the *reversibility* of the heat exchange in a cyclic process. Note that a reversible process proceeds with infinitesimal gradients within the system, and the direction can be reversed at any point by an infinitesimal change in external conditions.

$$dS = \frac{\delta Q_{rev}}{T} \qquad (2.13)$$

where δQ_{rev} is the infinitesimal amount of heat exchanged in a reversible process at temperature T.

Entropy plays a central role in evaluating the reversibility of processes, as discussed in the next chapter. For the transformation from point A to point B, the following relation is valid:

$$S_B - S_A = \int \frac{dQ_{AB}}{T} \geq 0 \qquad (2.14)$$

Equation 2.14 leads to the following possibilities:

- $\Delta S = 0$, the process is reversible and adiabatic.
- $\Delta S > 0$, the process is irreversible.
- $\Delta S < 0$, the process is impossible. The entropy of an isolated system never decreases.

Third law of thermodynamics: When a system approaches absolute zero, the entropy of the system approaches a minimum value. In other words, absolute zero is not attainable via a finite series of processes, or according to the Nernst–Simon statement: The entropy change associated with any isothermal reversible process of a condensed system approaches zero as the temperature approaches zero.

Basically, the third law allows the calculation of an absolute value for entropy. By definition, *the entropy of a perfect crystalline substance is zero at zero absolute temperature*. The entropy of a system that has a nondegenerate ground state vanishes at absolute zero. For a pure component ideal gas at temperature T, the entropy is given by

$$S_T = \int_0^{T_f} \frac{C_{P,s}}{T} dT + \frac{\Delta H_f}{T_f} + \int_{T_f}^{T_V} \frac{C_{P,l}}{T} dT + \frac{\Delta H_{vap}}{T_v} + \int_{T_v}^{T} \frac{C_{P,v}}{T} dT \qquad (2.15)$$

where T_f and T_v are the fusion and vaporization temperatures, respectively, whereas ΔH_f and ΔH_{vap} are the enthalpies of the respective phase transitions. Note that many thermodynamic variables are discontinuous across first-order phase transitions, as illustrated by Figure 2.5.

2.2.2 Free Energy Functions

Free energy functions (*Helmholtz free energy A* and *Gibbs free energy G*) are more appropriate for the computation of physical and chemical equilibrium than primary functions (*internal energy U* and *entropy S*). The definitions of A and G are as follows:

$$A = U - TS \quad \textit{Helmholtz free energy A} \qquad (2.16)$$

$$G = H - TS = U + PV - TS = A + PV \quad \textit{Gibbs free energy G} \qquad (2.17)$$

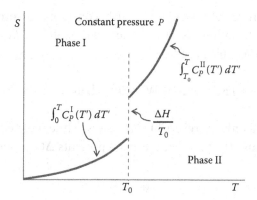

FIGURE 2.5   Discontinuity of thermodynamic variables across first-order phase transitions.

Note that the functions A and G are called *potentials* because they allow the calculation of the maximum work W_{max} only from initial and final states. For a reversible process in a closed system, the maximum work that can be produced only by a reversible transformation is given by

$$dU = TdS + \delta W_{max} \tag{2.18}$$

The *total work* W takes into account the *volumetric work* PV, as well as other forms of *useful work* $W^{\#}$, such as shaft work or electrochemical work. The volumetric work is accounted for as a change in the volume at constant pressure. For an infinitesimal change, the total work is $\delta W = -PdV + \delta W^{\#}$. Consequently, the following relation can express the maximum useful work:

$$\delta W_{max}^{\#} = dU - TdS + PdV \tag{2.19}$$

$$\delta W_{max}^{\#} = d(U - TS) \text{ for a process at } (T, V) \text{ constant} \tag{2.20}$$

Hence, the *Helmholtz free energy* $A = U - TS$ is a potential at (T, V) constant because it gives the maximum work for finite variation:

$$W_{max}^{\#} = \Delta A \tag{2.21}$$

In a similar way, the *Gibbs free energy* $G = H - TS$ is a potential at (P, T) constant because

$$W_{max}^{\#} = \Delta G \tag{2.22}$$

As most industrial processes take place at fairly constant pressure, Gibbs free energy is preferred in technical calculations. Note that the equation $dU = TdS + \delta W_{max}$ can be further extended to include irreversible processes:

$$\Delta U - T\Delta S + P\Delta V - W^{\#} \leq 0 \quad \text{Clausius inequality} \tag{2.23}$$

The equality sign is valid only for reversible processes, while the inequality applies for irreversible processes. Because the first three terms represents $\Delta G_{T,P}$, one can write

$$\Delta G_{T,P} \leq W^{\#} \tag{2.24}$$

If there is only volumetric work, then $W^{\#} = 0$, and therefore

$$\Delta G_{T,P} \leq 0 \tag{2.25}$$

A chemical process occurs spontaneously when $\Delta G_{T,P} < 0$, while at equilibrium $\Delta G_{T,P} = 0$. The first and the second laws of thermodynamics may be combined in the form of differential equations. For 1 mole of pure gaseous component in a closed system, the first law in a reversible process can be written as $dU = \delta Q_{rev} + \delta W_{rev}$. Replacing δQ_{rev} and considering only volumetric work ($\delta W_{rev} = -PdV$) leads to the expression for dU. Similar expressions can be obtained by differentiating the other state functions. The following four equations are *fundamental equations* in thermodynamics and the starting point to derive the network of thermodynamic properties (Dimian et al., 2014):

$$dU = TdS - PdV \tag{2.26}$$

$$dH = TdS + VdP \tag{2.27}$$

$$dA = -PdV - SdT \tag{2.28}$$

$$dG = VdP - SdT \tag{2.29}$$

2.2.2.1 Pressure and Temperature Dependency of Gibbs Free Energy

From the equation $dG = VdP - SdT$, at constant temperature one obtains

$$dG = VdP \text{ or } \left(\frac{\partial G}{\partial P}\right)_T = V \tag{2.30}$$

Therefore, the Gibbs free energy G depends strongly on pressure for gases, but only slightly for liquids. Note that G can be obtained experimentally from volumetric measurements. In case of an ideal gas, the integration of Equation 2.30 between P and $P_0 = 1$ atm at constant T leads to

$$G = G^0 + RT \ln(P/1 \text{ atm}) \quad (\text{constant } T) \tag{2.31}$$

This relation can be generalized for real gases by means of the fugacity concept, discussed further in this chapter. Similarly, at constant pressure but variable temperature, the following relations hold true:

$$dG = -SdT \text{ or } \left(\frac{\partial G}{\partial T}\right)_P = -S \tag{2.32}$$

Therefore, phase transitions give important changes in the slope of Gibbs free energy versus temperature because of the large entropy variation (see also Figure 2.5).

2.2.3 Phase Equilibrium Condition

2.2.3.1 Pure Components

The equilibrium condition for a pure species distributed between two phases α and β at the same and uniform P and T can be expressed as the equality of the Gibbs free energy in each phase:

$$[G^\alpha = G^\beta]_{P,T} \tag{2.33}$$

Equation 2.33 leads to an important thermodynamic relation that links the properties of phases at equilibrium. Assume an infinitesimal variation of temperature and pressure around the equilibrium, such that we can write $dG^\alpha = dG^\beta$. We may apply for each infinitesimal variation the equation $dG = VdP - SdT$ and obtain

$$\frac{dP}{dT} = \frac{S^\beta - S^\alpha}{V^\beta - V^\alpha} \tag{2.34}$$

From the Gibbs free energy definition and the equilibrium condition, it is easy to find that $S^\beta - S^\alpha = (H^\beta - H^\alpha)/T$. Replacing it in Equation 2.34 leads to the *Clapeyron equation*:

$$\frac{dP}{dT} = \frac{H^\beta - H^\alpha}{T(V^\beta - V^\alpha)} \text{ } Clapeyron \text{ } equation \text{ (valid for any phase transition)} \tag{2.35}$$

$$\left(\frac{dP}{dT}\right)_{eq} = \frac{\Delta H_{vap}}{T \Delta V} \text{ (at vapor} - \text{liquid phase transition)} \tag{2.36}$$

where $\Delta V = V^V - V^L$, V^V and V^L being the molar specific volumes of vapor and liquid, respectively. The relation can be extended for other phase equilibria, such as melting and sublimation. If the vapor is ideal, then $V^V - V^L \approx V^V = RT/P$, and this leads to the *Clausius–Clapeyron equation*:

$$\frac{d\ln P}{d(1/T)} = -\frac{\Delta H_{vap}}{R} \quad \text{or} \quad \frac{d\ln P}{d(1/T)} = -\frac{\Delta H_{vap}}{R} \quad \textit{Clausius – Clapeyron equation} \tag{2.37}$$

Hence, the enthalpy of vaporization can be determined from vapor pressure data. Conversely, the vapor pressure may be estimated more accurately if the enthalpy of vaporization is considered among experimental data. Details for estimating these two fundamental physical properties can be found in the monograph of Poling et al. (2001).

2.2.3.2 Mixtures

A composition dependency can be introduced in the analysis of any state function U, H, A or G. For multicomponent systems, the equation $dG = VdP - SdT$ may be formulated in extensive manner as

$$d(nG) = (nV)dP - (nS)dT \tag{2.38}$$

For a single-phase *closed* system without reaction and an infinitesimal variation in P and T,

$$d(nG) = \left[\frac{\partial(nG)}{\partial P}\right]_{T,n} dP + \left[\frac{\partial(nG)}{\partial T}\right]_{P,n} dT \tag{2.39}$$

$$\left[\frac{\partial(nG)}{\partial P}\right]_{T,n} = (nV) \quad \text{and} \quad \left[\frac{\partial(nG)}{\partial T}\right]_{P,n} = -(nS) \tag{2.40}$$

For a single-phase *open system*, the mixture composition may vary, and $nG = g(P, T, n_1, n_2, \ldots, n_i, \ldots)$:

$$d(nG) = \left[\frac{\partial(nG)}{\partial P}\right]_{T,n} dP + \left[\frac{\partial(nG)}{\partial T}\right]_{P,n} dT + \sum_i \left[\frac{\partial(nG)}{\partial n_i}\right]_{P,T,n_j} dn_i \tag{2.41}$$

Partial derivatives of the mixture Gibbs free energy function nG can be introduced with respect to the composition, which by definition are called *chemical potentials*:

$$\mu_i = \left[\frac{\partial(nG)}{\partial n_i}\right]_{P,T,n_j} \tag{2.42}$$

As a result, the generic relation takes the more practical form, which is fundamental for calculations regarding multicomponent mixtures that are designated in thermodynamics by *PVTx systems*:

$$d(nG) = (nV)dP - (nS)dT + \sum (\mu_i dn_i) \tag{2.43}$$

From this form, one can obtain the generalization of the phase equilibrium condition as follows:

$$\left[\mu_i^\alpha = \mu_i^\beta = \mu_i^\gamma = \ldots = \mu_i^\pi \right]_{P,T} \tag{2.44}$$

To apply this equation, one needs *models for chemical potentials*, as a function of pressure, temperature and composition, as discussed in the next section.

2.2.4 Gibbs–Duhem Equation

Chemical potential was introduced as a partial property of the Gibbs free energy to solve the phase equilibrium problem. Similar partial properties may be considered for other extensive properties (e.g. volume, enthalpy, entropy). A generalized approach would be useful. Considering that M represents the mean molar value of a property, for the whole system, $nM = f(T, P, n_1, n_2, n_3, \ldots)$. The derivation of the (nM) as a function of T, P, and composition gives

$$d(nM) = \left[\frac{\partial(nM)}{\partial T} \right]_{P,n} dT + \left[\frac{\partial(nM)}{\partial P} \right]_{T,n} dP + \sum \left[\frac{\partial(nM)}{\partial n_i} \right]_{P,T,n_j} dn_i \tag{2.45}$$

$$\bar{M}_i = \left[\frac{\partial(nM)}{\partial n_i} \right]_{P,T,n_j} \quad \text{(partial molar property)} \tag{2.46}$$

$$d(nM) = n \left(\frac{\partial M}{\partial T} \right)_{P,x} dT + n \left(\frac{\partial M}{\partial P} \right)_{T,x} dP + \sum \bar{M}_i dn_i \tag{2.47}$$

Because $n_i = x_i n$, $dn_i = x_i dn + n dx_i$, then

$$ndM + Mdn = n \left(\frac{\partial M}{\partial T} \right)_{P,x} dT + n \left(\frac{\partial M}{\partial P} \right)_{T,x} dP + \sum \bar{M}_i (n dx_i + x_i dn) \tag{2.48}$$

Grouping the terms and multiplying by n and dn leads to

$$\left[dM - \left(\frac{\partial M}{\partial T}\right)_{P,x} dT - \left(\frac{\partial M}{\partial P}\right)_{T,x} dP - \sum \bar{M}_i dx_i\right] n + \left[M - \sum x_i \bar{M}_i\right] dn = 0 \quad (2.49)$$

Equation 2.49 is valid only when both terms in brackets are zero; hence,

$$dM - \left(\frac{\partial M}{\partial T}\right)_{P,x} dT - \left(\frac{\partial M}{\partial P}\right)_{T,x} dP - \sum \bar{M}_i dx_i = 0 \quad (2.50)$$

$$M = \sum x_i \bar{M}_i \quad (2.51)$$

The last equation is important because it states that a property of a mixture may be calculated from partial molar properties weighted by molar fractions. By differentiation, one obtains

$$dM = \sum x_i d\bar{M}_i + \sum \bar{M}_i dx_i \quad (2.52)$$

Replacing dM into the previous equation leads to the following relation:

$$\left(\frac{\partial M}{\partial T}\right)_{P,x} dT + \left(\frac{\partial M}{\partial P}\right)_{T,x} dP - \sum (x_i d\bar{M}_i) = 0 \quad \textit{Gibbs–Duhem equation} \quad (2.53)$$

The relation of Equation 2.53 is valid for any thermodynamic property in a homogeneous phase. At constant temperature and pressure, it leads to the simplified Gibbs–Duhem equation, which is widely used, for instance, in deriving models for LACT coefficients.

$$\sum (x_i d\bar{M}_i) = 0 \quad \text{Simplified } \textit{Gibbs–Duhem equation} \quad (2.54)$$

2.2.5 Network of Thermodynamic Properties

Thermodynamic state functions (such as U, H, S, A, G) have the property that the variation between two states depends only on the *state variables* (P, V, T) characterizing the initial and the final state, but independent of the (*real* or *hypothetical*) path followed between these states. The differential of such a state function is *exact*. Considering the function $z = f(x,y)$, one may write

$$dz = \left(\frac{\partial z}{\partial x}\right)_y dx + \left(\frac{\partial z}{\partial y}\right)_x dy = M dx + N dy \quad (2.55)$$

The function z is an *exact* differential if the following relation holds:

$$\left(\frac{\partial M}{\partial y}\right)_x = \left(\frac{\partial N}{\partial x}\right)_y \tag{2.56}$$

Not only P, V, T, but also *entropy S* may be considered state variables. Useful relations between the state variables (P, V, T, S) and state functions (U, H, A, G) can be obtained as follows:

$$T = \left(\frac{\partial U}{\partial S}\right)_V = \left(\frac{\partial H}{\partial S}\right)_P \tag{2.57}$$

$$-P = \left(\frac{\partial U}{\partial V}\right)_S = \left(\frac{\partial H}{\partial V}\right)_T \tag{2.58}$$

$$V = \left(\frac{\partial H}{\partial P}\right)_S = \left(\frac{\partial G}{\partial P}\right)_T \tag{2.59}$$

$$-S = \left(\frac{\partial A}{\partial T}\right)_S = \left(\frac{\partial G}{\partial T}\right)_P \tag{2.60}$$

The derivatives of the state variables are interrelated by the *Maxwell equations* as follows:

$$\left(\frac{\partial T}{\partial V}\right)_S = -\left(\frac{\partial P}{\partial S}\right)_V \tag{2.61}$$

$$\left(\frac{\partial T}{\partial P}\right)_S = -\left(\frac{\partial V}{\partial S}\right)_P \tag{2.62}$$

$$\left(\frac{\partial P}{\partial T}\right)_V = \left(\frac{\partial S}{\partial V}\right)_T \tag{2.63}$$

$$\left(\frac{\partial V}{\partial T}\right)_P = -\left(\frac{\partial S}{\partial P}\right)_T \tag{2.64}$$

The assembly of these relations and the *fundamental equations* defines the *thermodynamic network*. These relations are the starting point for finding generalized computational methods for nonmeasurable thermodynamic properties from other measurable properties.

2.2.6 Fugacity

Fugacity f of a real gas is an effective pressure that replaces the true pressure in accurate chemical equilibrium calculations, and it is equal to the pressure of an *ideal gas* that has the same chemical potential as the *real gas*. The concept of fugacity was introduced initially to account for the nonideal behavior of real gases. Later, the concept was generalized to phase equilibrium calculations. At constant temperature, $dG = VdP$, and for an ideal gas, $dG^{ig} = V^{ig}dP = RT/P\,dP$, thus leading to

$$dG^{ig} = RTd\ln P \tag{2.65}$$

Equation 2.65 suggests a similar relation for a real fluid, but the pressure would be substituted by a more general property. By definition, the variation of Gibbs free energy can be linked with a thermodynamic property of a real fluid (called *fugacity*) by the following equation:

$$dG = RTd\ln f \tag{2.66}$$

Notably, fugacity has the meaning of a real pressure that a fluid would have when obeying different thermodynamic changes. Fugacity is in general different from the external (measurable) pressure. The reference state is 1 atm, where Gibbs free energy is a function only of temperature $G^0(T)$. The integration between the state of a real fluid and the reference state leads to the expression

$$G(T, P) = G^0(T) + RT\ln(f/1\text{ atm}) \text{ (Applicable to gases, liquids and solids)} \tag{2.67}$$

A better perception of the fugacity concept can be obtained by relating it to pressure:

$$d(G - G^{ig}) = dG^R = RTd\ln(f/P) = RT\,d\ln\phi \tag{2.68}$$

By introducing an important class of thermodynamic functions (*residual* or *departure functions*), the *residual Gibbs free energy* G^R may be defined. The *fugacity coefficient* ϕ is also introduced.

$$G^R(T, P) = G(T, P) - G^{ig}(T, P) \tag{2.69}$$

$$\phi = f/P \tag{2.70}$$

At low pressures, fugacity is the same as pressure: $f = P$ or $\phi = 1$. Another interpretation of fugacity may come by integration between a state of an ideal gas ($G^R = 0$ and $\phi = 1$) and a real state:

$$\frac{G^R}{RT} = \ln\phi \qquad (2.71)$$

Hence, the fugacity coefficient ϕ may be seen as a measure of the *residual Gibbs free energy*. Equation 2.71 shows also how to compute a nonmeasurable thermodynamic function (Gibbs free energy) via an indirect measurable physical property (fugacity). In practice, the opposite approach is used, so by appropriate modeling of G^R, the fugacity may be calculated in various conditions of temperature and pressure, reducing considerably the experimental efforts (Dimian et al., 2014).

The fugacity $\hat{f}_i$ of a component in a mixture is defined by the equation

$$d\mu_i = d\overline{G}_i = RTd\ln\hat{f}_i \;(\text{constant T}) \qquad (2.72)$$

Note that μ_i designates the chemical potential of the component i, identical with the partial Gibbs free energy $\left(\overline{G}_i\right)$. The circumflex sign emphasizes that $\hat{f}_i$ is a property in a mixture but not a partial property. The corresponding equation for an ideal gas is

$$d\mu_i^{ig} = d\overline{G}_i^{ig} = RTd\ln(Py_i)\,(\text{constant T}) \qquad (2.73)$$

Consequently, the *fugacity coefficient in a mixture* is defined as

$$\hat{\phi}_i = \frac{\hat{f}_i}{y_iP} \qquad (2.74)$$

$$\ln\hat{\phi}_i = \left(\overline{G}_i - \overline{G}_i^{ig}\right)/RT = \overline{G}_i^R/RT \qquad (2.75)$$

Now, the quantity $\left(\ln\hat{\phi}_i\right)$ is a partial property; hence, the following relations can be written:

$$\ln\hat{\phi}_i = \left[\frac{\partial(n\ln\phi)}{\partial n_i}\right]_{P,T} \qquad (2.76)$$

$$\ln\phi = \sum x_i \ln\hat{\phi}_i \;(\text{at constant } P,T) \qquad (2.77)$$

$$\sum x_i d\ln\hat{\phi}_i = 0 \;(\text{at constant } P,T) \qquad (2.78)$$

The above Equation 2.78 is a form of the Gibbs–Duhem equation with $\ln \hat{\phi}_i$ as a partial property. Fugacity is a property intensively used in engineering, the most important being the chemical equilibrium of gases at high pressures and VLE. Thus, the main problem is how to calculate fugacity from directly measurable quantities, such as pressure, volume, temperature and composition.

2.2.6.1 Fugacity Calculation for Pure Components

Considering a process at constant temperature, the following relation for Gibbs free energy can be formulated based on the previous equations (Dimian et al., 2014):

$$dG = RTd \ln f = VdP \tag{2.79}$$

By adding and subtracting $RTd\ln P$, one obtains

$$RTd \ln \frac{f}{P} = \left(V - \frac{RT}{P} \right) dP = d(G - G^{ig}) \tag{2.80}$$

The integration between zero pressure (ideal gas) and the actual pressure leads to

$$RT \ln \frac{f}{P} = \int_{0}^{P} \left(V - \frac{RT}{P} \right) dP = G - G^{ig} \tag{2.81}$$

An immediate result is that the fugacity can be calculated with any PVT relationship. Replacing here $Z = PV/RT$ (*compressibility factor*) leads to a formulation based on compressibility:

$$\ln \phi = \int_{0}^{P} (Z-1) \frac{dP}{P} \tag{2.82}$$

Changing the integration variable from P to V leads to another useful expression in practice:

$$\ln \frac{f(T,P)}{P} = \ln \phi = \frac{1}{RT} \int_{V=\infty}^{V=ZRT/P} \left[\frac{RT}{V} - P \right] dV - \ln Z + (Z-1) \tag{2.83}$$

Of interest is the fugacity dependency with pressure and temperature. From the definition, one obtains

$$RT \left(\frac{\partial \ln f}{\partial P} \right)_T = \left(\frac{\partial G}{\partial P} \right)_T = V \tag{2.84}$$

Consequently, the pressure dependency of fugacity is much higher for a gas than for a liquid because of the difference in phase volumes. The variation of fugacity with the temperature can be found by examining the derivative of the equation

$$\frac{\partial}{\partial T}\left(\ln\frac{f}{P}\right)_P = \frac{\partial}{\partial T}\left[\frac{G-G^{ig}}{RT}\right]$$

(2.85)

Taking into account that $G = H - TS$ and $(\partial G/\partial T)_P = -S$, the following relation holds true:

$$\frac{\partial}{\partial T}\left(\ln\frac{f}{P}\right)_P = \left[\frac{H-H^{ig}}{RT^2}\right]$$

(2.86)

2.2.6.1.1 Gases   The basic relation for gases is the following equation:

$$\ln\frac{f^V(T,P)}{P} = \ln\phi = \frac{1}{RT}\int_{V=\infty}^{V=Z^V RT/P}\left[\frac{RT}{V}-P\right]dV - \ln Z^V + (Z^V - 1)$$

(2.87)

For an *ideal gas*, $Z^V = 1$ and $f^V(T, P) = P$; hence, at low pressure the fugacity of a pure species equals the total pressure. However, at moderate pressures one may consider the *virial* EOS (described in Section 2.3.1). Using the form with the second coefficient $Z = 1 + B(T)/V$ leads to the following simple expression:

$$\ln\phi = \frac{BP}{RT}$$

(2.88)

2.2.6.1.2 Liquids   The fugacity of a pure liquid component is close but not identical to its vapor pressure. Assuming that the liquid is incompressible and integrating at constant temperature between the saturation pressure P^{sat} and the system pressure P gives the relation

$$RT\ln\frac{f^L}{f^{L,sat}} = V^L(P-P^{sat})$$

(2.89)

V^L is the liquid molar volume that may be taken as equal with the liquid volume in saturation conditions at P^{sat}. Replacing $f^{L,sat} = \phi^{sat}P^{sat}$ leads to following equation:

$$f^L = \phi^{sat}P^{sat}\exp\left(\frac{V^L(P-P^{sat})}{RT}\right)$$

(2.90)

The exponential term gives the influence of pressure on liquid fugacity, and it is known as the *Poynting correction*. At moderate pressures, the Poynting correction is negligible, but it becomes important at high pressures (e.g. super-critical fluids). At equilibrium, the component fugacity in liquid and vapor phases must be equal. It is also important to note that in calculating fugacity coefficients with a cubic EOS, the same expression holds for both vapor and condensed phase.

$$f^L = f^V \tag{2.91}$$

2.2.6.2 Fugacity Calculations for Mixtures

The computation of component fugacity in a mixture follows the same conceptual path as for pure species, but with a notable difference: The influence of composition must be accounted for. This aspect is taken into consideration by means of *mixing rules* for the parameters entering in the EOS model. The geometric mixing rules do not have a theoretical basis, but they are suitable for applications involving hydrocarbons for which the components are of comparable size, although still not accurate enough for mixtures with strong nonideal behavior. More complex mixing rules, as well as predictive methods for nonideality with EOS models, are presented in subsequent sections of this chapter.

2.3 EQUATIONS OF STATE

An *equation of state* is an equation correlating the fundamental state variables of a fluid (P, V, T), thus describing the state of matter under a given set of physical conditions. The simplest EOS is the *ideal gas law*, given by

$$PV = RT \tag{2.92}$$

The models based on EOS are the most used in simulations as they allow a full calculation of both thermodynamic properties and phase equilibrium with minimum data. EOS models are applied not only to nonpolar (hydrocarbon) mixtures, but also to mixtures containing species of various chemical structures (including water and polar components) or even to solutions of polymers.

2.3.1 Virial Family of Equations of State

The nonideality of a gas can be expressed by the compressibility factor Z, defined as

$$Z = PV/RT \tag{2.93}$$

2.3.1.1 Virial EOS

Virial EOS expresses the compressibility factor Z as a power series of molar volume or pressure:

$$Z = \frac{PV}{RT} = 1 + \frac{B}{V} + \frac{C}{V^2} + \ldots \tag{2.94}$$

$$Z = \frac{PV}{RT} = 1 + B'P + C'P^2 + \dots \tag{2.95}$$

where the *virial coefficients B, C*, or *B', C'* are functions only of temperature. The virial EOS has a theoretical basis in statistical mechanics: For example, the *B/V* term accounts for interactions between pairs of molecules, *C/V²* describes three molecule interactions, and so on. The form $Z = 1 + B/V$ (known as the *second virial coefficient*) has many applications as it is sufficient for technical computations, but it is restricted to gas phase calculations at moderate pressures up to 20 bar. The virial EOS is able to handle a variety of chemical classes, including polar species.

Hayden and O'Connell (1975) have proposed correlations for the second virial parameter. The method is predictive, as it considers only physical data (e.g. dipole moment, critical temperature, critical pressure, degree of association between the interacting components).

2.3.1.2 Benedict, Webb, and Rubin EOS

Proposed by Benedict, Webb, and Rubin (1940), the BWR-EOS is a more sophisticated form, a complex dependency of pressure versus volume (Poling, Prausnitz and O'Connel, 2001):

$$P = \frac{RT}{V} + \frac{B_0 RT - A_0 - C_0/T^2}{V^2} + \frac{bRT - a}{V^3} + \frac{\alpha a}{V^6} + \frac{c}{V^3 T^2}\left(1 + \frac{\gamma}{V^3}\right)\exp\frac{-\gamma}{V^2} \tag{2.96}$$

The constants A_0, B_0, C_0, a, b, c, α, β and γ are characteristic for a given fluid. The BWR-EOS and its recent modifications allow accurate calculations of physical properties of light gases and nonpolar or slightly polar components at medium pressure. The BWR-EOS can be used for both thermodynamic properties (e.g. enthalpy and entropy) and phase equilibrium. It is also suited for gas liquefaction and processes involving light hydrocarbons (e.g. rich methane and hydrogen mixtures). However, the BWR equation may contain sophisticated terms with a large number of parameters, mostly between 10 and 20, and needs substantial experimental data for tuning.

2.3.1.3 Benedict–Webb–Rubin–Lee–Starling EOS

The *Benedict–Webb–Rubin–Lee–Starling* (BWR-LS) EOS is an *extended virial-type EOS* formulation still used for special applications (e.g. gas processing and liquefaction), being one of the most accurate for hydrogen-rich hydrocarbon mixtures (Brulé et al., 1982). Remarkably, extended virial EOSs may calculate not only volumetric properties but also VLE.

2.3.1.4 Lee–Kesler Method

The *Lee–Kesler* (LK) method allows the estimation of the saturated vapor pressure at a given temperature for all components for which the critical pressure/temperature and acentric factor are known. The compressibility factor is described by the three-parameter corresponding states correlation:

$$Z = Z^{(0)} + (Z^{(r)} - Z^{(0)})\frac{\omega}{\omega^{(r)}} \qquad (2.97)$$

The contributions $Z^{(0)}$ and $Z^{(r)}$ are represented by generalized functions having as parameters the reduced temperature and pressure, obtained using a special form of the BWR-EOS. Mixture critical parameters and acentric factor are calculated by means of mixing rules, which do not have interaction parameters. Tables of values for hand calculations may be found in the work of Reid et al. (1987), while graphical representations are presented in Perry and Green (1997). The LK method can be used to compute phase properties (specific volume, enthalpy, entropy) for both vapor and liquid phases. It has been accepted as an accurate option for enthalpy and entropy of hydrocarbons and slightly polar components. The LK method has been extended to oil and petrochemical-type mixtures (e.g. hydrocarbons and alcohols with CO_2, H_2, CH_4 and H_2S), being known as the *Lee–Kesler–Ploecker* (LKP) method (Ploecker et al., 1978). Special mixing rules are designed to describe both symmetric (nonpolar) and symmetric (polar) molecules (Poling et al., 2001).

2.3.2 Cubic Equations of State

2.3.2.1 van der Waals EOS

Proposed by van der Waals (1873), the van der Waals EOS is a landmark equation, being one of the first to perform better than the ideal gas law. This is important historically as it is also the basis for other cubic EOS formulations of slightly greater complexity but much higher accuracy. It is called cubic EOS because it can be rewritten as a cubic function of the molar volume.

$$P = \frac{RT}{V-b} - \frac{a}{V^2} \text{ (van der Waals equation)} \qquad (2.98)$$

$$a = 27R^2T_c^2/64P_c \quad \text{and} \quad b = RT_c/8P_c \qquad (2.99)$$

where the term a/V^2 takes into account the increasing attractive forces between molecules at higher pressure, and b is a measure of the covolume of molecules at infinite pressure. The parameters a and b depend only on the critical pressure and critical temperature, as described previously.

2.3.2.2 Redlich and Kwong EOS

Proposed by Redlich and Kwong (1949), the Redlich and Kwong (RK) EOS is a modified form convenient for describing the properties of hydrocarbon mixtures. The RK-EOS model is represented by the relation

$$P = \frac{RT}{V-b} - \frac{a}{T^{1/2}V(V+b)} \qquad (2.100)$$

The RK-EOS introduced an important modification in the formulation of the van der Waals equation, namely, the temperature dependency of the a parameter. Note that the correction factor $\alpha(T)$ of the a parameter is often designated as the *alpha function*. For the RK-EOS, this factor is $\alpha(T) = 1/\sqrt{T}$.

$$a(T) = a_c(T_c, P_c)\alpha(T) \tag{2.101}$$

2.3.2.3 Soave–Redlich–Kwong EOS

The Soave–Redlich–Kwong (SRK) EOS was proposed much later by Soave (1972) as a new modification; the model incorporates another important molecular parameter, the *acentric factor* (ω). The accuracy of VLE calculations thus improved considerably. In the case of SRK-EOS, the alpha function is given by the following relation, where factor κ is a polynomial function of ω, $\kappa = f(\omega)$.

$$\alpha^{0.5} = 1 + \kappa\left(1 - T_r^{0.5}\right) \tag{2.102}$$

2.3.2.4 Peng and Robinson EOS

The PR-EOS is another cubic EOS proposed by Peng and Robinson (1976) that has proved similar accuracy as the SRK-EOS but has more robustness near the critical point. The PR-EOS is also generally superior in predicting the liquid densities of many materials, especially nonpolar ones. The PR-EOS has the following form, where the function α is the same as in the SRK-EOS but with different dependency of κ:

$$P = \frac{RT}{V-b} - \frac{a(T)}{V(V+b)+b(V-b)} \tag{2.103}$$

2.3.2.5 Cubic Plus Association EOS

The *cubic plus association* EOS (CPA-EOS) was proposed by Kontogeorgis et al. (1996). Strongly dipolar molecules (e.g. acetone) or quadrupolar molecules (e.g. CO_2) are treated using the concept of *pseudoassociation*, that is, as being able to act as associating compounds. The use of this concept in CPA provides good results for their mixtures with non-polar, polar and associating compounds. This approach renders the CPA-EOS applicable to such compounds without the need for extra terms to account for polar or quadrupolar interactions. In the CPA-EOS, the compressibility factor has a physical term Z^{ph} that includes all the usual repulsive and attractive forces, calculated from the cubic SRK-EOS or PR-EOS, while the association term Z^{assoc} accounts for specific interactions (self- and cross-association), and it is calculated from the Wertheim perturbation theory.

$$Z^{CPA} = Z^{ph} + Z^{assoc} \tag{2.104}$$

$$Z^{ph} = \frac{V}{V-b} - \frac{aV}{RT[V(V+b)+b(V-b)]} \tag{2.105}$$

$$Z^{assoc} = -\frac{1}{2}\left(1+\rho\frac{\partial(\ln g)}{\partial\rho}\right)\sum_i\sum_{A_i}x_i d(A_i)(1-X^{A_i}); \ X^{A_i} = \left(1+\rho\sum_i\sum_{B_j}x_j X^{B_j}\Delta^{A_iB_j}\right)^{-1} \quad (2.106)$$

where g is the hard sphere radial distribution function, A_i denotes association site A on component i, and X^{Ai} is the mole fraction of molecules i not bonded at site A, while $\Delta^{A_iB_j}$ is the association strength between sites A and B of the associating molecules.

Practically, the CPA-EOS reduces to PR-EOS with fitted T_c, P_c and ω parameters for mixtures that contain nonassociating compounds. It provides satisfactory correlation results for binary nonpolar mixtures as well as predictions for multicomponent mixtures using binary interaction parameters (BIPs). For mixtures containing associating compounds, the CPA-EOS has been applied with remarkable success in phase equilibrium calculations for a great variety of binary and multicomponent systems by many research groups (Kontogeorgis and Folas, 2010).

Cubic EOS methods are currently standard options in process simulators due to their simplicity and robustness in calculating phase equilibrium and enthalpy- or entropy-derived properties, while distinguishing between monophase and two-phase systems. New modifications and mixing rules extended the applications to all nonelectrolytic systems, including supercritical gases, hydrocarbon, polar components and small and large molecules, over a large range of pressures and temperatures. As further reading in this field, the review of Sandler et al. (1993) and the monograph on EOS edited by the International Union of Pure and Applied Chemistry Commission on Thermodynamics (Sengers et al., 2000) are recommended.

2.3.2.6 Nonpolar Mixtures

The origin of the cubic EOS goes back to the landmark equation of van der Waals, which corrects the ideal gas law by an *attraction term* on pressure and a *repulsion term* on volume. The van der Waals equation is explicit in pressure and implicit (cubic) in volume, and it contains two parameters, a and b, which can be expressed as a function of T_c and P_c:

$$\left(p+\frac{a}{V^2}\right)(V-b) = RT \quad (2.107)$$

Parameters a and b can be found from conditions set for the critical point, as (1) maximum pressure and (2) the inflection point of the isotherms. Both are expressed mathematically by the conditions

$$\left(\frac{\partial P}{\partial V}\right)_{T_c} = \left(\frac{\partial^2 P}{\partial V^2}\right)_{T_c} = 0 \quad (2.108)$$

Modified van der Waals EOSs may be written in a general manner as follows:

$$P = \frac{RT}{V-b} + \Delta \text{ with } \Delta = \frac{\theta}{V^2+\delta V+\varepsilon} \quad (2.109)$$

TABLE 2.1 General Expression for Some Cubic Equations of State

Author (Year)	θ	δ	ε	Δ
van der Waals (1873)	a	0	0	$\dfrac{a}{V^2}$
Redlich–Kwong (1949)	$a/\sqrt{T}$	b	0	$\dfrac{a/\sqrt{T}}{V(V+b)}$
Soave (1972)	$\theta_S(T)$	b	0	$\dfrac{\theta_S(T)}{V(V+b)}$
Peng–Robinson (1976)	$\theta_{PR}(T)$	2b	$-b^2$	$\dfrac{\theta_{PR}(T)}{V(V+b)+b(V-b)}$
Patel–Teja (1982)	$\theta_{PT}(T)$	b + c	$-cb$	$\dfrac{\theta_{PT}(T)}{V(V+b)+c(V-b)}$

Table 2.1 presents the most representative cubic EOS formulations. Note that an important feature is the temperature dependency of parameters, as shown in Table 2.2. Parameter a can be expressed as a product of its value at the critical point a_c and a dimensionless correction function $\alpha(T)$.

Note that a cubic EOS can be written as a third-degree polynomial of the compressibility factor, where the coefficients α, β and γ can be expressed as functions of the dimensionless parameters A and B (which depend on the particular cubic EOS):

$$Z^3 + \alpha Z^2 + \beta Z + \gamma = 0 \qquad (2.110)$$

TABLE 2.2 Parameters for SRK and PR Equations of State

Parameter	Soave–Redlich–Kwong	Peng–Robinson
u	1	2
w	0	−1
α	−1	$-1 + B$
β	$A - B - B^2$	$A - 3B^2 - 2B$
γ	$-AB$	$-AB + B^2 + B^3$
a	$\dfrac{0.42748R^2T_c^2}{P_cT^{0.5}}[1+\kappa(1-T_r^{0.5})^2]$	$\dfrac{0.45724R^2T_c^2}{P_cT^{0.5}}[1+\kappa(1-T_r^{0.5})^2]$
	$\kappa = 0.48 + 1.574\omega - 0.176\omega^2$	$\kappa = 0.37464 + 1.54226\omega - 0.26992\omega^2$
b	$\dfrac{0.0866RT_c}{P_c}$	$\dfrac{0.07780RT_c}{P_c}$

Note: Where $A = aP/(RT)^2$ and $B = bP/(RT)$.

Taking into account the general formulation, the fugacity coefficients for a pure species (calculated by SRK-EOS or PR-EOS) may be expressed by a single relation:

$$\ln\phi = (Z-1) - \ln(Z-B) + \frac{A}{B\sqrt{u^2 - 4w}} \ln\frac{2Z + B(u + \sqrt{u^2 - 4w})}{2Z + B\sqrt{u^2 - 4w}} \tag{2.111}$$

The same expression holds for vapor and liquid phases, but Z is different. As mentioned, a weakness of a cubic EOS consists of an inaccurate computation of liquid density. A significant improvement has been obtained by the introduction of a new parameter c_m on volume (Peneloux et al., 1982). Several modified EOSs have adopted the idea of volume translation (e.g. volume-translated Peng–Robinson or *VTPR*; Ahlers and Gmehling, 2001).

For SRK-EOS, the volume correction is: $V^L = V^L(\text{SRK}) - c_m$, while parameter c_m may be estimated as

$$c_m = 0.40768(0.29441 - Z_{RA})\frac{RT_c}{P_c} \tag{2.112}$$

$$Z_{RA} = 0.29056 - 0.08775\omega \text{ (Rackett compressibility factor)} \tag{2.113}$$

The *Rackett method* is notably one of the most accurate for calculating liquid densities (Reid et al., 1987):

$$V^{l,sat} = \frac{RT_c}{P_c} Z_{RA}^{[1+(1-T_r)^{2/7}]} \tag{2.114}$$

2.3.2.7 Polar Fluids

The first EOS dealing with highly polar components was proposed by Soave (1979, 1984). The SRK2-EOS modification consists of two adjustable parameters in the alpha function:

$$\alpha(T) = 1 + m(1 - T_r) + n(1/T_r - 1) \tag{2.115}$$

The parameters m and n, characteristic for each pure component, can be determined by regression of experimental vapor pressure data. The accuracy of SRK2 is typically improved by an order of magnitude with respect to the classical SRK over a broad range of chemical classes.

2.3.2.8 Peng–Robinson–Stryjek–Vera EOS

The Peng–Robinson–Stryjek–Vera (PRSV) EOS is a modified formulation proposed by Stryjek and Vera (1986) and is based on the PR-EOS. In the simplest modification, the factor α is still given by the original equation, but T dependency is

$$\alpha^{0.5} = k_0 + k_1 \left(1 - T_r^{0.5}\right)(0.7 - T_r) \tag{2.116}$$

The parameter k_0 is a general function of the acentric factor ω, while k_1 is a regressed parameter. The error in correlating vapor pressure is under 1%, typically between 0.2% and 0.3%. The accuracy of PRSV is by an order of magnitude better than PR-EOS, but the same as the *Antoine* equation, while SRK2 and PRSV are of comparable accuracy. Another form (*PRSV2*) is even more accurate, but it has three adjustable parameters k_0, k_2, k_3.

Other EOS formulations for polar components use variations of the alpha function, as for example:

- Mathias and Copeman (1983)

$$\alpha(T) = [1 + \kappa\left(1 - T_r^{0.5}\right) + c_2\left(1 - T_r^{0.5}\right) + c_3\left(1 - T_r^{0.5}\right)]^2 \tag{2.117}$$

- *Schwartzentruber, Renon and Watanasiri* (1990), also known as SR-POLAR. The parameter κ can be calculated with either SRK-EOS or PR-EOS. Parameters c_2 and c_3 and p_1, p_2, p_3 must be identified from experimental data.

$$\alpha(T) = [1 + \kappa\left(1 - T_r^{0.5}\right) - \left(1 - T_r^{0.5}\right)\left(p_1 + p_2 T_r + p_3 T_r^2\right)]^2 \tag{2.118}$$

The ability of a cubic EOS to accurately describe the physical properties and behavior of a fluid depends on both the alpha function and the mixing rules. There are several benefits of using cubic EOS formulations, as EOSs enable process simulations with a minimum of input data (for both phase equilibrium and energy balance purposes) and can handle mixtures of various chemical species, from nonpolar to very polar molecules (but the mixing rules should be carefully selected). Notably, EOS models allow a unified treatment of both supercritical and subcritical components, removing the inconvenience of using Henry constants and asymmetric convention *in computing K* values (described in the section about GLE).

However, there are also some drawbacks of using cubic EOS, as a cubic EOS cannot predict all the properties with equal accuracy (e.g. there are errors in estimating liquid volume, which may lead to errors in computing enthalpies), the VLE and LLE calculations at low pressure and highly nonideal mixtures are by far less accurate than LACT models, and mixtures with large asymmetry in shape and exhibiting strong interactions cannot be accurately handled (Dimian et al., 2014).

2.3.3 EOS Based on Molecular Modeling

Applications of molecular modeling techniques regard the thermodynamic properties of mixtures containing components with large differences in size, shape and associating bonds, such as mixtures involved in gas and oil production, polymer technology and bio-technologies. The theoretical basis can be found in specialized monographs, such as by Kontogeorgis and Folas (2010).

2.3.3.1 Statistical Associating Fluid Theory EOS

Statistical associating fluid theory (SAFT) EOS was proposed by Chapman et al. (1989), who considered the molecules as formed by *chains of equal spherical segments* connected by covalent bonds and associating by hydrogen bonding – instead of modeling the molecules as objects more or less deviating from a spherical shape, as for the case with classical cubic EOS. Note that a segment is not necessarily a well-defined molecular group, but rather a zone of molecular interactions. The SAFT model aims to handle complex mixtures (e.g. small molecules and long-chain molecules, such as monomers and polymers), as well as with associating species. The starting point is the residual Helmholtz free energy, which is described as the sum of three contributions:

$$A^{res} = A^{seg} + A^{chain} + A^{assoc} \tag{2.119}$$

where A^{seg} is segment-segment interactions, A^{chain} is covalent bonds between segments in chains, and A^{assoc} is the association through hydrogen bonding. The segment contribution can be decomposed as follows in terms for hard-sphere repulsion and attractive/dispersion interactions:

$$A^{seg} = m(A^{hs} + A^{disp}) \tag{2.120}$$

The parameter m is the number of segments in the molecule. The terms in the equation are described by complex equations based on statistical thermodynamics. The SAFT model requires fitting only three parameters from vapor pressure and liquid density data: segment number, segment volume and segment energy.

2.3.3.2 Perturbed-Chain Statistical Associating Fluid Theory EOS

Perturbed-chain statistical associating fluid theory (PC-SAFT), developed by Gross and Sadowski (2001), is a significant improvement of SAFT. The molecules are seen as chains of freely jointed spherical segments that can host association sites and polar groups. Unlike the SAFT model, where the reference fluid was formed by distinct spherical segments, in PC-SAFT the reference is the chain of segments. Thus, the perturbation theory applies to the chains of segments rather than to disconnected segments. The residual Helmholtz energy is expressed as a reference term of a hard-chain fluid A^{hc} plus contributions due to dispersion A^{disp} and association A^{assoc} as well as to dipolar A^{DD} and quadrupolar interactions A^{QQ}:

$$A^{res} = A^{hc} + A^{disp} + A^{assoc} + A^{DD} + A^{QQ} \qquad (2.121)$$

The key difference between SAFT and PC-SAFT consists of the way of dealing with the dispersion term (Gross and Sadowski, 2001). PC-SAFT similarly requires three pure-component parameters for nonassociating and nonpolar components: segment number, segment diameter and segment energy, which can be regressed from vapor pressure and densities data. For associating molecules, two additional parameters are needed to account for interactions between the association sites. Further refinement has been developed for multipolar contributions and copolymers, as well as for electrolytes. One-fluid mixing rules are used with a k_{ij} interaction parameter to be fitted on binary equilibrium data, although in some cases even setting $k_{ij} = 0$ gives good results (Dimian et al., 2014).

PC-SAFT can indeed solve difficult problems in phase equilibrium (e.g. for highly asymmetric mixtures of small and large molecules with CO_2 and H_2, for polymer and copolymer systems with light gases or solvents, for estimating the solubility of gases and solids in various solvents) (Tumaka et al., 2005). Nevertheless, the reader should be aware of numerical pitfalls that may lead to unrealistic results or the failure of simulation (Privat et al., 2010).

2.4 *PVT* BEHAVIOR AND RELATIONSHIPS

2.4.1 Phase Diagrams

The behavior of a pure species that can exist as solid, liquid or vapor can be conveniently illustrated in a pressure-temperature diagram, as shown in Figure 2.6. Three types of

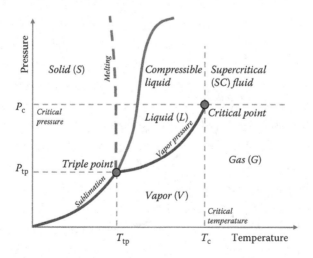

FIGURE 2.6   Generalized phase diagram.

two-phase equilibrium are possible: solid-liquid, vapor-liquid and solid-vapor. The *phase rule* gives the dependence of the degree of freedom F versus number of components C and phases at equilibrium P: $F = C + 2 - P$. At the *triple point*, the degree of freedom is zero ($F = 1 + 2 - 3 = 0$), so pressure and temperature cannot be used to modify the equilibrium; all three phases coexist. If only two phases can be found at equilibrium ($F = 1 + 2 - 2 = 1$), then either pressure or temperature can vary (Dimian et al., 2014).

Vapor-liquid equilibrium is basically the most important equilibrium in process engineering. The two phases coexist up to a point where it is difficult to make a distinction between vapor and liquid. This is known as the *critical point* and is characterized by the *critical parameters* P_c, T_c and V_c. When $P > P_c$ and $T > T_c$, then the state of the fluid is *supercritical*, meaning that the fluid behaves as a single-phase superdense gas or light liquid. Consequently, small variations in the parameters, pressure or temperature lead to large variations in density. The change in diffusion properties is also remarkable; for example, supercritical water behaves as an organic solvent.

When the pressure is represented as a function of the molar specific volume of a pure component at constant temperature, the isotherms have the aspect given in Figure 2.7 (Dimian et al., 2014), where the state variables (reduced values) are scaled by reference to the critical point:

$$T_r = \frac{T}{T_c}; \quad P_r = \frac{P}{P_c}; \quad V = \frac{V}{V_c} \tag{2.122}$$

Note that at the critical point the difference between the molar specific volumes of liquid and vapor approaches zero, suggesting that the energy needed for phase transition (enthalpy of vaporization) should reduce to zero. This illustrates in a simple manner the link between the PVT behavior of a fluid and the energy implied in its physical changes.

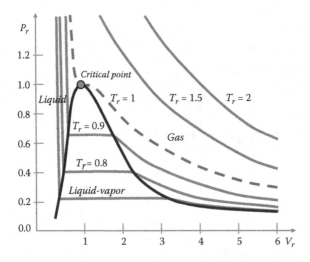

FIGURE 2.7   Pressure-volume curves using reduced variables.

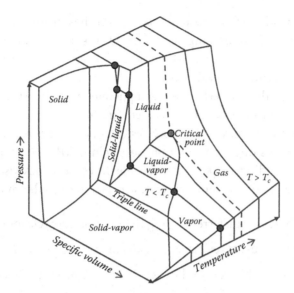

FIGURE 2.8   Generalized *PVT* diagram (three dimensional).

Figure 2.8 presents the generic *PVT* relationship for a pure component in a three-dimensional (3-D) diagram, which captures both the volumetric behavior and the temperature evolution of the phase boundaries at equilibrium. Based on this, the 2-D representation for different cases is rather straightforward.

2.4.2 Property Charts

Thermodynamic diagrams are valuable tools in designing processes based on thermodynamic calculations, such as heat pumps and refrigeration. These tools allow the designer to determine key properties (e.g. enthalpy, entropy, specific volume) and execute by simple graphical representations some essential calculations (e.g. adiabatic transformations). Large-scale charts are available for technically important substances, such as water, ammonia, freons, or CO_2. The most common property diagrams are *T-S* (temperature-entropy), *P-H* (pressure-enthalpy) and *H-S* (enthalpy-entropy).

Charts, tables and diagrams are available in handbooks (Green and Perry, 2008) or specialized publications (e.g. compilations of the US Bureau of Standards; *International Tables of the Fluid State*, Pergamon Press). These representations combine experimental data and correlation methods. Note that generic simulation methods could successfully replace the use of specialized charts and tables if the accuracy is properly managed by the selection of the most appropriate method, validation over the range of interest, and calibration of parameters wherever necessary.

2.4.3 Thermodynamic Cycles

A thermodynamic cycle consists of a linked sequence of processes that involve transfer of heat and work into and out of the system – while varying pressure, temperature, and other state variables within the system – with the eventual return of the system to its initial state.

Through a cycle, the working fluid (system) may convert heat from a warm source into useful work and dispose of the remaining heat to a cold sink, thereby acting as a heat engine. The cycle may be reversed and use work to move heat from a cold source and transfer it to a warm sink, thereby acting as a heat pump.

Power cycles are cycles that convert some heat input into mechanical work output, while *heat pump cycles* transfer heat from low to high temperatures using mechanical work as the input. Cycles composed entirely of quasi-static processes can operate as power or heat pump cycles by controlling the process direction. On a pressure-volume (*P-V*) diagram or *T-S* diagram, the clockwise direction indicates a power cycle, while the counterclockwise a heat pump cycle. Note that the following thermodynamic processes are possible in a thermodynamic cycle:

- *Adiabatic*: No heat is transferred during this part of the cycle ($\delta Q = 0$).
- *Isothermal*: The process takes place at a constant temperature (T = constant, $\delta T = 0$).
- *Isobaric*: Pressure in this part of the cycle remains constant (P = constant, $\delta P = 0$).
- *Isochoric*: The process takes place at constant volume (V = constant, $\delta V = 0$).
- *Isentropic*: The process is one of constant entropy (S = constant, $\delta S = 0$).

The most common power cycles used to model *internal combustion engines* (ICEs) are the Otto cycle (gasoline engines) and Diesel cycle (diesel engines), while cycles for *external combustion engines* include the Brayton cycle (gas turbines) and Rankine cycle (steam turbines). As discussed previously, Carnot showed that the maximum work from a heat engine is given by a cycle formed by two reversible adiabates and two isotherms (Figure 2.4). For example, the well-known Otto cycle used in ICEs – and illustrated in Figure 2.9 – consists of

- 1 → 2: Isentropic expansion: constant entropy (S), while $P\downarrow$, $V\uparrow$ and $T\downarrow$
- 2 → 3: Isochoric cooling: constant volume (V), while $P\downarrow$, $S\downarrow$ and $T\downarrow$
- 3 → 4: Isentropic compression: constant entropy (S), while $P\uparrow$, $V\downarrow$ and $T\uparrow$
- 4 → 1: Isochoric heating: constant volume (V), while $P\uparrow$, $S\uparrow$ and $T\uparrow$

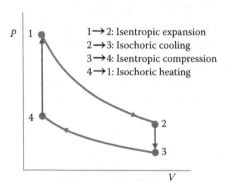

FIGURE 2.9   Thermodynamic cycle: Otto cycle (heat engine).

In practice, simple idealized thermodynamic cycles are usually made of four thermodynamic processes (e.g. isothermal, isobaric, isochoric, isenthalpic, isentropic or reversible adiabatic). Some examples of thermodynamic cycles and their constituent processes are given in Table 2.3.

Heat pump cycles are used to model heat pumps and refrigeration machines (both move heat from a cold space to a warm space). A heat pump is practically a heat engine that is operating in reverse. As heat cannot spontaneously flow from a colder location to a hotter one, work is required to achieve this task. The most common heat pump cycles are the vapor-compression (VC) cycle (systems using refrigerants that change phase), absorption-refrigeration (AR) cycle (refrigerant is absorbed in a liquid rather than compressed), gas cycles (reversed Brayton, Hampson–Linde cycle) and Stirling cycle types. More details about the heat pump cycles are given in the dedicated subsequent chapters.

The *vapor-compression cycle* (shown in the *T-S* diagram of Figure 2.10) is used the most in heat pumps and refrigeration machines. A working fluid (e.g. ammonia) enters the compressor as vapor that is isentropically compressed and exits as superheated vapor, which goes through a condenser that first cools and removes the superheat and then condenses the vapor into a liquid by removing more heat at constant pressure and temperature. The liquid goes through an expansion (throttle) valve where pressure decreases, causing flash evaporation and autorefrigeration of the liquid. The resulting mixture of liquid and vapor, at a lower temperature and pressure, travels then through an evaporator, where it is completely vaporized by cooling the space being refrigerated. The resulting vapor restarts the thermodynamic cycle.

The *absorption-refrigeration cycle* is similar to the VC cycle except for the method of raising the pressure of vapor: The compressor is replaced by an absorber that dissolves the refrigerant in a suitable liquid; a liquid pump raises the pressure, and a generator drives off the vapor from the high-pressure liquid on heat addition. In the AR cycle, a suitable combination of refrigerant and absorbent is used, such as ammonia-water and water-LiBr.

TABLE 2.3   Examples of Thermodynamic Cycles and Constituent Processes

Thermodynamic Cycle	Process 1–2: Compression	Process 2–3: Heat Addition	Process 3–4: Expansion	Process 4–1: Heat Removal
Power cycles with external combustion				
Bell Coleman (Joule-Brayton cycle)	Isentropic	Isobaric	Isentropic	Isobaric
Carnot (heat engine)	Isentropic	Isothermal	Isentropic	Isothermal
Ericsson (second Ericsson cycle)	Isothermal	Isobaric	Isothermal	Isobaric
Rankine (steam engine)	Isentropic	Isobaric	Isentropic	Isobaric
Stirling (Stirling engine)	Isothermal	Isochoric	Isothermal	Isochoric
Power cycles with internal combustion				
Otto (gasoline engines)	Isentropic	Isochoric	Isentropic	Isochoric
Diesel (diesel engine)	Isentropic	Isobaric	Isentropic	Isochoric
Power cycles with external combustion				
Joule-Brayton (jet engines)	Isentropic	Isobaric	Isentropic	Isobaric

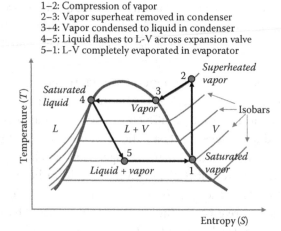

1–2: Compression of vapor
2–3: Vapor superheat removed in condenser
3–4: Vapor condensed to liquid in condenser
4–5: Liquid flashes to L-V across expansion valve
5–1: L-V completely evaporated in evaporator

FIGURE 2.10   Temperature-entropy (*T-S*) diagram of the vapor-compression cycle.

The *gas cycle* uses a gas (e.g. air) as the working fluid, such that when compressed and expanded no phase change occurs. As there is no intended condensation and evaporation, components corresponding to the condenser and evaporator in a VC are replaced by the hot and cold gas-to-gas heat exchangers. The gas cycle is less efficient than the VC cycle because the ideal cycle has a lower efficiency than Carnot. Note that the gas cycle works on the Joule–Brayton cycle (instead of the reverse Rankine cycle), so the working fluid does not receive and reject heat isothermally.

The *Stirling cycle* heat engine can be driven in reverse, using a mechanical energy input to drive the heat transfer in a reversed direction (as for example in a thermoacoustic heat pump or refrigerator). Several design configurations are possible for the Stirling cycle, some of them requiring rotary or sliding seals that can introduce trade-offs between the frictional losses and refrigerant leakage.

2.4.4 Departure Functions

The calculation of properties in computer simulations is largely based on generalized methods using the *PVT* relationship. Consider the calculation of enthalpy variation for a fluid when going from (T_1, P_1) to (T_2, P_2), as shown in Figure 2.11 (Dimian et al., 2014). Note that, as with any thermodynamic function, the variation is independent of the path followed.

A first one could be an isothermal compression at T_1 followed by an isobaric heating at P_2:

$$H_2 - H_1 = \int_{P_1}^{P_2} \left(\frac{\partial H}{\partial P} \right)_{T_1} dP + \int_{T_1}^{T_2} \left(\frac{\partial H}{\partial T} \right)_{P_2} dT \tag{2.123}$$

A second path could be an isobaric heating at P_1, followed by an isothermal compression at T_2. Other paths are also possible, and although all routes are feasible, they are not practical as the partial derivatives of enthalpy with pressure and temperature should be available at

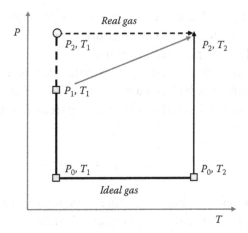

FIGURE 2.11 Paths for enthalpy computation.

different isotherms or isobars. One other way could be at sufficiently low pressure P_0, where the gas is ideal.

$$H_2 - H_1 = \int_{P_1}^{P_0} \left(\frac{\partial H}{\partial P} \right)_{T_1} dP + \int_{T_1}^{T_2} C_p^0 dT + \int_{P_0}^{P_2} \left(\frac{\partial H}{\partial P} \right)_{T_2} dP \qquad (2.124)$$

$$\Delta H = (H^0 - H_{P_1})_{T_1} + \int_{T_1}^{T_2} C_p^0 dT - (H^0 - H_{P_2})_{T_2} \qquad (2.125)$$

The first and the third term consider the enthalpy variation between a *real fluid* (at given P and T), and an *ideal gas state* (at the same T, but a reference pressure P_0). These may be designated by the function $(H - H^0)_{P,T}$ called the *departure enthalpy*. Generally, the *departure function* for a property M is

$$(M - M^0) = M(P, T) - M^{ig}(P_0, T) \qquad (2.126)$$

Applying this to enthalpy, one may write the equation in a more convenient way:

$$\Delta H = \int_{T_1}^{T_2} C_p^0 dT + \{(H_{P_2} - H^0)_{T_2} - (H_{P_1} - H^0)_{T_1}\} \qquad (2.127)$$

Therefore, the enthalpy change between (T_1, P_1) and (T_2, P_2) may be calculated from the variation for an ideal gas plus the variation of the departure function, which accounts for nonideality. The key benefit of departure functions is that they can be evaluated with a *PVT* relationship, including the corresponding states principle. Furthermore, the use of departure functions leads to a unified framework of computational methods, both for thermodynamic properties and phase equilibrium.

Residual property is a concept derived using a similar treatment. By definition, a residual property M^R is the difference between a molar value of any extensive thermodynamic property (*V, U, H, S, G*) of a real fluid M and its counterpart as an ideal fluid M^{ig} at the same *P* and *T*.

$$M^R(P, T) = M(P, T) - M^{ig}(P, T) \tag{2.128}$$

The difference between residual and departure functions is the quantity $M^{ig}(P,T) - M^{ig}(P_0,T)$. Thus, residual and departure functions are identical for *U* and *H*, but slightly different for *S* and *G*. Hence, for the last function the difference is $G^R = (G - G^0) - R \ln(P/P_0)$, where P_0 is the reference pressure.

2.4.4.1 Evaluation of Departure Functions

For example, the Helmholtz free energy *A* can be calculated at constant temperature as follows:

$$A - A^0 = -\int_{V^0}^{V} P dV \tag{2.129}$$

The reference ideal gas state is taken at the same *T* but at pressure P_0, where the volume has value V^0. As previously, one may consider a path passing by a very low pressure (infinite volume):

$$A - A^0 = -\int_{\infty}^{V} P dV - \int_{V^0}^{\infty} P dV \tag{2.130}$$

Adding and subtracting $\int_{\infty}^{V} (RT/V)dV$ and replacing $P = RT/V$ in the second integral leads to $(A - A^0)$. Other departure functions for *S, H, U, G* can be similarly calculated in a straightforward manner.

$$A - A^0 = -\int_{\infty}^{V} \left(P - \frac{RT}{V} \right) dV - RT \ln \frac{V}{V^0} \tag{2.131}$$

$$S - S^0 = -\frac{\partial}{\partial T}(A - A^0)_V = \int_{\infty}^{V} \left(\left(\frac{\partial P}{\partial T} \right)_V - \frac{R}{V} \right) dV + R \ln \frac{V}{V^0} \tag{2.132}$$

$$H - H^0 = (A - A^0) + T(S - S^0) + RT(Z - 1) \qquad (2.133)$$

$$U - U^0 = (A - A^0) + T(S - S^0) \qquad (2.134)$$

$$G - G^0 = (A - A^0) + RT(Z - 1) = -\int_{\infty}^{V}\left(P - \frac{RT}{V}\right)dV + RT(Z - 1) - RT\ln\frac{V}{V_0} \qquad (2.135)$$

The value of $(A - A^0)$, $(G - G^0)$ and $(S - S^0)$ depend on the choice of the reference state V_0, but in contrast, $(U - U^0)$ and $(H - H^0)$ are independent. The fugacity coefficient may be found starting from

$$RT\ln(f/P) = G(T, P) - G^{ig}(T, P) \qquad (2.136)$$

The right term is the residual Gibbs energy. Further development leads to the fugacity coefficient:

$$G(T, P) - G^{ig}(T, P) = G(T, P) - G^{ig}(T, P_0) - [G^{ig}(T, P) - G^{ig}(T, P_0)] \qquad (2.137)$$

$$G(T,P) - G^{ig}(T,P) = (G - G^0) - RT\ln\frac{P}{P^0} = (G - G^0) - RT\ln Z - RT\ln\frac{V_0}{V} \qquad (2.138)$$

$$\ln\frac{f}{P} = -\frac{1}{RT}\int_{\infty}^{V}\left(P - \frac{RT}{V}\right)dV + (Z - 1) - \ln Z \qquad (2.139)$$

2.4.4.2 Equations of State
Any EOS can be used to generate analytical expressions for the residual or departure functions. In the case of PR-EOS, the results for enthalpy and entropy are as follows, while the equation $f^L = f^V$ has already presented the result for fugacity:

$$H(T,P) - H^0(T,P) = RT(Z - 1) + \frac{T\left(\dfrac{da}{dT}\right) - a}{2\sqrt{2}b}\ln\left[\frac{Z + (1+\sqrt{2})B}{Z + (1-\sqrt{2})B}\right] \qquad (2.140)$$

$$S(T,P) - S^0(T,P) = R\ln(Z - B) + \frac{\left(\dfrac{da}{dT}\right)}{2\sqrt{2}b}\ln\left[\frac{Z + (1+\sqrt{2})B}{Z + (1-\sqrt{2})B}\right] \qquad (2.141)$$

where

$$\frac{da}{dT} = -0.45724 \frac{R^2 T_c^2}{P_c} \kappa \left(\frac{\alpha}{TT_c} \right)^{1/2} \; ; \quad b = \frac{0.077796 RT_c}{p_c}; \quad B = \frac{bp}{RT} \qquad (2.142)$$

2.4.4.3 Corresponding States

The *principle of corresponding states* was the first attempt towards a universal method for calculating thermodynamic properties, and it is expressed as *the equilibrium properties that depend on intermolecular forces are related to critical properties in a universal way*. In the *two-parameters* formulation (van der Waals, 1873), the compressibility factor is a function only of the reduced temperature and pressure:

$$Z = f(T_r, P_r) \qquad (2.143)$$

Further improvement in generalizing the correlation of the compressibility factor was obtained with a *three-parameter* formulation, where the third parameter is the *acentric factor* ω, which is a measure of the shape and size of a molecule. A mathematical definition is based on the slope of the vapor pressure curve P^v of a pure component versus temperature at the reduced value $T_r = 0.7$:

$$\omega = 1.0 - \log(P^v/P_c)_{T_r=0.7} \qquad (2.144)$$

According to Pitzer and Curl (1957), the compressibility factor Z may be expressed by

$$Z = Z^0(T_r, P_r) + \omega Z^1(T_r, P_r) \qquad (2.145)$$

Hence, the computation of Z is split in two parts: Z^0 is the term for an ideal spherical molecule, and Z^1 is the deviation function accounting for non-sphericity. This formulation is exploited by many generalized predictive methods for physical properties, as for example the LK method.

Lee and Kesler (1975) found an accurate representation for the compressibility of both gases and liquids by combining BWR-EOS with the corresponding states law. Departure functions for enthalpy, entropy, fugacity coefficient and heat capacity were generated, and tables are given in Reid et al. (1987), whereas graphs are presented in Green and Perry's (2008) handbook. The method is similar to that developed for compressibility. As an example, the enthalpy departure function is given by

$$\left(\frac{H^0 - H}{RT_c} \right) = \left(\frac{H^0 - H}{RT_c} \right)^{(0)} + \omega \left(\frac{H^0 - H}{RT_c} \right)^{(1)} \qquad (2.146)$$

The LK method is considered more accurate for enthalpy and entropy than those based on cubic EOS, so it is the preferred choice in cryogenics, refrigeration or gas processing.

2.5 VAPOR/GAS-LIQUID EQUILIBRIUM

Vapor-liquid equilibrium is a condition where a liquid and its vapor phase are in equilibrium with each other, and the rate of evaporation equals the rate of condensation on a molecular level such that there is no net vapor-liquid interconversion. A substance at VLE is generally referred to as a saturated fluid – for pure chemical substances this means at the boiling point. Equilibrium is practically reached in a relatively closed location if a liquid and its vapor are allowed to stand in contact with each other with no (or only gradual) interference from the outside. Figure 2.12 shows some examples of *T-xy* and *xy* diagrams used in VLE simulations (Kiss, 2013).

2.5.1 Ideal Solution Concept

The *ideal solution* concept is based on the assumption that there are no interactions between molecules, only information regarding pure component properties and mixture composition. The *Lewis and Randall rule* allows obtaining the component fugacity in an ideal solution by multiplying the pure species fugacity at given T and P by its molar fraction. For the vapor phase, one may write

$$\hat{f}_i^V = f_i^{0,V} y_i \tag{2.147}$$

where $\hat{f}_i^V$ and $f_i^{0,V}$ are the fugacities of species i in a mixture and as a pure component, respectively. Likewise, in an ideal liquid solution, the component fugacity $\hat{f}_i^L$ is given by multiplying the pure liquid fugacity $f_i^{0,L}$ by its molar fraction x_i:

$$\hat{f}_i^L = f_i^{0,L} x_i \tag{2.148}$$

At equilibrium, the component fugacities in both phases are equal, such that

$$\hat{f}_i^V = \hat{f}_i^L \tag{2.149}$$

Combining these equations gives the following expression for *ideal K values*:

$$K_i^{id}(EoS) = \frac{y_i}{x_i} = \frac{f_i^{0,L}}{f_i^{0,V}} \tag{2.150}$$

Ideal K values (*VLE ratio*) can be defined as the ratio of fugacity of the pure components in liquid and vapor phase, which depend only on T and P but not on composition. A parallel can be made with *ideal mixtures,* defined by the *Raoult–Dalton law,* where $K_i^{id} = P_i/P$.

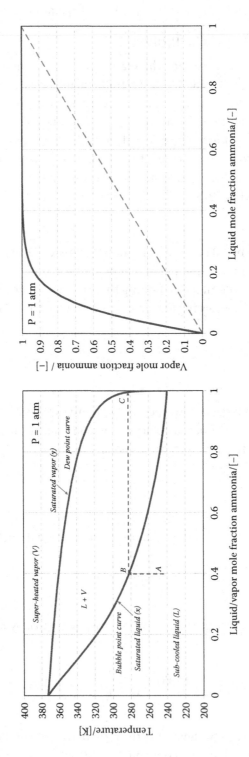

FIGURE 2.12 Examples of *T-xy* and *xy* diagrams used in VLE: *T-xy* diagram of ammonia-water (left), *xy* diagram of ammonia-water (right).

TABLE 2.4   Analogy between Ideal Gas and Ideal Solution

Property	Ideal Gas	Ideal Solution
Volume	$V^{ig} = \sum y_i V_i$	$V^{id} = \sum x_i V_i$
Enthalpy	$H^{ig} = \sum y_i H_i^{ig}$	$H^{id} = \sum x_i H_i^{id}$
Entropy	$S^{ig} = \sum_i y_i S_i^{ig} - R \sum_i y_i \ln y_i$	$S^{id} = \sum_i x_i S_i^{id} - R \sum_i x_i \ln x_i$
Gibbs free energy	$G^{ig} = \sum y_i G_i^{ig} + RT \sum y_i \ln y_i$	$G^{id} = \sum y_i G_i^{id} + RT \sum x_i \ln x_i$
Chemical potential	$\mu_i^{ig} = \mu_i^0 + RT \ln y_i$	$\mu_i^{id} = G_i^0 + RT \ln x_i$

An important analogy exists between the concepts of ideal solutions and ideal gases. Table 2.4 (Dimian et al., 2014) presents formulas for averaging main properties: Some are additive (e.g. V, H), while others (e.g. S, G) need correction for composition.

2.5.2 Equation-of-State Approach

A cubic EOS model can calculate fugacity for both phases at equilibrium; thus, a correlation linking fugacity and composition is needed. The derivation starts with variation of the chemical potential:

$$d\hat{\mu}_i = \left(\frac{\partial V}{\partial n} \right)_{T,P,n_j} dP = RTd \ln \hat{f}_i \tag{2.151}$$

In view of referring to an ideal state, the term $RTd(\ln p_i)$ may be subtracted from both sides:

$$RTd \ln \frac{\hat{f}_i}{p_i} = \left(\frac{\partial V}{\partial n_i} \right)_{T,P,n_j} dP - RTd \ln p_i \tag{2.152}$$

Because $d \ln y_i = 0$, then $RTd \ln p_i = RTd \ln P + RTd \ln y_i = RTd \ln P$, and thus

$$RTd \ln \frac{\hat{f}_i}{p_i} = \left[\left(\frac{\partial V}{\partial n} \right)_{T,P,n_j} - \frac{RT}{P} \right] dP \tag{2.153}$$

The inversion property of functions allows changing the variation of volume by pressure:

$$\left(\frac{\partial V}{\partial n_i} \right)_{T,P,n_j} dP = - \left(\frac{\partial P}{\partial n_i} \right)_{T,V,n_j} dV \tag{2.154}$$

For the vapor phase, the fugacity coefficient is given by $\hat{\phi}_i^V = \dfrac{f_i(T,P,y_i)}{Py_i}$. Combining equations leads to

$$\ln\hat{\phi}_i^V = \frac{1}{RT}\int_V^\infty\left[\left(\frac{\partial P}{\partial N_i}\right)_{T,V,N_j} - \frac{RT}{V}\right]dV - \ln Z^V \tag{2.155}$$

For the liquid phase, a similar expression is obtained:

$$\ln\hat{\phi}_i^L = \frac{1}{RT}\int_V^\infty\left[\left(\frac{\partial P}{\partial N_i}\right)_{T,V,N_j} - \frac{RT}{V}\right]dV - \ln Z^L \tag{2.156}$$

Using the previous equations requires knowledge of pressure dependency on composition, which is described via *mixing rules*. Replacing the fugacity coefficients in equation $\hat{f}_i^V = \hat{f}_i^L$ leads to

$$\hat{\phi}_i^V y_i P = \hat{\phi}_i^L x_i P \tag{2.157}$$

Real K values are given by the ratio of fugacity coefficients in liquid-vapor phases at equilibrium:

$$K_i = \frac{y_i}{x_i} = \frac{\hat{\phi}_i^L}{\hat{\phi}_i^V} \tag{2.158}$$

Models based on EOS give good results for most hydrocarbon systems, but the extension to polar components and aqueous mixtures needs *modified EOSs* with appropriate mixing rules.

Mixing rules are relations used to average the parameters of an EOS by taking into account the mixture composition. The accuracy of pure component properties is a necessary, but not sufficient, condition to ensure accurate mixture properties. The type of mixing rules and quality of the BIPs used in these relations are of equal importance. Having more than one adjustable parameter and by considering temperature dependency, the mixing rules may handle suitably non-ideal mixtures. SR-POLAR (Schwartzentruber et al., 1989) is such an option:

$$a_m = \sum_i\sum_j x_i x_j (a_i a_j)^{0.5}[1 - k_{a,ij} - l_{ij}(x_i - x_j)] \tag{2.159}$$

$$b_m = \sum_i\sum_j x_i x_j \frac{b_i + b_j}{2}(1 - k_{b,ij}) \tag{2.160}$$

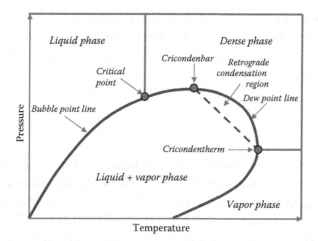

FIGURE 2.13   Phase envelope of a gas: The *cricondenbar* is the maximum pressure above which no gas can be formed regardless of the temperature, while the *cricondentherm* is the maximum temperature above which liquid cannot be formed regardless of the pressure.

where the attraction parameter a is averaged via two interaction parameters $k_{a,ij}$ and l_{ij}, while the covolume b is averaged via another parameter $k_{b,ij}$. The interaction parameters are temperature dependent.

Phase envelopes are useful thermodynamic analysis tools for hydrocarbon mixtures (see Figure 2.13). The phase equilibrium is explored over a large range of pressures and temperatures (up to the critical point) by bubble and dew point series, as well as by variable vapor fractions. EOS models (e.g. SRK, PR) are mostly employed. Phase envelopes are intensively used in the exploration of oil and gas fields, as well as in designing pipelines for natural gas transportation.

Liquid-liquid equilibrium (*LLE*) or the formation of two liquid phases at equilibrium occurs when the mixture exhibits high non-ideality in the liquid phase (e.g. aqueous mixtures containing partially miscible components). The LLE can be treated conveniently using the LACT concept, but an EOS model is also able to deal with this problem, provided that the algorithms are adapted accordingly and that the BIPs are regressed from LLE experimental data.

2.5.3 Liquid Activity Coefficient Approach

By definition, the VLE of ideal mixtures is described by the *Raoult–Dalton law* $Py_i = P_i x_i$, where P is the total pressure, P_i is the vapor pressure at equilibrium temperature, while x_i and y_i are the composition of liquid and vapor phase, respectively. The *ideal K values* are given by the ratio of vapor pressure to total pressure, meaning that the interactions between components are negligible in vapor/liquid phases:

$$K_i^{id} = \frac{y_i}{x_i} = \frac{P_i}{P}$$

(2.161)

For non-ideal mixtures, the interactions between species (typically stronger in liquid than vapor) should be accounted for. The starting point is the equality of fugacities in each phase: $\hat{f}_i^V = \hat{f}_i^L$. Computing the vapor phase fugacity can follow the method based on fugacity coefficient $\hat{\phi}_i^V$:

$$\hat{f}_i^V = y_i P \hat{\phi}_i^V \tag{2.162}$$

For determining $\hat{\phi}_i^V$, an appropriate PVT relationship should be selected. Any EOS can be used (e.g. virial EOS or cubic EOS: RK, SRK, PR). When the species in the vapor phase give association (e.g. hydrofluoric acid [HF] or short carboxylic acids), the fugacity calculation has to be modified accordingly. The methods of Hayden and O'Connell (1975) and Nothnagel et al. (1973) are based on virial EOS, while the VPA/IK-CAPE (Vapor Phase Association/Industrie Konsortium CAPE [Industrial Cooperation on Computer Aided Process Engineering]) EOS was also proposed to tackle di-, tetra- and hexamerization simultaneously. More details about the chemical theory of VLE can be found in the book by Prausnitz et al. (1980).

The route for computing the fugacity in a liquid phase is based on the *LACT coefficient* concept, which is theoretically founded by the *excess Gibbs energy* G^{ex} as the difference between the actual value and that corresponding to an ideal solution in the same conditions. Thus, the definition is

$$G^{ex}(T, P, x_i) = G(T, P, x_i) - G^{id}(T, P, x_i) \tag{2.163}$$

By multiplying with n (total moles of mixture) and differentiating with respect to composition n_i,

$$\left[\frac{\partial(nG^{ex})}{\partial n_i}\right]_{P,T,n_j} = \left[\frac{\partial(nG)}{\partial n_i}\right]_{P,T,n_j} - \left[\frac{\partial(nG^{id})}{\partial n_i}\right]_{P,T,n_j} \tag{2.164}$$

$$\bar{G}_i^{ex} = \bar{G}_i - \bar{G}_i^{id} \tag{2.165}$$

The notation $\bar{G}_i$ designates the partial Gibbs energy of the component i in the mixture. By definition $d\bar{G}_i = d\mu_i = RTd\ln\hat{f}_i$. The state of *pure liquid* can be chosen as a reference, for which $\bar{G}_i = G_i(T, P)$ and $\hat{f}_i = f_i^{0,L}$. Integrating at constant T and P between the reference and actual states gives

$$\bar{G}_i - G_i = RT\ln\frac{\hat{f}_i}{f_i^{0,L}} \tag{2.166}$$

Assuming an ideal solution, the component fugacity can be described by $\hat{f}_i = f_i^{0,L} x_i$, hence:

$$\bar{G}_i^{id} - G_i = RT \ln x_i \tag{2.167}$$

By combining these equations, the next expression can be formulated for excess Gibbs energy:

$$\bar{G}_i^{ex} = \bar{G}_i - \bar{G}_i^{id} = RT \ln \frac{\hat{f}_i}{x_i f_i^{0,L}} \tag{2.168}$$

By definition, the *activity coefficient* γ_i indicates the deviation of the component *activity* from its actual measured concentration. If a molar fraction is chosen as the composition variable, then $\gamma_i = a_i/x_i$.

Conversely, the component activity is defined by the ratio of component fugacity in the mixture to its value as a pure liquid at T and P of the system, which by combinations leads to real K values:

$$a_i = \frac{\hat{f}_i}{f_i^{0,L}} \tag{2.169}$$

$$\gamma_i = \frac{\hat{f}_i}{x_i f_i^{0,L}} \tag{2.170}$$

$$\hat{f}_i^L = \gamma_i x_i f_i^{0,L} \tag{2.171}$$

$$y_i P \phi_i^V = \gamma_i x_i f_i^{0,L} \tag{2.172}$$

$$K_i = \frac{\gamma_i f_i^{0,L}}{P \phi_i^V} \tag{2.173}$$

Equation 2.173 for real K values combines the fugacity description for the vapor phase with LACT modeling, but it needs the pure liquid fugacity $f_i^{0,L}$ as a function of more accessible properties (Dimian et al., 2014). The following relation was already demonstrated:

$$f_i^{0,L} = P_{s,i} \phi_{s,i}^V \int_{P_i}^{P} \frac{\bar{V}_i}{RT} dP \tag{2.174}$$

where $P_{s,i}$ is the vapor pressure at saturation temperature T, while $\phi_{s,i}^V$ is the fugacity coefficient of the saturated vapor (for an ideal vapor, $\phi_{s,i}^V$ is 1). The last term taking into account the effect of the pressure between the saturation state of component i and the value of the system is known as the *Poynting correction*. The computation uses the variation of partial molar liquid volume of the component. At low and moderate pressures, the Poynting correction is negligible. At low pressures, the fugacity of pure liquid is practically given by the vapor pressure, $f_i^{0,L} \approx P_i$, and thus the equation becomes

$$y_i P \phi_i^V = \gamma_i x_i P_i \tag{2.175}$$

Equation 2.175 finds large applications in the simulation of distillation-based separations involving highly non-ideal mixtures. The accuracy of the equation for the vapor pressure also is important. High-quality results are obtained by employing an *extended Antoine-type equation*:

$$\ln P_i = C_1 + \frac{C_2}{C_3 + T} + C_4 T + C_5 \ln T + C_6 T^{C_7} \tag{2.176}$$

In case of hydrocarbon mixtures with hydrogen and methane, the method of *Chao-Seader* (Chao and Seader, 1961) is often used due to its proven accuracy. The fugacity coefficients may be calculated by a corresponding states formulation, as the sum of two contributions, for spherical molecule and for deviation from sphericity (where $\upsilon_i^{(0)}$ and $\upsilon_i^{(1)}$ are semiempirical functions of T_r and P_r):

$$\ln \frac{f_i^{0,L}}{P} = \ln \upsilon_i^{(0)} + \omega_i \ln \upsilon_i^{(1)} \tag{2.177}$$

Excess Gibbs free energy and LACT coefficients are linked. A key relation is obtained, expressing that the LACT coefficient is a measure of excess Gibbs free energy of a component in a mixture:

$$\bar{G}_i^{ex} = RT \ln \gamma_i \tag{2.178}$$

Because $\bar{G}_i^{ex}$ is a partial property of G^{ex}, then $\ln \gamma_i$ is a partial property as well. The total excess Gibbs free energy is obtained by summation over the composition:

$$G^{ex}/RT = \sum_i x_i \ln \gamma_i \tag{2.179}$$

Hence, the total excess Gibbs energy can be easily determined from experimental values of activity coefficients, as will be shown in the next section. Because $\ln \gamma_i$ is a partial property, the following relation may be written as a form of the *Gibbs–Duhem equation* (Dimian et al., 2014):

$$\left(\sum_i x_i d\ln\gamma_i = 0\right)_{P,T} \tag{2.180}$$

2.5.4 Gas-Liquid Equilibrium

Gas-liquid equilibrium and VLE display not only similarities but also key differences. Considering a component i in a gaseous mixture dissolved in a solvent (at equilibrium and constant temperature), if the solution is diluted and no chemical reaction takes place, then the *Henry law* is applicable, which links the solute partial pressure p_i to its molar fraction in the liquid phase x_i:

$$H_{iA} = p_i/x_i \tag{2.181}$$

The Henry constant H_{iA} is characteristic for a pair component i–solvent A and depends strongly on temperature, while having pressure dimension. GLE may be described by equilibrium constant:

$$K_i = \frac{y_i}{x_i} = \frac{H_{iA}}{P} \tag{2.182}$$

Note that the Henry law is not valid at a higher solute concentration. Also at higher pressure, the influence of pressure on GLE should be accounted for. The first approach is based on the *asymmetric definition* of the K values, while the second one takes advantage of the ability of cubic EOS to handle simultaneously sub- and super-critical components (Dimian et al., 2014).

2.5.4.1 Asymmetric Definition of Equilibrium Constants

The approach is illustrated in Figure 2.14, left (Dimian et al., 2014), where the fugacity of a solute in liquid $\hat{f}_i^L$ is plotted as a function of its molar fraction x_i. The Henry constant

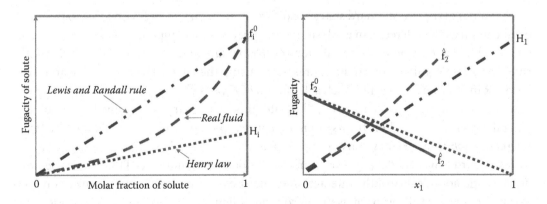

FIGURE 2.14 Asymmetric definition of reference states (left); asymmetric definition of GLE (right).

H_i may be defined as the limit of component fugacity $\hat{f}_i^L$ at infinite dilution (left corner of the diagram):

$$H_i(T,P) = \lim_{x_i \to 0} \frac{\hat{f}_i^L(T,P,x_i)}{x_i} \tag{2.183}$$

The temperature dependency of H_i may be expressed similarly with the vapor pressure, as follows:

$$\ln H_i = A + B/T + C \ln T + DT^2 \tag{2.184}$$

The pressure dependency of H_i can be described by a similar relation with liquid fugacity:

$$H_i(T,P) = H_i(T,P_{ref}) \exp\left[\int_{P_{ref}}^{P} \frac{V_i^L}{RT} dP \right] \tag{2.185}$$

Typically, the reference pressure is $P_{ref} = 1$ atm. In the correction term (analogue to *Poynting factor*), V_i^L is the solute liquid molar volume available in tables or estimated by various methods (Poling et al., 2001). The Poynting correction is small at low pressures, but necessary at high pressures. Another reference state for the solute i is its pure liquid fugacity f_i^0. This state is a virtual one because in practice $x_i \ll 1$. If the actual liquid mixture has as reference an ideal solution obeying the *Lewis–Randall rule*, the reference state f_i^0 can be defined as the limit of component fugacity at $x_i \to 1$:

$$\lim_{x_i \to 1} \frac{\hat{f}_i^L(T,P,x_i)}{x_i} = f_i^0(T,P) \tag{2.186}$$

The values of the two standard states are different, such that $H_i \neq f_i^0$. An ideal solution is often chosen as the reference in analysis, so the fugacity of a component is $\hat{f}_i^{id}(HL) = x_i H_i$ and $\hat{f}_i^{id}(LR) = x_i f_i^0$, where *HL* stands for the *Henry law* and *LR* for the *Lewis–Randall* rule. The phase equilibrium can be expressed by using the Henry law for a solute and the Lewis–Randall rule for solvent – such a definition of K values being considered *asymmetric*. Figure 2.14, right (Dimian et al., 2014), illustrates this for a binary system involving a solute (1) and solvent (2). Based on the *asymmetric convention*, the K values are defined by two references: *solute* by a Henry constant ($x_1 \to 0$) and *solvent* by a liquid fugacity ($x_2 \to 1$).

A standard LACT model (e.g. van Laar, non-random two liquid [NRTL]) can be used to describe the non-ideality of the interaction solute-solvent. The asymmetric convention for K values is a powerful approach useful in solving difficult problems, such as absorption in solvents of gases containing both sub-critical and super-critical components, absorption of sour gases (NH_3, H_2S, SO_2, CO_2) in water or aqueous solutions, or design of gas-liquid processes with very concentrated solutions (Dimian et al., 2014).

2.5.4.2 Cubic EOS Approach

Cubic EOSs – particularly those modified for handling polar species – are suitable for the simulation of processes at higher pressures. A major advantage is that all components (sub- and super-critical) are treated in the same manner. The key problems here are the selection of appropriate mixing rules and using good-quality interaction parameters. If regression is necessary, sufficient data should be available. After the conversion of solubility data into *PT-xy* data, the regression of BIPs can follow the usual way. Predictive EOS (e.g. predictive Soave–Redlich–Kwong [PSRK]) can be employed for estimating the solubility of some important technical gases by the extension of UNIFAC (UNIQUAC [universal quasi-chemical] functional group activity coefficients) with groups as for NH_3, CO_2, CH_4, O_2, Ar, N_2, H_2S, H_2 and CO (Dimian et al., 2014).

2.6 LIQUID ACTIVITY MODELS

2.6.1 Modeling Excess Gibbs Free Energy

For a binary mixture, the activity coefficients and the excess Gibbs free energy may be computed with the relations $y_i P \phi_i^V = \gamma_i x_i P_i$ and $G^{ex}/RT = \sum_i x_i \ln \gamma_i$. Any form of dependency for excess Gibbs energy can be used if it respects the Gibbs–Duhem equation. Two types of series expansion have proved to be the most suitable: power series and *Legendre functions*. The power expansion consists of the following expression or the next alternative form known as the *Redlich–Kister expansion*:

$$\frac{G^{ex}}{RT} = x_1 x_2 \left(a + b x_1 + c x_1^2 + \ldots \right) \tag{2.187}$$

$$\frac{G^{ex}}{RT} = x_1 x_2 [A + B(x_1 - x_2) + C(x_1 - x_2)^2 + \ldots] \, (\text{Redlich–Kister expansion}) \tag{2.188}$$

2.6.2 Correlation Models for Liquid Activity

Classical models (e.g. *Margules* and *van Laar*) are described first as correlation models for liquid activity, as they are simple and typically used in spreadsheet calculations. They are still applicable when accuracy is not crucial or for fitting local models in dynamic simulation. Then, accurate models suited for computer simulation (e.g. *Wilson, NRTL, UNIQUAC*) are described. Only the formulas for binary mixtures are given here, as extension to multicomponent mixtures is given in textbooks (Smith et al., 2005).

2.6.2.1 Margules

A direct application of power series (Redlich–Kister expansion) is the *Margules* LACT model. If $B = C = 0$, then the result is a *two-suffix* Margules *equation* because G^{ex} is a second-degree function. The equation has only one constant, but it may be used for qualitative computations:

$$\frac{G^{ex}}{RT} = Ax_1x_2 \qquad (2.189)$$

If both A and B are non-zero, then a *three-suffix equation* is obtained:

$$\frac{G^{ex}}{RT} = x_1x_2(A_{21}x_1 + A_{12}x_2) \qquad (2.190)$$

There is also an accurate four-suffix formula that may be employed to test the thermodynamic consistency of VLE data. Another expansion accurate for data correlation is

$$\frac{G^{ex}}{RT} = x_1x_2[p_0 + p_1(2x_1 - 1) + p_2(6x_1^2 - 6x_1 + 1) + \ldots] \qquad (2.191)$$

The terms in parentheses are *Legendre polynomials*. Their advantage over the ordinary polynomial expansion is that they are orthogonal, meaning that the values of p_k found by regressing experimental data are independent of each other, and accuracy can be adapted to the number of data points. The Margules model can be used for both VLE and VLLE, with the most used form given by

$$\ln\gamma_1 = x_2^2[A_{12} + 2(A_{21} - A_{12})x_1] \quad \ln\gamma_2 = x_1^2[A_{21} + 2(A_{12} - A_{21})x_2] \qquad (2.192)$$

2.6.2.2 van Laar

The van Laar model is simple, gives good results, and can be used for both VLE and VLLE. The model has two adjustable BIPs; the following equations are used:

$$\ln\gamma_1 = \frac{B_{12}}{[1 + (B_{12}x_1/B_{21}x_2)]^2} \quad \ln\gamma_2 = \frac{B_{21}}{[1 + (B_{21}x_2/B_{12}x_1)]^2} \qquad (2.193)$$

$$B_{12} = \ln\gamma_1\left[1 + \frac{x_2\ln\gamma_2}{x_1\ln\gamma_1}\right]^2 \quad \text{and} \quad B_{21} = \ln\gamma_2\left[1 + \frac{x_1\ln\gamma_1}{x_2\ln\gamma_2}\right]^2 \qquad (2.194)$$

The van Laar model allows the direct calculation of the BIPs from single data, as the azeotropic point, where $\gamma_1 = P/P_1(T_{az})$ and $\gamma_2 = P/P_2(T_{az})$.

2.6.2.3 Wilson

The Wilson model (Wilson, 1964) uses the *local composition* concept, related to the segregation caused by different interaction energies between pairs of molecules. Thus, the probability of finding a molecule of species 1 surrounded by molecules of species 2, relative to the probability of being surrounded by the same species 1, is given by the following expression:

$$\frac{x_{12}}{x_{11}} = \frac{x_2 \exp(-\lambda_{12}/RT)}{x_1 \exp(-\lambda_{11}/RT)} \tag{2.195}$$

Quantities λ_{12} and λ_{11} denote interaction energies between molecules, while x_{12} has the meaning of a *local composition*. Also, a *local volume* fraction of the component 1 can be formulated as

$$\zeta_1 = \frac{v_{1L}x_{11}}{v_{1L}x_{11} + v_{2L}x_{12}} \tag{2.196}$$

Quantities v_{1L} and v_{2L} are the molar liquid volumes of the two components. The following relation can describe the excess Gibbs free energy of the mixture:

$$\frac{G^{ex}}{RT} = x_1 \ln\frac{\zeta_1}{x_1} + x_2 \ln\frac{\zeta_2}{x_2} \tag{2.197}$$

Binary interaction constants may be defined as

$$\Lambda_{12} = \frac{v_{2L}}{v_{1L}}\exp\left[-\frac{(\lambda_{12}-\lambda_{11})}{RT}\right] \quad \text{and} \quad \Lambda_{21} = \frac{v_{1L}}{v_{2L}}\exp\left[-\frac{(\lambda_{21}-\lambda_{22})}{RT}\right] \tag{2.198}$$

Energies of interaction are $\lambda_{12} = \lambda_{21}$, but $\lambda_{11} \neq \lambda_{22}$. After substitution, excess Gibbs free energy is

$$\frac{G^{ex}}{RT} = -x_1 \ln(x_1 + \Lambda_{12}x_2) - x_2 \ln(x_2 + \Lambda_{21}x_1) \tag{2.199}$$

The following equations for the activity coefficients are eventually obtained:

$$\ln\gamma_1 = -\ln(x_1 + x_2\Lambda_{12}) + x_2\left(\frac{\Lambda_{12}}{x_1 + x_2\Lambda_{12}} - \frac{\Lambda_{21}}{x_2 + x_1\Lambda_{21}}\right) \tag{2.200}$$

$$\ln\gamma_2 = -\ln(x_2 + x_1\Lambda_{21}) - x_1\left(\frac{\Lambda_{12}}{x_1 + x_2\Lambda_{12}} - \frac{\Lambda_{21}}{x_2 + x_1\Lambda_{21}}\right) \qquad (2.201)$$

The Wilson model describes accurately the VLE of strong non-ideal mixtures but is not convenient for LLE. If infinite-dilution γ_i^∞ values are available from measurements, the interaction parameters can be found easily by solving the following algebraic system (note that the parameters obtained from γ_i^∞ may be used over the whole concentration range):

$$\ln\gamma_1^\infty = 1 - \ln\Lambda_{12} - \Lambda_{21} \quad \text{and} \quad \ln\gamma_2^\infty = 1 - \ln\Lambda_{21} - \Lambda_{12} \qquad (2.202)$$

2.6.2.4 Non-Random Two Liquids
The NRTL model was developed by Renon and Prausnitz (1968) as an extension of the local composition concept that accounts for the non-randomness of interactions. The following expression for G^{ex} is valid:

$$\frac{G^{ex}}{RT} = x_1 x_2\left(\frac{\tau_{21}G_{21}}{x_1 + x_2 G_{21}} + \frac{\tau_{12}G_{12}}{x_2 + x_1 G_{12}}\right) \qquad (2.203)$$

Quantities τ_{ij} express the differences in interaction energies, $\tau_{ji} = (g_{ji} - g_{ii})/RT$. The parameters G_{ij} take into account the nonrandomness of interactions $G_{ij} = \exp(-\alpha\tau_{ij})$. The parameter α is adjustable, but it can be fixed to a specific value, such as 0.20 (saturated hydrocarbon with polar nonassociated species); 0.30 (nonpolar compounds, water, nonassociated species); 0.40 (saturated hydrocarbon and homologue perfluorocarbons) or 0.47 (alcohol and other self-associated nonpolar species). The binary activity coefficients can be calculated using the following equations:

$$\ln\gamma_1 = x_2^2\left[\frac{\tau_{21}G_{21}^2}{(x_1 + x_2 G_{21})^2} + \frac{\tau_{12}G_{12}}{(x_2 + x_1 G_{12})^2}\right] \qquad (2.204)$$

$$\ln\gamma_2 = x_1^2\left[\frac{\tau_{12}G_{12}^2}{(x_2 + x_1 G_{21})^2} + \frac{\tau_{21}G_{21}}{(x_1 + x_2 G_{12})^2}\right] \qquad (2.205)$$

The three-parameter formulation requires the regression of τ_{12}, τ_{21} and α. Even if τ_{12} and τ_{21} account implicitly for the effect of temperature, it is possible to introduce an explicit dependency, such as $\tau_{ij} = a_{ij} + b_{ij}/T$. NRTL is somewhat sensitive in computation, but it describes accurately both VLE and VLLE of high non-ideal mixtures (Dimian et al., 2014).

2.6.2.5 Universal Quasi-Chemical

The UNIQUAC model was developed by Abrams and Prausnitz (1975). In contrast to Wilson and NRTL (where the local volume fraction is used), in UNIQUAC the primary variable is the *local surface area* fraction θ_{ij}. Each molecule is characterized by two structural parameters: r, the relative number of segments of the molecule (*volume parameter*), and q, the relative surface area (*surface parameter*). Values of these parameters are obtained in some cases by employing statistical mechanics. A special form of UNIQUAC for mixtures containing alcohols introduces a third surface parameter q' that notably increases the accuracy (Prausnitz et al., 1980).

The excess Gibbs free energy is the result of two contributions. The first one, called the *combinatorial part*, represents the influence of the structural parameters as size (parameter r) and shape (area parameter q). The second one, called the *residual part*, accounts for the energy of interactions between segments. In the case of a binary mixture, the expression for the excess Gibbs free energy is

$$\frac{G_C^{ex}}{RT} = x_1 \ln \frac{\phi_1}{x_1} + x_2 \ln \frac{\phi_2}{x_2} + \frac{z}{2}\left(q_1 x_1 \ln \frac{\theta_1}{\phi_1} + q_2 x_2 \ln \frac{\theta_2}{\phi_2} \right) \text{(Combinatorial part)} \quad (2.206)$$

$$\frac{G_R^{ex}}{RT} = -q_1 x_1 \ln(\theta_1 + \theta_2 \tau_{21}) - q_2 x_2 \ln(\theta_2 + \theta_1 \tau_{12}) \,\text{(Residual part)} \quad (2.207)$$

The parameters in these equations have the following significance: average segment fraction, $\phi_1 = \dfrac{x_1 r_1}{x_1 r_1 + x_2 r_2}$; average surface area fraction, $\theta_1 = \dfrac{x_1 q_1}{x_1 q_1 + x_2 q_2}$; lattice coordination number z (set equal to 10) and binary interaction energy, $\tau_{ji} = \exp\left(-\dfrac{u_{ji} - u_{ii}}{RT} \right)$.

Size and area parameters can be computed from two other quantities: the van der Waals area A_W and the volume V_W. This information is usually stored in databases of pure components. UNIQUAC has only two adjustable parameters (τ_{12} and τ_{21}), but this number increases to four if temperature dependency is also considered. UNIQUAC is just as accurate as Wilson, but it may be applied to LLE. Despite apparent complexity, the UNIQUAC model is robust in computations (Dimian et al., 2014).

Concerning multicomponent mixtures, note that only BIPs are sufficient for accurate VLE simulations, but LLE computations might need ternary data. As the quality of the BIPs is crucial for the reliability of results, special attention must be given to parameter regression.

2.6.3 Predictive Methods for Liquid Activity

Predictive methods allow the treating of non-ideal VLE without knowledge of BIPs fitted from experimental data. These models are used mainly for exploratory purposes, although recent developments have steadily increased their accuracy. Several approaches are summarized here: *regular solution* theory, which requires only information about pure components; *UNIFAC*, which is a group contribution method (GCM) using indirectly experimental data and conductor-like screening model (*COSMO*), which is based on quantum mechanics theory and allows ab initio computations of properties.

2.6.3.1 Regular Solution Theory

The *regular solution theory* concept is based on two assumptions: molecules of equal size and no heat of mixing. Scatchard and Hildebrand (Sandler, 2006) arrived at the following simple relation:

$$\ln \gamma_i = \frac{\upsilon_{i,L}\left(\delta_i - \sum \Phi_i \delta_i\right)^2}{RT} \tag{2.208}$$

where γ_i is the activity coefficient, and Φ_i is the volume fraction of component i, which can be determined from the component molar liquid volumes υ_{iL} as follows:

$$\Phi_i = \frac{x_i \upsilon_{i,L}}{\sum x_i \upsilon_{i,L}} = \frac{x_i \upsilon_{i,L}}{V_L} \tag{2.209}$$

The *solubility parameter* δ_i can be determined by the relation

$$\delta_i = \left[\frac{\Delta H_i^{vap} - RT}{\upsilon_{i,L}}\right]^{1/2} \tag{2.210}$$

The key information in this relation is the enthalpy of vaporization ΔH_i^{vap} and the partial liquid volume. Note that the interactions are neglected.

2.6.3.2 UNIQUAC Functional Group Activity Coefficients

UNIFAC is an extension of the UNIQUAC model in which the interaction parameters are estimated by means of group contributions. The molecules are decomposed in characteristic structures (functional groups and sub-groups), while some small molecules are taken separately for better accuracy. Then, the parameters involved are determined as follows: Molecular volume and area parameters in the combinatorial part are replaced by

$$r_i = \sum_k v_k^{(i)} R_k; \quad q_i = \sum_k v_k^{(i)} Q_k \tag{2.211}$$

where $v_k^{(i)}$ is the number of functional groups of type k in the molecule i, while R_k, Q_k are *volume* and *area parameters* of the functional group, respectively. The residual term is replaced by

$$\ln \gamma_i^R = \sum_k v_k^{(i)} \left(\ln \Gamma_k - \ln \Gamma_k^{(i)} \right) \tag{2.212}$$

Here, Γ_k is the residual activity coefficient of functional group k in the actual mixture. $\Gamma_k^{(i)}$ is the residual activity coefficient of functional group k in a reference mixture that contains only i molecules. The activity coefficient for the group k in molecule i depends on the molecule i in which the k group is situated. The following relation gives the activity coefficient of each group:

$$\ln \Gamma_k = Q_k \left[1 - \ln \left(\sum_m \theta_m T_{mk} \right) - \sum_m \frac{\theta_m T_{km}}{\sum_n \theta_n T_{nm}} \right] \tag{2.213}$$

where θ_m is the area fraction of group m, and X_m is the mole fraction of group m in the solution:

$$\theta_m = \frac{X_m Q_m}{\sum_n X_m Q_m} \quad \text{and} \quad X_m = \sum_i v_m^{(i)} x_j / \sum_j \sum_n \left(v_n^{(j)} x_j \right) \tag{2.214}$$

There are also group interaction parameters T_{mk} expressed by $T_{mk} = \exp(-a_{mk}/T)$, where $a_{mk} \neq a_{km}$, but $a_{mk} = 0$ when $m = k$.

In the UNIFAC decomposition, the computation is managed by defining main groups, with each one further broken down in sub-groups. For example, alkanes are described by the main group CH_2 with four sub-groups (CH_3, CH_2, CH, C), while alkenes are handled by the main group $C=C$ with five sub-groups, alcohols by the main group $-OH$ with three sub-groups and so on. A number of small and polar molecules are counted as both group and sub-group (e.g. H_2O, CH_3OH, acetonitrile, formic acid, $CHCl_3$, CCl_4, CS_2, furfural, dimethylformamide [DMF], acrylonitrile). Each sub-group is characterized by R_k and Q_k values. The R and Q parameters of a molecule are obtained by summing the contribution of sub-groups. The compilation of 46 groups and 106 sub-groups with recent values are given in a book by Sandler (2006). With respect to the parameters a_{mn} and a_{nm}, the procedure considers that the sub-groups within a main group have the same interaction parameters with other main groups but are zero for interactions inside their own main group. As key limitations, UNIFAC is unable to distinguish between isomers and ignores the proximity effects of strong polar groups (e.g. a CH_2 group has different charge and dipole moment in alkanes vs. alcohols, depending on its location).

Note that there are several versions of the UNIFAC model available: *standard UNIFAC*, proposed by Fredenslund et al. (1975, 1977) and completed by Gmehling et al. (2004), *UNIFAC-Lyngby* modification (Larsen et al., 1987) and *UNIFAC-Dortmund* modification proposed by Gmehling and co-workers (Weidlich and Gmehling, 1987; Wittig et al., 2003) and presently developed by a consortium implemented around the Dortmund Database (http://www.ddbst.com). These variants differ theoretically by the formulas used for the residual and combinatorial parts, as well as for the temperature dependency of interaction parameters. The computation of LLE uses different values of the interaction parameters. However, the major differences are in the extension of groups and subgroups, as well as in the quality of regressed parameters. UNIFAC–Dortmund provided good results in a large number of applications. Although the precision is about half that of UNIQUAC, the UNIFAC–Dortmund modified version is significantly better than the original UNIFAC.

2.6.3.3 Joback–Reid

The *Joback–Reid* method is another important GCM as proposed by Joback and Reid (1987) that predicts 11 pure component thermodynamic properties from molecular structure only: normal boiling point; melting point; critical temperature, pressure or volume; heat of formation; Gibbs energy of formation; heat capacity; heat of vaporization; heat of fusion and liquid dynamic viscosity. The Joback–Reid method uses the basic structural information of a chemical molecule (e.g. list of simple functional groups), adds parameters to these functional groups, and calculates physical properties as a function of the sum of group parameters. Unlike UNIFAC, the Joback–Reid method assumes no interactions between groups and thus uses only additive contributions. The main advantage of using only simple group parameters is the small number of parameters needed.

2.6.3.4 Models Based on Quantum Mechanics

Models emerged as new types of thermodynamic tools based on quantum mechanics (Sandler, 2003). COSMO is a powerful tool that allows the prediction of thermophysical properties of pure components, as well as phase equilibrium (VLE and SLE), without any experimental data. The basic idea is that a molecule (rather than being divided into a set of functional groups) is deconstructed into a collection of very small surface elements, for which the charge density on each is computed using quantum electrostatic calculations. The unique characteristic of each molecule is its *sigma profile*, which represents the charge density versus likelihood of occurrence. The excess Gibbs energy at any composition is computed from the sigma profiles of each molecule by statistical-mechanical analysis. Because each molecule has a unique sigma profile, the method does not suffer from the limitations of GCMs mentioned previously. The model requires few adjustable parameters for each component, such as molecule volume and sigma profiles. Unlike UNIFAC (for which the interactions take place between groups with BIPs regressed from experimental data), in COSMO-SAC (segment activity coefficient) the interactions are accounted only from the description of molecules.

COSMO-RS (real solvent) was the first model developed by Klamt (1995, 2005). Another variant, COSMO-SAC, was proposed later by Sandler and co-workers (Lin and Sandler, 2002; Sandler, 2003; Mullins et al., 2006; Wang et al., 2007; Hsieh et al., 2010). A sigma profile database for 1,432 chemicals was proposed, generated with only 10 elements: hydrogen, carbon, nitrogen, oxygen, fluorine, phosphorus, sulphur, chlorine, bromine and iodine.

The COSMO-like methods are still in development, but the last modifications are competitive with UNIFAC for describing VLE and LLE (Wang et al., 2007; Hsieh et al, 2010). The advantage of almost independency of experimental data makes these methods attractive for new applications dealing with complex molecules. Recent achievements include the prediction of properties for large molecules and asymmetric mixtures (Wang et al., 2007), high-pressure VLE (Constantinescu et al., 2005), solubility of pharmaceuticals (Bouillot et al., 2011), as well as solvent design for specific LLE and solid-liquid equilibrium (Shah & Yadav, 2011, 2012).

2.7 COMBINED EOS AND LACT MODELS

2.7.1 G^{ex} Mixing Rules

Huron and Vidal (1979) obtained the first significant advance in combining the approaches based on EOSs and LACT, starting with the following equality:

Excess Gibbs free energy of mixing (EOS) = Excess Gibbs free energy of mixing (LACT)

Theoretically, high-dense phase or infinite pressure should be considered to make the two models compatible. Hence, the previous condition becomes

$$G_{EoS}^{ex}(T,P=\infty,x_i)=G^{ex}(T,P=\infty,x_i) \tag{2.215}$$

The basic assumption is that G^{ex} may be replaced by the Helmholtz free energy A^{ex}:

$$G^{ex} = A^{ex} + PV^{ex} \cong A^{ex} \tag{2.216}$$

where V^{ex} is the excess volume. Because for a liquid phase A^{ex} is practically independent of pressure, it follows that $V^{ex} = 0$, and further

$$V^{ex} = V - \sum x_i V_i = b - \sum x_i b_i = 0 \tag{2.217}$$

The next equation gives a *mixing rule* for the parameter b_m as follows:

$$b_m = \sum x_i b_i \tag{2.218}$$

The b_m equation is identical with the *one-fluid van der Waals mixing rule*. Further, a mixing rule for the parameter a_m can be obtained, leading to the relation

$$\frac{a_m}{RTb_m} = \sum x_i \frac{a_i}{RTb_i} - \frac{G^{ex}}{CRT} \qquad (2.219)$$

In Equation 2.219, C is a constant depending on the particular type of EOS employed, while the excess Gibbs free energy G^{ex} can be determined accurately by means of LACT models. In the mixing rules of Huron and Vidal, the parameters in the LACT model depend on pressure and thus must be regressed again from experimental data.

Wong and Sandler (1992) reviewed the approach and proposed thermodynamic consistent mixing rules. The first mixing rule sets a constraint on both parameter a and parameter b as follows:

$$b_m - \frac{a_m}{RT} = \sum \sum x_i x_j \left(b_{ij} - \frac{a_{ij}}{RT} \right) \text{ where } b_{ij} - \frac{a_{ij}}{RT} = \frac{1}{2}\left[\left(b_{ii} - \frac{a_{ii}}{RT} \right) + \left(b_{jj} - \frac{a_{jj}}{RT} \right) \right] \qquad (2.220)$$

The second mixing rule remains unchanged. Therefore, combining the previous equations leads to

$$\frac{a_m}{RT} = Q\frac{D}{1-D}; \quad b_m = \frac{Q}{1-D} \qquad (2.221)$$

$$Q = \sum \sum x_i x_j \left(b_{ij} - \frac{a_{ij}}{RT} \right); \quad D = \sum x_i \frac{a_i}{b_i RT} + \frac{G^{ex}(x_i)}{CRT} \qquad (2.222)$$

The mixing rules proposed by Wong and Sandler allow the extrapolation of BIPs for LACT models from low to high pressure, such that a large amount of existing experimental information can be reused. When UNIFAC is used to express G^{ex}, the mixing rules of Wang and Sandler transform the EOS modeling in predictive methods for non-ideal mixtures.

2.7.2 Predictive EOS-G^{ex} Models

The development of G^{ex} *mixing rules* based on the calculation of excess Gibbs free energy by LACT methods incited the proposal of EOS offering prediction capabilities. Holderbaum and Gmehling (1991) proposed *predictive Soave–Redlich–Kwong* (PSRK), in which the alpha function is described by the Mathias and Copeman equation (accurate for polar components), while the excess Gibbs free energy is calculated by UNIFAC. The advantage is the extension of EOS models to mixtures of polar components without knowledge of experimental binary interaction coefficients. There are also weaknesses resulting from the original combined methods, such as inaccuracy in describing liquid densities, enthalpies and strong non-ideal behavior (Dimian et al., 2014).

The limitations of PSRK have been largely addressed by the VTPR EOS (Schmid and Gmehling, 2012), which is based on these assumptions: translation of volume by the correction proposed by Peneloux, predictive alpha function and G^{ex} *mixing rules* described by group contributions. The VTPR EOS can be applied successfully to systems containing (non-)polar, (a)symmetric, sub- or super-critical compounds for V-L, G-L and S-L calculations.

Jaubert et al. (2010) proposed the *predictive Peng–Robinson* (PPR78) using as support the Peng–Robinson EOS (1978 form), in which the k_{ij} BIPs are correlated by group contributions. The PPR78 model was fitted for accurate prediction of the phase equilibrium of petroleum fluids.

The *universal mixing rule* combined with PR and UNIFAC (UMR-PRU) was proposed by Voutsas et al. (2004) as an *EOS-G^{ex}* model applicable to all types of system asymmetries, including mixtures containing polymers. Its application in phase equilibrium and heat of mixing predictions, utilizing the existing parameters of the UNIFAC model, led to successful results. Also, UMR-PRU successfully predicts the dew points of synthetic and real natural gas mixtures. The performance of UMR-PRU is better than SRK, PR, and PC-SAFT, especially at the higher dew point pressures.

2.8 PROPERTY MODEL SELECTION AND PARAMETER REGRESSION

2.8.1 Selection of Property Models

The selection of an appropriate physical property method is a crucial issue, as improper modeling is one of the major causes of failure in designing chemical processes. The 'garbage in, garbage out' aphorism states that the results cannot be better than the quality of models and associated data. In practice, there are several types of physical property data that an engineer can use, such as various collections of properties and experimental data or more often process simulators that have guiding tools for selecting property models and methods. Figure 2.15 provides a convenient scheme for quick selection of property models (Kiss, 2013; Dimian et al., 2014). The decision strategy is based on distinction between polar, non-polar and electrolyte mixtures and on the operating pressure.

The first decision level is separating the mixtures in non-polar and polar components. Non-polar components typically include not only hydrocarbons but also inorganic gases (CO_2, N_2, O_2, H_2S, H_2) and CH_4. Because the processing conditions are usually above the critical coordinates, these components are super-critical, so models based on EOS and simple mixing rules are suitable. Cubic EOS, such as PR and SRK, are very flexible and reasonably accurate. The estimation of phase density improves when volume-translation modification is used (e.g. VTPR and VTSRK). More accurate for handling light gases in higher concentration are virial EOSs, such as Benedict–Webb–Rubin–Starling (BWRS) and BWR–Lee–Starling (BWR-LS), as well as the models based on the principle of corresponding state, such as LK and LKP. LK and LKP are recommended when accurate enthalpy

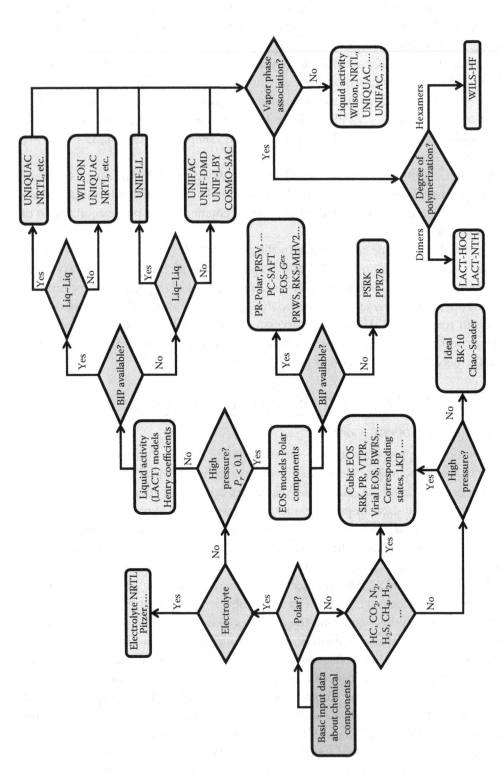

FIGURE 2.15 Property methods selection scheme. UNIF-LL, UNIFAC for liquid-liquid systems; UNIF-LBY, UNIFAC Lyngby-modified; UNIF-DMD, UNIFAC Dortmund-modified; WILS-HF, Wilson with HF hexamerization model.

calculation is required. The next selection condition is the operating pressure, with higher pressure meaning a reduced pressure $P_r = P/P_c > 0.1$ for the light-key component (or typically over 10 bars). At lower pressure, the hydrocarbon mixtures behave closely to ideal systems, while at higher pressures the standard EOS models are suitable. Specific accurate methods for direct K value computation can be used, such as Chao–Seader (for aliphatic hydrocarbons, including hydrogen) and BK-10 (for aromatics and pseudocomponents). The option of an ideal behavior is employed typically for preliminary computations or for comparison purposes.

In case of polar components, a selection separates electrolytes and non-electrolytes. The electrolytes require special methods described elsewhere (e.g. Electrolyte NRTL).

Polar non-electrolytes (e.g. most refrigerants) are further treated based on the operating pressure. At higher pressures, EOS models (modified for polar components) are recommended. Parameters for single components are required and BIPs for the mixing rules. Several modified EOSs for polar components are available, such as Schwartzentruber and Renon (SR-Polar) and PRSV or PC-SAFT as a new model for innovative applications. For weakly and moderately polar components, the combined treatment EOS-G^{ex} can be applied, such as PR with Wong–Sandler (PRWS) mixing rules or SRK with Michelsen–Huron–Vidal (SRK-MHV2) mixing rules. In prediction mode, the following models can be used: PSRK (BIP by UNIFAC in SRK), UMR-PRU, VTPR and PPR78 (BIP by GCM in PR). The combined EOS-G^{ex} models can handle directly super-critical components, as well as the GLE at higher solute concentration. For lower pressures, LACT models are most accurate. When the mixture contains super-critical components in low amounts, these are treated as Henry components. Further, the selection splits based on the occurrence of a second liquid phase. If BIPs are known from regression of experimental data, the most accurate methods for VLE computations are Wilson, UNIQUAC, NRTL and van Laar (options valid also for VLLE, except Wilson). For handling gas solubility at higher concentrations, LACT modeling by the asymmetric convention is better than Henry description.

For the prediction of phase equilibrium, the UNIFAC-based models are the most employed, in three variants: classical, modified Dortmund and modified Lyngby. Note that VLE and LLE make use of different group parameters. If the mixture contains components that may form strong associations in vapor phase, such as HF and dicarboxylic acids, then the vapor phase should consider specific models (e.g. Hayden–O'Connell [HOC], Nothnagel). The impressive development of predictive methods based on quantum mechanics (COSMO-SAC) should be taken advantage of – even if the accuracy is still limited, they can be used for exploring the phase equilibrium of special mixtures for which few or even no experimental data are available (Dimian et al., 2014).

2.8.2 Regression of Parameters

The regression of parameters of property models from experimental data is a systematic activity when accuracy is required, and the quality of experimental data is essential. A large amount of data is needed, but experiments are quite expensive. Simulations can help to reduce the experimental effort. Industrial data may be used for calibrating thermodynamic models for design or operation purposes (Dimian, 1994). Several types of equilibrium data can be treated (Dimian et al., 2014):

- *PT-xy data* contain the maximum information. Isothermal *P-xy* and isobaric *T-xy* data are mostly available, the last one being preferred because it incorporates the temperature effect.

- Bubble pressure (P-x) or temperature (T-x) is the easiest information to obtain experimentally (e.g. industrial data [temperature and liquid composition in distillation or flash vessels]).

- Azeotropic points are available for a great number of binary mixtures. The information can be extrapolated via LACT models over the whole concentration range, but the accuracy is not guaranteed. Several collections of azeotrope data are accessible (Gmehling et al., 2004).

- Infinite dilution activity coefficients can be measured accurately by chromatography. Studies claim that extrapolation is reliable over the whole composition range for homogeneous mixtures.

- Reciprocal solubility information is required for handling the LLE.

- *Solubility of gases in liquids.* At low gas concentration, the phase equilibrium may be treated by means of Henry coefficients. The thermodynamic treatment is more subtle when dealing with concentrated solutions or with species giving interactions in the liquid phase.

2.8.3 Thermodynamic Consistency

The quality of experimental data can be assessed before regression by a test for *thermodynamic consistency*. The treatment is based on the Gibbs–Duhem equation written as (Dimian et al., 2014)

$$\frac{\Delta H_m}{RT^2} dT + \frac{\Delta V_m}{RT} dP + x_1 d\ln\gamma_1 + x_2 d\ln\gamma_2 = 0 \tag{2.223}$$

ΔH_m and ΔV_m are molar enthalpy and molar volume of mixing, respectively. The heat of mixing is negligible in most cases. In an isobaric system, the second term also vanishes, so the equation becomes

$$x_1 d\ln\gamma_1 + x_2 d\ln\gamma_2 = 0 \tag{2.224}$$

Applying Equation 2.224 over the experimental range gives the following relation:

$$\int_{x_i}^{x_f} \ln\frac{\gamma_1}{\gamma_2} dx_1 = 0 \,(\text{Thermodynamic consistency test}) \tag{2.225}$$

The preceding test is necessary but not sufficient because the ratio of activity coefficients eliminates the effect of pressure, which should be checked for correctness. The calculated value of pressure is

$$P* = \frac{x_1 \gamma_1^* f_1^{0,L}}{\phi_1} + \frac{x_2 \gamma_2^* f_2^{0,L}}{\phi_2} \tag{2.226}$$

where the asterisk marks the estimated values. Consequently, it is required to use a pre-correlation of the activity coefficients γ_i^*, for example, by *Legendre polynomials* or a *four-suffix Margules*-type model. The correction for vapor non-ideality ϕ_i should be included. Equally important is the accuracy of liquid fugacities $f_i^{0,L}$. The parameters entering in precorrelating γ_i^* can be obtained by minimizing the following objective function (Dimian et al., 2014):

$$Q = \sum_{i=1}^{n} (P - P*)_i^2 \tag{2.227}$$

The next step is computing the vapor molar fractions y_i^* from the estimated quantities:

$$y_1^* = \frac{x_1 \gamma_1^* f_1^{0,L}}{P* \phi_1} \tag{2.228}$$

An accurate consistency test is passed if two conditions are satisfied (Gess et al., 1991):

- Average deviation $\Delta \bar{y} = \frac{1}{n} \sum_{i=1}^{n} |(y - y^*)_i|$ below 0.01.
- Deviation $\Delta y = y_i - y_i^*$ evenly distributed, as measured by the bias $\frac{1}{n} \sum_{i=1}^{n} \Delta y$.

2.8.4 Statistical Methods

The regression of parameters from experimental data can follow two statistical techniques: least square quadratic (LSQ) and maximum likelihood (ML).

2.8.4.1 Least Square Regression

The LSQ method consists of finding the set of parameters minimizing the square error between predicted and experimental values. Regressions for LACT and EOS models are encountered. In the LACT approach, the phase equilibrium equations for a binary system are

$$y_1 P \phi_1 = x_1 \gamma_1 f_1^{0,L} \text{ and } y_2 P \phi_2 = x_2 \gamma_2 f_2^{0,L} \tag{2.229}$$

Directly measured variables are x, y, P and T, while γ may be seen as an indirectly measurable quantity. Among the different formulations of objective function, the following options are recommended:

- Bubble pressure (often referred to as Baker's method, BM):

$$Q = \sum_{i=1}^{n} \left(P_i - P_i^*\right)^2 \tag{2.230}$$

- Vapor composition and bubble pressure:

$$Q = \sum_{i=1}^{n} \left\{ \left(y_1 - y_1^*\right)^2 + \left(y_2 - y_2^*\right)^2 + \left(\frac{P - P^*}{P}\right)^2 \right\} \tag{2.231}$$

An objective function formulated in terms of K values has provided good results for fitting interaction parameters from EOS models. Experimental values are $K_i = y_i/x_i$, while calculated values are $K_i^* = \phi_i^V / \phi_i^L$.

$$Q = \sum_{i=1}^{n} \left\{ \left[\frac{K_1 - K_1^*}{K_1} \right] + \left[\frac{K_2 - K_2^*}{K_2} \right] \right\} \tag{2.232}$$

2.8.4.2 Maximum Likelihood Approach

The ML approach makes the basic assumption that all measured and calculated variables are subject to random errors. The errors are normally distributed and independent, while data errors may be characterized by a variance σ^2, either globally or individually. The numerical method consists of finding a suitable set of parameters that maximizes the *likelihood* between true values and measured data. Several types of objective function may be considered. One of the most efficient consists of minimizing the deviations in T, P, x, y weighted by the variance of errors:

$$S = \sum_{i=1}^{n} \left\{ \frac{\left(P_i^o - P_i^e\right)^2}{\sigma_{P,i}^2} + \frac{\left(T_i^o - T_i^e\right)^2}{\sigma_{T,i}^2} + \frac{\left(x_i^o - x_i^e\right)^2}{\sigma_{x,i}^2} + \frac{\left(y_i^o - y_i^e\right)^2}{\sigma_{y,i}^2} \right\} \tag{2.233}$$

Superscripts o and e designate observed and estimated values, while $\sigma_{X,i}^2$ represents the standard deviation of measurements in pressure, temperature and compositions. A problem in the ML approach is the estimation of experimental errors, seldom reported in original articles. Recommended values are $\Delta T = 1$ K, $\Delta P = 133$ Pa (1 mm Hg), $\Delta x = 0.005$, $\Delta y = 0.015$. In the ML approach, the equilibrium equations are considered constraints of the optimization algorithm (Dimian et al., 2014).

Note that the ML method should produce better estimates than LSQ if the errors are random and not systematic. When the errors are systematic, graphical representation is the simplest way to detect them. With good-quality data, the two approaches should give similar good results, but if the data are inaccurate, the regression procedure plays an important role. The quality of data regression should be checked systematically by inspecting the deviation of individual points with respect to model prediction and the statistical significance of parameters (Dimian et al., 2014). Clearly, *the choice of a model is not as important as the procedure used to obtain the parameters from limited and inaccurate experimental data* (Prausnitz et al., 1980).

2.8.5 Evaluation of Models

The monograph of Gess et al. (1991) contains a comprehensive evaluation of different models. The evaluation studied 104 binaries covering a large variety of chemical classes. The components were assigned to the following categories for this study: non-polar, weakly polar, strongly polar, and aqueous. Typical systems have been formed by combining components with different polarity from different chemical classes. They have been completed with mixtures containing carboxylic acids and immiscible systems, and special attention was paid to the quality of the VLE data. The authors proposed a simple method to distinguish between ideal and non-ideal systems. This consists of correlating *excess Gibbs free energy* with a simple *one-suffix Margules equation* of the form $G^{ex}/RT = Ax_1x_2$. Theoretically, $A = 0$ for non-ideal mixtures, but practically this limit may be set to $A = 0.6$. Note that ideal systems may be formed not only from non-polar components but also from the combination of some polar components, such as alcohols.

The following are the conclusions of the study (Dimian et al., 2014):

- For slightly non-ideal systems, there was little difference between models.

- UNIQUAC seemed to be the best for non-polar/weakly polar, weakly polar/weakly polar and weakly polar/strongly polar combinations.

- Wilson was best for non-polar/strongly polar mixtures, but UNIQUAC and Margules were most accurate at times.

- In case of strongly polar/strongly polar systems, no conclusion was drawn. Simple models (van Laar and Margules) sometimes behaved better than sophisticated models. UNIQUAC/Margules are recommended.

- For aqueous miscible systems, UNIQUAC and Wilson behaved the best.

- EOS models were in general less accurate than LACT models, especially for strongly polar components. For non-polar/weakly polar combinations, the EOS models were in some cases among the best.

- Huron–Vidal mixing rules produced better results than geometrical mixing rules. For non-polar/non-polar systems, there was no difference. Modified EOS polar and WS mixing rules were not evaluated.

TABLE 2.5 Recommendations on Liquid Activity Models

	Non-Polar (NP)	Weakly Polar (WP)	Strongly Polar (SP)
Non-Polar	All models	–	–
Weakly polar	UNIQUAC	UNIQUAC	–
Strongly polar	Wilson	UNIQUAC	None
Aqueous miscible	–	–	UNIQUAC
Aqueous immiscible		NRTL or UNIQUAC	

- At high pressure, EOS models were better, while at low pressure the LACT models were superior.

- Immiscible systems can be handled by NRTL, UNIQUAC, van Laar and Margules.

Table 2.5 summarizes the recommendations regarding LACT models (Dimian et al., 2014). UNIQUAC gives good results in most cases, but for mixtures of strong polar molecules, all LACT models should be checked. Wilson is a good choice for homogeneous organic mixtures, while NRTL and UNIQUAC are recommended for immiscible systems.

2.9 CONCLUDING REMARKS

Thermodynamics plays a central role in mastering the heat pump cycles within chemical processes. This chapter provided a review of the fundamental concepts in thermodynamics, the network that links the properties of a fluid (enthalpy, entropy, Gibbs free energy and fugacity) with the primary measurable state parameters (temperature, pressure, volume, concentration). The key consequence is that a comprehensive computation of physical properties is possible with a suitable *PVT* model and using only limited fundamental data, such as critical coordinates and ideal gas heat capacity.

Generalized methods for calculating the thermodynamic properties of fluids are based on the concept of departure functions, which are the difference between the property of a real fluid and ideal fluid at given pressure and temperature. A complete set of thermodynamic properties can be determined (e.g. enthalpy, entropy, Gibbs free energy and fugacity). The integration of the closed-form equations for different functions requires the availability of an accurate *PVT* relationship.

The main types of *PVT* representation used are based on EOS and corresponding states principle. EOS models are standard tools currently, and cubic EOS types are particularly beneficial because they offer a consistent computation of both thermodynamic properties and phase equilibrium. However, there is no single EOS that could offer accurate prediction of thermodynamic properties of all types of components, from non-polar and low molecules up to polar species and polymers.

Fugacity is a key concept in phase equilibrium, and its calculation implies either EOSs (pure components and mixtures) or LACT coefficients (non-ideal liquid mixtures). Classical cubic EOS models (such as SRK and PR) are suitable only for processes involving hydrocarbons. Several modified cubic EOSs with supplementary adjustable parameters are available for handling large, asymmetric and polar molecules, while the estimation of volumetric properties is improved using VTPR models. Non-analytical EOS models based on molecular modelling are able to handle systems with large molecular asymmetry (polymers, biomolecules). Many thermodynamic options and routes of methods are possible when performing a simulation, so the choice should be guided by the model compatibility with the physical situation and the required accuracy, as well as by the (experimental) availability of specific parameters.

As further reading, a good introduction in thermodynamics is the book by Elliott and Lira (2012), the main concepts in phase equilibrium were clearly explained by Smith et al. (2005), while modeling in phase equilibrium was treated by Sandler et al. (1993) with recent developments in the monograph of Kontogeorgis and Folas (2010). An in-depth presentation of thermodynamics applied in simulation can be found in the book by Gmehling et al. (2012), while Sandler's (2006) work remains the reference work in chemical engineering thermodynamics.

LIST OF SYMBOLS

A	Helmholtz free energy	J/mol
a	Activity	mol/L
C_P	Constant-pressure heat capacity	J/mol K
C_V	Constant-volume heat capacity	J/mol K
f	Fugacity	kPa
F	Degrees of freedom	
G	Gibbs free energy	J/mol
H	Enthalpy	J/mol
h	Height	m
H_{iA}	Henry constant	kPa
k	Interaction parameter	
K	Equilibrium constant	
l	Interaction parameter	
m	Number of segments in molecule	
M	Mean molar value of a property (or a generic property)	
n	Component number	–
P	Pressure	kPa
Q	Heat	J/mol
R	Universal/ideal gas constant	J/mol K
S	Entropy	J/mol K
T	Temperature	K
U	Internal energy	J/mol
V	Volume (molar)	m^3/mol
W	Work	J/mol
x	Liquid molar fraction	–
y	Vapor molar fraction	–
Z	Compressibility coefficient	–

Greek Symbols

κ	Polynomial function of ω	–
ΔH	Enthalpy of phase transition	J/mol
ϕ	Fugacity coefficient/average segment fraction	–
γ	Activity coefficient	–
η	Efficiency	–
λ	Interaction energy	
μ	Chemical potential	
τ	Differences in interaction energy	
ω	Acentric factor	–

Superscripts

0	Reference/standard condition
assoc	Association
DD	Dipolar interaction
disp	Dispersion
e	Experimental/estimated value
ex	Excess
hc	Hard chain
hs	Hard sphere
id	Ideal
ig	Ideal gas
iL	Ideal liquid
L	Liquid
o	Observed value
ph	Physical
QQ	Quadrupolar interaction
r	Reduced
R|res	Residual
sat	Saturated
seg	Segment
V|vap	Vapor

Subscripts

0	Reference state
ad	Adiabatic
c	Critical/cold
C	Combinatorial/cold
eq	Equilibrium
f	Fusion
h	Hot
H	Hot
i	Component *i*

in	Inlet conditions
j	Component *j*
l	Liquid
max	Maximum
n	Number of components
out	Outlet conditions
r	Reduced
R	Residual
ref	Reference
rev	Reversible
s	Shaft/solid/saturated
T	At temperature T
v|*vap*	Vapor/vaporization
vap	Vaporization

Abbreviations

AR	Absorption refrigeration
BIP	Binary interaction parameter
BK-10	Braun K-10
BM	Baker's method
BRW	Benedict–Webb–Rubin
BWR-LS	Benedict–Webb–Rubin–Lee–Starling
BWRS	Benedict–Webb–Rubin–Starling
COSMO	Conductor-like screening model
COSMO-RS	COSMO real solvent
COSMO-SAC	COSMO segment activity coefficient
CPA	Cubic plus association
EOS	Equation of state
GCM	Group contribution method
G	Gas
GLE	Gas-liquid equilibrium
HC	Hydrocarbon
HL	Henry's law
HP	Heat pump
HOC	Hayden–O'Connell
ICE	Internal combustion engine
L	Liquid
LACT	Liquid activity (model)
LLE	Liquid–liquid equilibrium
LK	Lee–Kesler
LKP	Lee–Kessler–Ploecker
LR	Lewis–Randall
LSQ	Least square method

ML	Maximum likelihood approach
NP	Non-polar
NRTL	Non-random two liquid
PC-SAFT	Perturbed-chain statistical associating fluid theory
PPR78	Predictive Peng–Robinson (1978 form)
PR	Peng–Robinson
PRSV	Peng–Robinson–Stryjek–Vera
PRU	Peng–Robinson UNIFAC
PRWS	Peng–Robinson with Wong–Sandler mixing rules
PSRK	Predictive Soave–Redlich–Kwong
PT	Patel–Teja
PVT	Pressure-volume-temperature
RK	Redlich–Kwong
S	Solid
SAFT	Statistical associating fluid theory
SC	Supercritical
SLE	Solid-liquid equilibrium
SP	Strong polar
SRK	Soave–Redlich–Kwong
SRK-MHV2	Soave–Redlich–Kwong with Michelsen–Huron–Vidal mixing rules
SR-POLAR	Schwartzentruber, Renon, Watanasiri
UMR	Universal mixing rule
UNIFAC	UNIQUAC functional-group activity coefficients
UNIQUAC	Universal quasi-chemical
V	Vapor
VC	Vapor compression
VLE	Vapor-liquid equilibrium
VLLE	Vapor-liquid-liquid equilibrium
VPA/IK-CAPE	Vapor Phase Association/Industrie Konsortium CAPE (Industrial Cooperation on Computer Aided Process Engineering)
VTPR	Volume-translated Peng–Robinson
VTSRK	Volume-translated Soave–Redlich–Kwong
WP	Weakly polar
WS	Wong-Sandler

REFERENCES

Abrams D. S., Prausnitz J. M., Statistical thermodynamics of liquid mixtures: A new expression for the excess Gibbs energy of partly or completely miscible systems, *AIChE Journal*, 21 (1975), 116–128.

Ahlers J., Gmehling J., Development of an universal group contribution equation of state. I. Prediction of liquid densities for pure compounds with a volume translated Peng-Robinson equation of state, *Fluid Phase Equilibria*, 191 (2001), 177–188.

Benedict M., Webb G. B., Rubin L. C., An empirical equation for thermodynamic properties of light hydrocarbons and their mixtures: I. Methane, ethane, propane, and n-butane, *Journal of Chemical Physics*, 8 (1940), 334–345.

Bouillot B., Teychené S., Biscans B., An evaluation of thermodynamic models for the prediction of drug solubility in organic solutes, *Fluid Phase Equilibria*, 309 (2011), 36–52.

Brulé M. R., Lin C. T., Lee L. L., Starling K. E., Multiparameter corresponding states correlation of coal-fluid thermodynamic properties, *AIChE Journal*, 28 (1982), 616–637.

Carlson E. C., Don't gamble with physical properties for simulations, *Chemical Engineering Progress*, 10 (1996), 35–46.

Chao K. C., Seader J. D., A general correlation of vapor-liquid equilibria in hydrocarbon mixtures, *AIChE Journal*, 7 (1961), 598–605.

Chapman G., Gubbins K., Jackson G., Radocz M., SAFT: Equation-of-state solution model for associating fluids, *Fluid Phase Equilibria*, 52 (1989), 31–38.

Constantinescu D., Klamt A., Geana D., Vapor-liquid equilibrium prediction at high pressures using activity coefficients at infinite dilution from COSMO-type methods, *Fluid Phase Equilibria*, 231 (2005), 231–238.

Dimian A. C., Use process simulation to improve plant operation, *Chemical Engineering Progress*, 90 (1994), 58–66.

Dimian A. C., Bildea C. S., Kiss A. A., *Integrated design and simulation of chemical processes*, 2nd edition, Elsevier, New York, 2014.

Elliott J. R., Lira C. T., *Introductory chemical engineering thermodynamics*, 2nd edition, Prentice Hall, Englewood Cliffs, NJ, 2012.

Fredenslund A., Gmehling J., Rasmussen P., *Vapor-liquid equilibria using UNIFAC*, Elsevier, New York, 1977.

Fredenslund A., Jones R. L., Prausnitz J. M., Group-contribution estimation of activity coefficients in non-ideal liquid mixtures, *AIChE Journal*, 21 (1975), 1086–1099.

Gess M. A., Danner P. R., Nagvekar M., *Thermodynamic analysis of vapor-liquid equilibria: Models and a standard database*, DIPPR, AIChE, New York, 1991.

Gmehling J., Li J., Fischer K., Further development of the PSRK model for the prediction of gas solubilities and vapor-liquid-equilibria at low and high pressures, *Fluid Phase Equilibria*, 141 (1997), 113–127.

Gmehling J., Kolbe B., Kleiber M., Rarey J., *Chemical thermodynamics for process simulation*, Wiley-VCH, Weinheim, Germany, 2012.

Gmehling J., Menke J., Krafczyk J., Fischer K., *Azeotropic data*, 3 volumes, Wiley-VCH, Weinheim, Germany, 2004.

Green D. W., Perry R. H., *Chemical engineer's handbook*, 8th edition, McGraw-Hill, New York, 2008.

Gross J., Sadowski G., Perturbed-chain SAFT: An equation of state based on a perturbation theory for chain molecules, *Industrial and Engineering Chemistry Research*, 40 (2001), 1244–1260.

Hayden J. G., O'Connell J. P., A generalized method for predicting second virial coefficients, *Industrial and Engineering Chemical Process Design and Development*, 14 (1975), 209–216.

Holderbaum T., Gmehling J., PSRK: A group-contribution equation of state based on UNIFAC, *Fluid Phase Equilibria*, 70 (1991), 251–265.

Hsieh C-M., Sandler S., Lin S-T., Improvement of COSMO-SAC for vapor-liquid and liquid-liquid equilibrium prediction, *Fluid Phase Equilibria*, 297 (2010), 90–97.

Huron M. J., Vidal J., New mixing rules in simple equations of state for representing vapor-liquid equilibrium of strongly non-ideal mixtures. *Fluid Phase Equilibria*, 3 (1979), 255–271.

Jaubert J. N., Privat R., Mutelet F., Predicting the phase equilibria of synthetic petroleum fluids with the PPR78 approach, *AIChE Journal*, 56 (2010), 3225–3235.

Joback K. G., Reid R. C., Estimation of pure-component properties from group-contributions, *Chemical Engineering Communications*, 57 (1987), 233–243.

Kiss A. A., *Advanced distillation technologies—Design, control and applications*, Wiley, Chichester, UK, 2013.

Klamt A., Conductor-like screening model for real solvents: A new approach to the quantitative calculation of solvation phenomena, *The Journal of Physical Chemistry*, 99 (1995), 2224–2235.

Klamt A., *COSMO-RS, from quantum chemistry to fluid phase thermodynamics and drug design*, Elsevier, New York, 2005.

Kontogeorgis G. M., Folas G., *Thermodynamic models for industrial applications*, Wiley, New York, 2010.

Kontogeorgis G. M., Voutsas E. C., Yakoumis I. V., Tassios D. P., An equation of state for associating fluids, *Industrial and Engineering Chemistry Research*, 35 (1996), 4310–4318.

Kyle B. G., *Chemical and process thermodynamics*, 3rd edition, McGraw-Hill, 1999.

Larsen B. L., Rasmunsen P., Fredenslund A., A modified UNIFAC group contribution model, *Industrial and Engineering Chemistry Research*, 26 (1987), 2274–2286.

Lee B. I., Kesler M. G., A generalized thermodynamic correlation based on three-parameter corresponding states, *AIChE Journal*, 21 (1975), 510–527.

Lin S. T., Sandler S., A priori phase equilibrium prediction from a segment contribution solvation model, *Industrial and Engineering Chemistry Research*, 41 (2002), 899–913.

Mathias P. M., Copeman T. W., Extension of the Peng Robinson equation of state to complex mixtures: Evaluation of the various forms of the local composition concept, *Fluid Phase Equilibria*, 13 (1983), 91–108.

Moran M., Shapiro H. N., Boettner D. D., Bailey M. B., *Fundamentals of engineering thermodynamics*, 7th edition, Wiley, New York, 2011.

Mullins E., Oldland R., Liu Y. A., Wang S., Sandler S. I., Chen C-C., Zwolak M., Seavey K. C., Sigma-profile database for using COSMO-based thermodynamic methods, *Industrial and Engineering Chemistry Research*, 45 (2006), 4389–4415.

Nothnagel K-H., Abrams D. S., Prausnitz J. M., Generalized correlation for fugacity coefficients at moderate pressures, *Industrial and Engineering Chemical Process Design and Development*, 12 (1973), 25–35.

Patel N. C, Teja A. S., A new cubic equation of state for fluids and fluid mixtures, *Chemical Engineering Science*, 37 (1982), 463–473.

Peneloux A., Rauzy E., Freeze R., A consistent correction for Redlich-Kwong-Soave volumes, *Fluid Phase Equilibria*, 8 (1982), 7–23.

Peng D. Y., Robinson D. B., A new two-constant equation of state, *Industrial and Engineering Chemistry: Fundamentals*, 15 (1976), 59–64.

Perry R. H., Green D. W., *Chemical engineer's handbook*, 7th edition, McGraw-Hill, New York, 1997.

Pitzer K. S., Curl R. F., Jr., The volumetric and thermodynamic properties of fluids. III. Empirical equation for the second virial coefficient, *Journal of the American Chemical Society*, 79 (1957), 2369–2370.

Ploecker U., Knapp H., Prausnitz J. M., Calculation of high-pressure vapor-liquid equilibria from a corresponding-states correlation with emphasis on asymmetric mixtures, *Industrial and Engineering Chemical Process Design and Development*, 17 (1978), 324–332.

Poling B. E., Prausnitz J. M., O'Connel J. P., *Properties of gases and liquids*, 5th edition, McGraw-Hill, New York, 2001.

Prausnitz J. M., Anderson T. F., Grens E. A., Eckert C. A., Hsieh R., O'Connell J. P., *Computer calculations for multi-component vapor-liquid and liquid-liquid equilibria*, Prentice-Hall, Englewood Cliffs, NJ, 1980.

Privat R., Gani R., Jaubert J. N., Are safe results obtained when the PC-SAFT equation of state is applied to ordinary pure chemicals? *Fluid Phase Equilibria*, 295 (2010), 76–92.

Redlich O., Kwong J. N. S., On the thermodynamics of solutions, *Chemical Reviews*, 44 (1949), 233–244.

Reid R. C., Prausnitz J. M., Poling B. E., *The properties of gases and liquids*, 4th edition, McGraw-Hill, New York, 1987.

Renon H., Prausnitz J. M., Local compositions in thermodynamic excess functions for liquid mixtures, *AIChE Journal*, 14 (1968), 135–144.

Sandler, S. I., Quantum mechanics: A new tool for engineering thermodynamics, *Fluid Phase Equilibria*, 210 (2003), 147–160.

Sandler S. I., *Chemical, biochemical and engineering thermodynamics*, 4th edition, Wiley, New York, 2006.

Sandler S. I., Orbey H., Lee B., Equations of state, in Sandler S. I. (Ed.), *Models for thermodynamics and phase equilibria calculations*, Dekker, New York, 1993, 87–186.

Schmid B., Gmehling J., Revised parameters and typical results of the VTPR group contribution equation of state, *Fluid Phase Equilibria*, 317 (2012), 110–126.

Schwartzentruber J., Renon H., Watanasiri S., Development of a new cubic equation of state for phase equilibrium calculations, *Fluid Phase Equilibria*, 52 (1989), 127–134.

Schwartzentruber J., Renon H., Watanasiri S., K-values for non-ideal systems: An easier way, *Chemical Engineering*, March (1990), 118–124.

Sengers J. V., Kayser R. F., Peters C. J., White H. J., *Equations of state for fluids and fluid mixtures*, Commission I.2 on Thermodynamics of the International Union of Pure and Applied Chemistry (IUPAC), Elsevier, Amsterdam, 2000.

Shah M. R., Yadav G. D., Prediction of liquid-liquid equilibria for biofuel applications using the COSMO-SAC method, *Industrial and Engineering Chemistry Research*, 50 (2011), 13066–13075.

Shah M. R., Yadav G. D., Prediction of liquid-liquid equilibria using the COSMO-SAC model, *Journal of Chemical Thermodynamics*, 49 (2012), 62–69.

Smith J. M., Van Ness H. C., Abbott M., *Introduction in chemical engineering thermodynamics*, 7th edition, McGraw-Hill, New York, 2005.

Soave G., Equilibrium constants from a modified Redlich-Kwong equation of state, *Chemical Engineering Science*, 27 (1972), 1197–1203.

Soave G., Application of a cubic equation of state to vapor-liquid equilibria of systems containing polar compounds, *Institute of Chemical Engineering Symposium Series*, No. 56 (1979), 1.2/1.

Soave G., Improvement of the Van Der Waals equation of state, *Chemical Engineering Science*, 39 (1984), 357–369.

Stryjek R., Vera J. H., PRSV: An improved Peng-Robinson equation of state for pure compounds and mixtures, *The Canadian Journal of Chemical Engineering*, 64 (1986), 323–333.

Tumaka F., Gross J., Sadowski G., Thermodynamic modeling of complex systems using PC-SAFT, *Fluid Phase Equilibria*, 228 (2005), 89–98.

van der Waals J. D., Over de continuiteit van den gas- en vloeistoftoestand (On the continuity of the gas and liquid state), PhD thesis, Leiden, the Netherlands, 1873.

Voutsas E., Magoulas K., Tassios D., Universal mixing rule for cubic equations of state applicable to symmetric and asymmetric systems: Results with the Peng-Robinson equation of state, *Industrial and Engineering Chemistry Research*, 43 (2004) 6238–6246.

Wang S., Sandler S. I., Chen C. C., Refinement of COSMO-SAC and the applications, *Industrial and Engineering Chemistry Research*, 46 (2007), 7275–7288.

Weidlich U., Gmehling J., A modified UNIFAC model, *Industrial and Engineering Chemistry Research*, 26 (1987), 1372–1381.

Wilson G. M., Vapor-liquid equilibrium. XI. A new expression for the excess free energy of mixing, *Journal of the American Chemical Society*, 86 (1964), 127–130.

Wittig R., Lohmann J., Gmehling J., Vapor-liquid equilibria by UNIFAC group contribution. 6. Revision and extension, *Industrial and Engineering Chemistry Research*, 42 (2003), 183–188.

Wong H., Sandler S. I., A theoretically correct mixing rule for cubic equations of state, *AIChE Journal*, 38 (1992), 671–680.

CHAPTER 3

Entropy Production and Exergy Analysis

3.1 INTRODUCTION

Entropy production plays a key role in the thermodynamics of irreversible processes, and it determines the performance of thermal machines such as heat engines, heat pumps, refrigerators, air conditioners and power plants. Carnot recognized early the importance of avoiding irreversible processes, thus reducing the entropy production (then called *uncompensated transformations*).

Exergy analysis is a powerful technique for evaluating and enhancing the efficiency of processes, systems and devices, as well as improving the ecoefficiency performance. Consequently, exergy analysis is used increasingly in many sectors, in particular by the chemical process industry, with the aim of improving the energy efficiency and sustainability.

This chapter introduces and discusses evaluation criteria for vapor compression cycles based on analysis with the first and second laws of thermodynamics. Although other cycles can be applied for attaining heat pump effects, vapor compression cycles are most common. The most simple vapor compression cycle is discussed here as an example. Industrial plants mostly require complex types of vapor compression cycles, but similar analysis can be developed for such complex cycles. Sahin et al. (2001), Sahin and Kodal (2002) and Chen (1999) presented thermoeconomic optimization studies of two-stage refrigeration systems, including the effect of the major irreversibilities and economic parameters.

3.2 FIRST AND SECOND LAW EFFICIENCY AND SUSTAINABILITY CRITERIA

Figure 3.1 shows schematically a vapor compression refrigeration/heat pump cycle. The cycle operates counterclockwise. A refrigerant undergoes a number of processes in a closed loop. Starting from point 1, low-pressure refrigerant vapor is compressed to high-pressure state 2. This requires the compressor shaft power $\dot{W}_{comp}$. In the condenser, heat is rejected to a heat sink. In industrial plants, the heat sink is mostly a

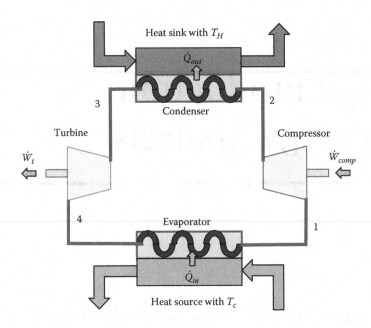

FIGURE 3.1   Vapor compression heat pump.

process stream that needs to be heated. Due to heat rejection, the refrigerant first cools to saturated vapor conditions, then it condenses until all refrigerant becomes liquid at saturated conditions. Eventually, the refrigerant is slightly subcooled as it leaves the condenser with state 3. Generally, a throttling device substitutes the turbine shown in the figure. In that case, no power is recovered from the expansion process, that is, $\dot{W}_t = 0$. After the throttling device, the refrigerant is again at low pressure and has a state in the two-phase region. In the evaporator, heat is absorbed from the heat source until the refrigerant attains vapor-saturated conditions. Eventually, the vapor is superheated until state 1. In industrial plants, the heat source is mostly a waste stream that needs to be cooled to environmental temperature. Sometimes, the heat can be taken from other process steps where cooling is needed. In that case, the cycle has a dual function: refrigerating machine and heat pump.

The rate of heat removal from the heat source, heat rejection to the heat sink and the power input to the compressor are determined from their definitions:

$$\dot{Q}_{in} = \dot{m}_{ref} \times (h_1 - h_4) \tag{3.1}$$

$$\dot{Q}_{out} = \dot{m}_{ref} \times (h_2 - h_3) \tag{3.2}$$

$$\dot{W}_{comp} = \dot{m}_{ref} \times (h_2 - h_1) \tag{3.3}$$

The enthalpy of state 2 follows from the definition of isentropic compressor efficiency if this efficiency is known:

$$\eta_{is} = \frac{h_{2s} - h_1}{h_2 - h_1} \tag{3.4}$$

The (first law) performance of refrigeration cycles and heat pumps is expressed in terms of the coefficient of performance (COP), which is defined as

$$COP_{hp} = \frac{objective\_of\_the\_cycle}{required\_input} \tag{3.5}$$

The objective of a heat pump cycle is to deliver heat to the high-temperature heat sink:

$$COP_{hp} = \frac{\dot{Q}_{out}}{\dot{W}_{comp} - \dot{W}_t} \tag{3.6}$$

Note that we also can define refrigeration cycle efficiency. In this case, the purpose of the cycle is to remove heat from the low-temperature heat source:

$$COP_r = \frac{\dot{Q}_{in}}{\dot{W}_{comp} - \dot{W}_t} \tag{3.7}$$

The COP will be a function of the operating conditions of the system and of the refrigerant used in the cycle. Figure 3.2 shows for a number of refrigerants the single-stage vapor

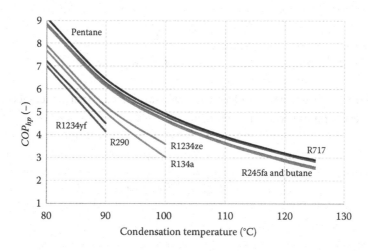

FIGURE 3.2 Coefficient of performance of heat pump cycle as a function of the condensation temperature for an evaporation temperature of 55°C, no subcooling, no superheating and compressor isentropic efficiency of 70%.

compression cycle COP for an evaporating temperature of 55°C. The COP has been calculated by filling Equations 3.2 and 3.3 in Equation 3.6 and calculating the different enthalpy values with REFPROP (Lemmon et al., 2013). All further presented refrigerant data have been calculated with this program. As the condensation temperature increases, the COP significantly decreases, but it remains mostly above 3. Pentane (R601) and ammonia (R717) show the largest values for the COP, followed by butane (R600) and R245fa. Figure 3.2 also shows that fluids with a lower critical temperature show lower COP_{hp} values: R1234ze, R134a, propane (R290) and R1234yf show significantly lower performance for these operating conditions. By using complex system designs, it is possible to attain higher COPs at higher temperatures. The preferred fluids for high-temperature heat pump applications are thus pentane, ammonia, butane and R245fa. Among these fluids, only ammonia operates at too high pressure levels for the considered conditions (23 to 109 bar); the other fluids operate at pressures lower than 25 bar. Butane is often preferred because it has the highest saturation pressure at room temperature (1.0 bar at −0.8°C), while pentane shows subatmospheric equilibrium pressures for temperatures below 35.7°C. R245fa becomes subatmospheric at 14.8°C and is a global warming gas (with a global warming potential [GWP] = 1030 kg CO_2 eq.).

The ideal refrigeration/heat pump cycle is the reversed Carnot cycle. This cycle consists of two reversible isothermal and two isentropic processes. This cycle is totally reversible, and as such it is sustainable and has the largest performance for a given set of sink and source conditions. Thus, it can be derived that

$$COP_{hp\_Carnot} = \frac{T_H}{T_H - T_C} \qquad (3.8)$$

and

$$COP_{r\_Carnot} = \frac{T_C}{T_H - T_C} \qquad (3.9)$$

These performance values serve as a (second law) standard against which actual cycles can be compared, resulting in second law efficiency:

$$\eta_{2nd\,law\_hp} = \frac{COP_{hp}}{COP_{hp\_Carnot}} \qquad (3.10)$$

and

$$\eta_{2nd\,law\_r} = \frac{COP_r}{COP_{r\_Carnot}} \qquad (3.11)$$

Because the Carnot heat pump cycle would be sustainable, these values also indicate the sustainability level of the system. Generally, practical heat pump systems show second law efficiencies below 50%. Figure 3.3 shows values of the second law efficiency for single-stage heat pump cycles with different refrigerants (considering the conditions of Figure 3.2). Pentane and ammonia (NH₃) appear again as the refrigerants with the largest potential to reach high second law efficiencies. Butane shows slightly lower second law efficiencies, but as already mentioned, it is preferred due to its more favorable operating pressure levels. Media such as R1234yf are clearly less-efficient refrigerants, being limited by their critical temperature. For larger systems, higher efficiencies can be expected due to higher compressor isentropic efficiencies. A global indication of the quality of a refrigeration plant (or heat pump) can be obtained by calculating its second law efficiency. If this efficiency shows a value below 40%, the cycle performance is poor but the cause of malfunction cannot be identified. Only a second law analysis of the cycle allows for this identification; this topic is addressed in a further section.

When evaluating the sustainability of a refrigeration/heat pump cycle, not only its energetic performance plays a role but also aspects such as contribution to ozone layer depletion, GWP, safety and toxicity may determine the sustainability of the cycle. In recent years, the total equivalent warming impact (TEWI) of the medium used in the refrigeration/heat pump plant has been adopted to evaluate its sustainability. This includes the direct global warming impact due to emissions of the refrigerant and the indirect global warming impact due to CO_2 emission during the production of the electrical energy needed to drive the cycle. In this way, it also includes the energetic performance of the plant.

$$TEWI = GWP \times refrigerant\_content \times (leak\_rate \times n + end\_life\_leak\_percentage)$$

$$+ \frac{\dot{Q}_{out}}{COP_{hp}} \times number\_operating\_hours \times \frac{CO_2\_emission}{kWh} \tag{3.12}$$

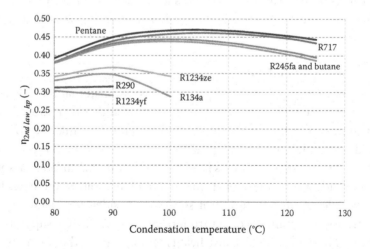

FIGURE 3.3 Second law (Carnot) efficiency of a single-stage heat pump cycle with temperature driving forces of 5 K for both sink and source heat exchangers (under conditions of Figure 3.2).

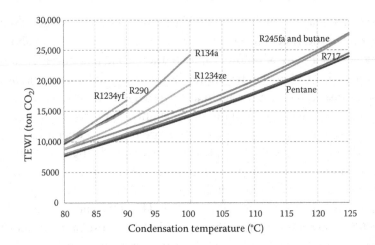

FIGURE 3.4 Total equivalent warming impact (TEWI) of heat pump systems with different refrigerants. Assumed operating conditions: refrigerant content 1000 kg, leak rate 5%, end life leak percentage 10%, operational lifetime 15 years, number of operating hours 8000 h/yr, energy conversion factor 0.58 kg CO_2/kWh, heating capacity 1.0 MW.

In Equation 3.12, *GWP* represents the global warming potential of the refrigerant in CO_2 equivalents, *n* is the expected number of operating years, and the number of operating hours includes all operating hours during the life of the system. Figure 3.4 shows the TEWI as a function of the condensation temperature for an industrial heat pump plant operating under the conditions of Figures 3.2 and 3.3. The assumptions under which the figure applies are given in the figure caption. Pentane and ammonia are again the best refrigerants with the lowest environmental impact. Refrigerant R134a has a 70% larger environmental impact than pentane.

Again, this evaluation method allows comparison of alternatives, but it does not allow for an identification of the reason why a plant shows too high TEWI values. A second law analysis of the cycle is needed if this identification is required, as described in a further section.

3.3 SECOND LAW ANALYSIS

In the derivation of COP_{hp_Carnot} and COP_{r_Carnot}, it was assumed that the heat sink and heat source temperatures are constant, even though heat is added or removed. This implies sink and heat sources of infinite size. In practical systems, the sink and source consist of a flow of a fluid that undergoes a temperature change as heat is exchanged. In heat pump systems, the heat sink is mostly a process flow, possibly indirectly through a water flow. The heat sink is frequently a liquid flow that undergoes a temperature increase. In industrial plants, waste heat flows are common heat sources. Consider, for instance, a heat pump in which a water waste stream is cooled from 62.5°C to 57.5°C by evaporating butane at 55°C, as illustrated in Figure 3.5. Because heat is removed from the water stream, its temperature is lower at the outlet (T_{source_out}) of the heat exchanger than at the inlet (T_{source_in}).

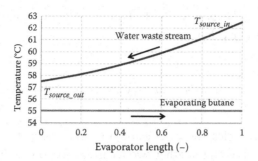

FIGURE 3.5   Heat source (waste stream of water) with a gliding temperature.

As shown further in this discussion, also the refrigerant temperature may change as the refrigerant flow passes the heat exchangers. This is here less relevant because COP_{hp_Carnot} and COP_{r_Carnot} refer to the ideal reversible reference process based on the external temperatures imposed on the cycle.

The inlet and outlet water stream temperatures of the heat pump follow from the available waste stream amount and conditions. If a different (complex) cycle, refrigerant or heat exchanger arrangement is used, the requirement for the source conditions remains.

Similar considerations can be drawn for the heat rejection side. Heat sink and heat source conditions give a constant basis for the comparison of options, but their conditions have to consider their gliding character. Touber (1996) derived that, in the case of gliding sink and source temperatures, the COP of the ideal reversible vapor compression heat pump cycle is obtained when the thermodynamic averaged sink and source temperatures are filled in COP_{hp_Carnot} and COP_{r_Carnot}. These temperatures cause the same entropy production in the condenser and evaporator as the gliding temperatures. The thermodynamic averaged sink and source temperatures are given by

$$\bar{T}_H = \frac{\dot{Q}_{out}}{\dot{m}_{sink} c_{p\_sink} \ln \dfrac{T_{sink\_out}}{T_{sink\_in}}} = \frac{T_{sink\_out} - T_{sink\_in}}{\ln \dfrac{T_{sink\_out}}{T_{sink\_in}}} \qquad (3.13)$$

and

$$\bar{T}_C = \frac{\dot{Q}_{in}}{\dot{m}_{source} c_{p\_source} \ln \dfrac{T_{source\_out}}{T_{source\_in}}} = \frac{T_{source\_out} - T_{source\_in}}{\ln \dfrac{T_{source\_out}}{T_{source\_in}}} \qquad (3.14)$$

Exergy or specific flow availability (Moran and Shapiro, 2007) is the work that would be delivered by a reversible process that would bring the flow in equilibrium with the environmental conditions. This process consists of an isentropic expansion to environment pressure and an isothermal expansion to the entropy state of the environment. The exergy of a flow with state n can be calculated from

$$ex_n = (h_n - h_o) - T_o(s_n - s_o) + \frac{c_n^2}{2} + gz_n \qquad (3.15)$$

In vapor compression cycles, the kinetic and potential energy terms are generally negligibly small in comparison with the other two terms, so that a simplified form of the equation can be used:

$$ex_n = (h_n - h_o) - T_o(s_n - s_o) \qquad (3.16)$$

When analysing vapor compression heat pump cycles, the environment conditions must be taken as the thermodynamic averaged source temperature, $T_0 = \bar{T}_C$. Generally, the environment pressure is the atmospheric pressure ($p_0 = 101.3$ kPa), while the environment enthalpy h_0 and entropy s_0 must be calculated for the medium under consideration. For the refrigerant state points, this implies that

$$h_o = f(T_o, p_o) \qquad (3.17)$$

and

$$s_o = f(T_o, p_o) \qquad (3.18)$$

must be calculated from the thermodynamic data tables or software (i.e. Lemmon et al., 2013) for the refrigerant. Assuming constant specific heat values for the heat sink and heat source media and neglecting pressure differences between air and surroundings, the exergy values for the sink and source media can be approached with

$$ex_{sink} = c_{p\_sink}(T_{sink} - T_o) - T_o c_{p\_sink} \ln \frac{T_{sink}}{T_o} \qquad (3.19)$$

$$ex_{source} = c_{p\_source}(T_{source} - T_o) - T_o c_{p\_source} \ln \frac{T_{source}}{T_o} \qquad (3.20)$$

3.4 EXERGY ANALYSIS OF HEAT PUMP CYCLES

The rate of availability destruction or exergy loss Ex_{loss} for each of the components of a vapor compression refrigeration cycle allows for an identification of the sources of malfunction within the cycle. When the specific exergy values for each state are known, a simple exergy balance will allow for the calculation of the exergy loss of a component. An analysis of the critical design components then becomes possible. In the following sections, each main component is analysed separately, following the state numbering given in Figure 3.1.

3.4.1 Heat Pump Cycle Components

Figure 3.6 shows the conditions around the condenser. The condenser is considered externally adiabatic, such that there is no heat transfer with the surroundings. The refrigerant coming from the compressor with state 2 is desuperheated and condensed to state 3 at the condenser outlet. In countercurrent flow, the heat sink medium (process fluid) is heated from T_{sink_in} to T_{sink_out}. Between the two media, the heat flow $\dot{Q}_{out}$ is exchanged. The exergy loss in the condenser is obtained from

$$Ex_{loss\_cond} = \dot{m}_{ref}(ex_2 - ex_3) + \dot{m}_{sink}(ex_{sink\_in} - ex_{sink\_out}) \qquad (3.21)$$

The useful exergy effect of the cycle is the exergy flow increase of the heat sink medium:

$$Ex_{rise\_sink\_flow} = \dot{m}_{sink}(ex_{sink\_out} - ex_{sink\_in}) \qquad (3.22)$$

Similar considerations can be made for the evaporator. The situation is schematically shown in Figure 3.7.

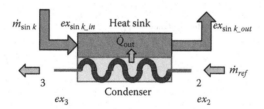

FIGURE 3.6   Conditions around the condenser.

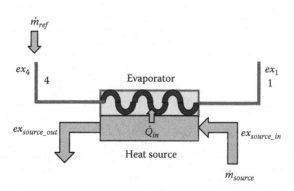

FIGURE 3.7   Conditions around the evaporator.

An analysis of Figure 3.7 gives, for the exergy loss in the evaporator,

$$Ex_{loss\_evap} = \dot{m}_{ref}(ex_4 - ex_1) + \dot{m}_{source}(ex_{source\_in} - ex_{source\_out}) \qquad (3.23)$$

Again, the compressor is considered to be externally adiabatic, which is a reasonable assumption for larger compressors without heat removal (Figure 3.8). Screw compressors for refrigeration/heat pump applications are generally oil injected. In the analysis of such compressors, the exergy flows associated with the oil flow must also be included in the analysis. The exergy loss in the compressor becomes

$$Ex_{loss\_comp} = \dot{m}_{ref}(ex_1 - ex_2) + \dot{W}_{comp} \qquad (3.24)$$

Similarly for the expansion/throttling device,

$$Ex_{loss\_exp} = \dot{m}_{ref}(ex_3 - ex_4) - \dot{W}_t \qquad (3.25)$$

Generally, there will be no work delivery to the surroundings, so that $\dot{W}_t = 0$.

3.4.2 Exergy Efficiency of the Cycle

The exergy efficiency of a vapor compression heat pump cycle is defined as the ratio between the objective of the cycle (useful exergy effect in condenser) and exergy input to the process:

$$\eta_{hp\_2nd\ law} = \frac{Ex_{rise\_\sin k\_flow}}{\dot{W}_{comp} - \dot{W}_t} \qquad (3.26)$$

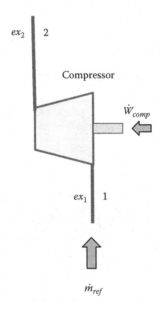

FIGURE 3.8   Conditions around the compressor.

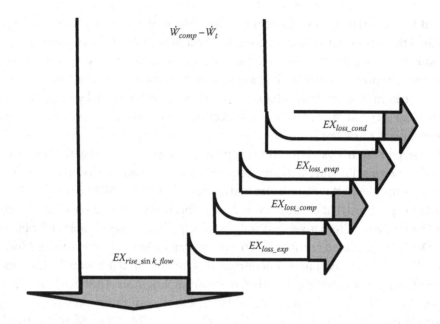

FIGURE 3.9 Sankey diagram illustrating how the different losses are visualized.

The difference between the exergy input $(\dot{W}_{comp} - \dot{W}_t)$ and the exergy rise of the sink flow is equal to the sum of exergy losses per component. This allows for the construction of Sankey diagrams that visualize the contribution of the different components to the total exergy losses of the system, as illustrated in Figure 3.9.

3.4.3 Application Example

A vapor compression heat pump system, using butane (R600) as the working fluid, operates under the conditions listed in Table 3.1. The condenser is a water-cooled condenser in

TABLE 3.1 Operating Conditions of Example Heat Pump Cycle

Medium	Variable	Value
Sink	$\dot{Q}_{out}$	5.8 MW
	T_{sink_in}	117.5°C
	$\dot{m}_{sink}$	273.4 kg/s
	C_{p_sink}	4.2427 kJ/kgK
Source	$\dot{Q}_{in}$	3.33 MW
	T_{source_in}	62.5°C
	$\dot{m}_{source}$	159.2 kg/s
	C_{p_source}	4.1845 kJ/kgK
Refrigerant	$T_{condensation}$	125.0°C
	$T_{condenser\_out}$	123.0°C
	$T_{evaporator\_in}$	45.0°C
	$T_{evaporator\_out}$ (saturation)	43.0°C
	$T_{evaporator\_out}$	57.5°C
	$T_{compressor\_out}$	135.0°C

which heat is rejected to a pressurized water flow. The heat source is a water waste stream flow released by an industrial plant. The specific heat of both water flows can be assumed to be constant. The refrigerant passes the condenser with negligible pressure drop but undergoes a significant pressure drop as it passes the evaporator. Consider all components and refrigerant lines to be externally adiabatic and kinetic and potential energy effects to be negligible. Make an exergy analysis of the heat pump system and compare this analysis with the results of a first law analysis.

The first step in the analysis is locating the principal operating states of the butane cycle in *T-s* and *p-h* diagrams. Pressure-enthalpy diagrams are most widely used in the refrigeration field, but for convenience the *T-s* diagram is also drawn here. REFPROP, a refrigerant properties calculation program that includes a variety of refrigerants (Lemmon et al., 2013), has been used for this analysis. Alternatively, saturated and superheated vapor tables for butane may be used. Data for a confined number of refrigerants can be found in the work of Moran and Shapiro (2007). An extensive number of refrigerants can be found in a work by the American Society of Heating, Refrigerating and Air-Conditioning Engineers (ASHRAE, 2009).

The pressure at the high-pressure side of the cycle is the saturation pressure corresponding to the condensation temperature (125°C), or $p_2 = p_3 = 24.17$ bar. State 2 is fixed by p_2 and $T_2 = T_{compressor\_out} = 135°C$: $h_2 = 776.3$ kJ/kg and $s_2 = 2.594$ kJ/kgK. State 3 is fixed by p_2 and $T_3 = T_{condenser\_out} = 123°C$: $h_3 = 537.3$ kJ/kg and $s_3 = 1.994$ kJ/kgK. The refrigerant state change in the condenser from stage 2 to stage 3 is given in Figure 3.10.

The pressure at the low pressure side of the cycle changes through the evaporator, and at the inlet of the evaporator equals the saturation pressure corresponding to the inlet saturation temperature (45.0°C), or $p_4 = 4.342$ bar and, at the outlet of the evaporator, the saturation pressure corresponding to the outlet saturation temperature (43.0°C), or $p_1 = 4.112$ bar. State 1 is fixed by p_1 and $T_1 = T_{evaporator\_out} = 57.5°C$: $h_1 = 674.7$ kJ/kg and $s_1 = 2.522$ kJ/kgK. The expansion through the valve is a throttling process implying that $h_4 = h_3 = 537.3$ kJ/kg. The entropy of state 4 follows from the local vapor quality:

$$s_4 = s_{L\_4.342} + \frac{h_3 - h_{L\_4.342}}{h_{V\_4.342} - h_{L\_4.342}} \tag{3.27}$$

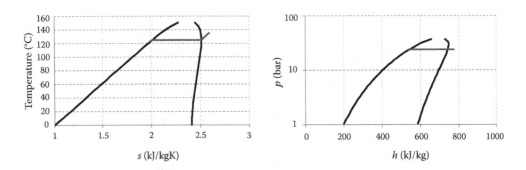

FIGURE 3.10 State change in condenser in *T-s* diagram (left) and *p-h* diagram (right). The line represents part of the 24.17-bar isobar in both diagrams.

with the index V indicating saturated vapor conditions and L saturated liquid conditions. From Equation 3.27, $s_4 = 2.039$ kJ/kgK. The refrigerant state change in the evaporator from stage 4 to stage 1 is given in Figure 3.11. The processes in the expansion valve and compressor can now be obtained by connecting state 3 to state 4 and state 1 to state 2, respectively, as shown in Figure 3.12.

The sink medium outlet temperature can be obtained from

$$T_{sink\_out} = T_{sink\_in} + \frac{\dot{Q}_{out}}{\dot{m}_{sink}c_{p\_sink}} \qquad (3.28)$$

Using the data from Table 3.1, $T_{sink\_in} = 117.5 + 273.15 = 390.65$ K and $T_{sink\_out} = 395.65$ K (122.5°C). The thermodynamic averaged sink temperature follows from Equation 3.13: $\bar{T}_H = 393.14$ K. Similarly, the source medium outlet temperature can be obtained from

$$T_{source\_out} = T_{source\_in} - \frac{\dot{Q}_{in}}{\dot{m}_{source}c_{p\_source}} \qquad (3.29)$$

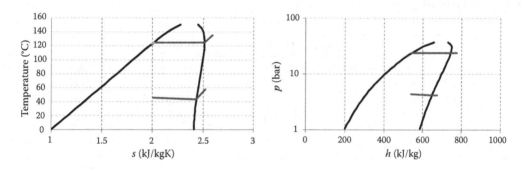

FIGURE 3.11 State change in evaporator in T-s diagram (left) and p-h diagram (right). The line represents the process in the evaporator in both diagrams, considering a pressure drop from $p_4 = 4.342$ bar to $p_1 = 4.112$ bar.

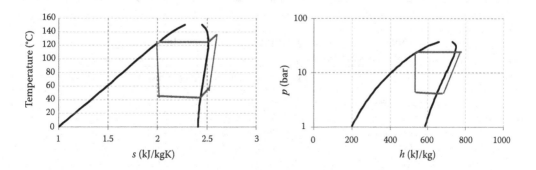

FIGURE 3.12 State change in cycle in T-s diagram (left) and p-h diagram (right).

With the data from Table 3.1, $T_{source\_in} = 62.5 + 273.15 = 335.65$ K and $T_{source\_out} = 330.65$ K (57.5°C). The thermodynamic averaged source temperature follows from Equation 3.14: $\overline{T}_C = 333.14$ K. The first law of thermodynamics gives

$$\dot{W}_{comp} = \dot{Q}_{out} - \dot{Q}_{in} \tag{3.30}$$

Equations 3.6 and 3.8 allow for calculation of the COP of the heat pump cycle $COP_{hp} = 2.35$ and of the reversible heat pump cycle operating under similar external conditions $COP_{hp\_Carnot} = 6.55$, respectively. Using Equation 3.10, the calculated second law efficiency is 36%.

In case of isentropic compression the discharge conditions are fixed by p_2 and $s_{2s} = s_1 = 2.522$ kJ/kgK: $h_{2s} = 747.4$ kJ/kg. The isentropic efficiency of the compressor follows then from Equation 3.4: $\eta_{is} = 0.715$, which is a common value for commercial refrigeration compressors.

Application of Equation 3.16 for the refrigerant states and Equations 3.19 and 3.20, respectively, for sink and source media gives the results listed in Table 3.2.

Equations 3.21 through 3.25 can now be used to evaluate the exergy losses in each component. The mass flow of refrigerant can be evaluated from, for example, Equation 3.2. The results are shown in Table 3.3. It is worth noting that the evaporator and compressor give the largest contribution.

TABLE 3.2   Overview of State Conditions and Calculated Exergy Values

State	T (K)	h (kJ/kg)	s (kJ/kgK)	ex (kJ/kg)
1		674.69	2.5220	63.68
2s		747.35	2.5220	136.34
2		776.31	2.5938	141.38
3		537.27	1.9943	102.06
4		537.27	2.0391	87.14
sink_in	390.65			18.91
sink_out	395.65			22.15
source_in	335.65			0.039
source_out	330.65			0.039
0	333.14			

TABLE 3.3   Overview of Exergy Losses in Application Example Components

	Exergy Loss (kW)	Exergy Loss (%)
Ex_{loss_cond}	69	2.8
Ex_{loss_exp}	362	14.7
Ex_{loss_evap}	569	23.1
Ex_{loss_comp}	580	23.5
$Ex_{rise\_sink\_flow}$	885	35.9
$Ex_{input} = \dot{W}_{comp}$	2466	100

When Equation 3.26 is filled in, it appears that the exergy efficiency of the cycle corresponds to the second law efficiency calculated previously: 36%. In addition to the global quality of the performance of the cycle given by the second law efficiency, Table 3.3 shows that an exergy analysis allows for identification of the components that give the largest contribution to deviation from ideal behavior. In this case, it is the compressor followed by the evaporator. In the next section, the causes of irreversibility per component are discussed.

3.4.4 Impact of Malfunction

The evaporator operates close to environmental conditions so that heat transfer to the surroundings is limited. The condenser is installed in a plant, and its purpose is to exchange heat with a process stream. The refrigerant side pressure drop in the evaporator is nonnegligible but is already accounted for in the conditions of state 1, at the outlet of the evaporator. The main cause of irreversibility in the heat exchangers is heat transfer between sink and condensing medium and between source and evaporating medium.

Figure 3.13 illustrates the conditions around the evaporator, as listed in Table 3.1. It appears that the temperature driving force for heat transfer is rather large. In this case, the malfunction of the superheating control associated with a wrong design of the refrigerant distribution is leading to this large temperature driving force. Assuming that the maintenance personnel can fix this malfunction, the situation schematically illustrated in Figure 3.14 could result. Now, states 4 and 1 have different conditions. The pressure at the low-pressure side of the cycle is, at the inlet of the evaporator, the saturation pressure corresponding to the inlet saturation temperature (52.0°C), or $p_4 = 5.22$ bar, and, at the outlet of the evaporator, the saturation pressure corresponding to the outlet saturation temperature (50°C), or $p_1 = 4.96$ bar. State 1 is fixed by p_1 and $T_1 = T_{evaporator\_out} = 57.5$°C: $h_1 = 671.3$ kJ kg and $s_1 = 2.488$ kJ/kgK. The expansion through the valve is a throttling process, implying that $h_4 = h_3 = 537.3$ kJ/kg. The entropy of state 4 follows from the local vapor quality Equation 3.27: $s_4 = 2.057$ kJ/kgK. Assuming that the isentropic efficiency remains the same ($\eta_{is} = 0.715$), also the state at the compressor discharge changes: $h_2 = 758.0$ kJ/kg and $s_2 = 2.549$ kJ/kgK.

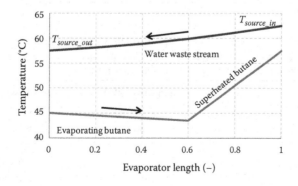

FIGURE 3.13 Temperature change of refrigerant and source medium (water waste stream) through the evaporator.

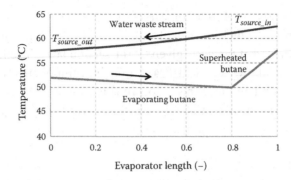

FIGURE 3.14 Temperature change of refrigerant and source medium (waste water stream) through the evaporator after correction of refrigerant distribution and control malfunction.

The effects of the correction of this malfunction are given in Table 3.4. The effect of the correction is quite clear both for first and for second law performance. The COP increases from 2.35 to 2.54 and the second law efficiency from 36% to 39%.

After this modification of the system, the compressor and the expansion valve appear now to be the main causes of nonideal system behavior. The compressor in this plant already has a reasonable isentropic efficiency. Nevertheless, it would be possible to substitute the compressor by a compressor with a higher isentropic efficiency of 85%. If this compressor would be installed, then an exergy analysis of the resulting system gives the results

TABLE 3.4   Overview of Exergy Losses in Application
Example Components after Correction of the Malfunction

	Exergy Loss (kW)	Exergy Loss (%)
Ex_{loss_cond}	63	2.8
Ex_{loss_exp}	551	24.2
Ex_{loss_evap}	247	10.8
Ex_{loss_comp}	533	23.4
$Ex_{rise\_sin k\_flow}$	885	38.8
$Ex_{input} = \dot{W}_{comp}$	2279	100

TABLE 3.5   Overview of Exergy Losses in Application Example
Components after Correction of the Malfunction and Substitution
of Compressor by One with 85% Isentropic Efficiency

	Exergy Loss (kW)	Exergy Loss (%)
Ex_{loss_cond}	61	3.0
Ex_{loss_exp}	586	28.5
Ex_{loss_evap}	263	12.8
Ex_{loss_comp}	258	12.6
$Ex_{rise\_sin k\_flow}$	885	43.1
$Ex_{input} = \dot{W}_{comp}$	2053	100

listed in Table 3.5. The expansion valve is a simple device that cannot easily be improved to recover part of the expansion work.

A relatively small step in isentropic efficiency of the compressor has a significant impact in the system performance: The COP goes from 2.54 to 2.83 (about 11% improvement) and the second law efficiency from 39% to 43% (same relative improvement).

3.5 EXERGY ANALYSIS OF HEAT PUMP PLANT

In the previous section, the exergy losses in heat pump cycles were considered. The exergy entering the cycle in the condenser also leaves the cycle in the condenser so that the net exergy entering the system is the exergy input through the compressor shaft.

When a heat pump plant is considered, globally there are three main exergy inflows: electric energy input for the compressor drive, the condenser fans or pump(s) and the evaporator fans or pump(s). The useful effect of the process corresponds with the exergy increase of the fluid flow passing the condenser. This exergy increase corresponds to the exergy reduction of the same fluid flow due to the heat required by the process plant. The exergy input to the condenser and evaporator fans or pumps is lost. Part of the exergy input is lost directly due to electric motor losses. The rest is lost due to mixing and friction of the external fluids. A small amount of the exergy input to the compressor drive is usefully applied to remove the heat load. The rest is lost in the different components of the heat pump system.

Mostly, plants are designed to handle extreme operating conditions. The result is that most of the time plants operate under part-load conditions. If the plant switches on and off, then a dynamic analysis of the plant is needed to be able to quantify the impact of the different exergy inputs as a function of time. An example of possible exergy loss distribution is given in Figure 3.15 (Infante Ferreira et al., 1999). The figure applies for the integrated exergy flows of a refrigeration plant for a period of 24 hours. The integration is needed to quantify the effect of frost formation and continuous operation of the evaporator fans. The continuous operation of the evaporator fans is needed to guarantee the product quality. The useful effect is only 9% of the total exergy input. The largest losses take place in the air cooler fans (41%), electric motor of the compressor drive (16%), condenser (14%), compressor (7%) and condenser fans (5%).

Some researchers propose that the required input in Equation 3.5 should include the auxiliary energy inputs to the system: to drive evaporator and condenser fans or pumps and required control systems. A so-called coefficient of system performance (COSP) is then obtained. In that case, the results obtained with Equation 3.10 are significantly lower. If the power usage is not simultaneous, the ratios of operating hours must be considered. Figure 3.15 illustrates that when studying the exergy losses, an integrated approach is needed to allow for the identification of the main causes of exergy losses. For this plant, application of two-speed evaporator fans that operate at high speed as the refrigeration cycle is in operation and at low speed if only air circulation is required to guarantee an acceptable product quality could lead to significant exergy (and energy) performance enhancement of the system.

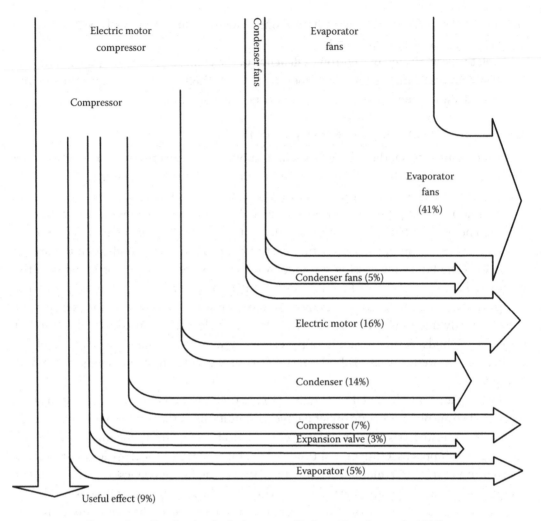

FIGURE 3.15 Exergy loss distribution in fruit storage (Infante Ferreira et al., 1999).

3.6 ENTROPY PRODUCTION ANALYSIS OF HEAT PUMP CYCLE

Most researchers make use of exergy to quantify the losses in refrigeration and heat pump cycles. However, quantification of the entropy production per component of the system can alternatively be used for the same purpose. In this section, the entropy production approach to analyse heat pump cycles is discussed.

3.6.1 Entropy Production and Heat Pump COP

According to the second law of thermodynamics, the total entropy production for a cycle with only $\dot{Q}_{in}$ and $\dot{W}_{comp}$ as input and $\dot{Q}_{out}$ as output is given by

$$\frac{\dot{Q}_{out}}{\overline{T}_H} - \frac{\dot{Q}_{in}}{\overline{T}_C} = \sum \Delta \dot{S} \qquad (3.31)$$

or

$$\frac{\dot{Q}_{out}\bar{T}_C - \dot{Q}_{in}\bar{T}_H}{\bar{T}_H} = \bar{T}_C \sum \Delta \dot{S} \tag{3.32}$$

So that

$$\dot{Q}_{out}\left[\frac{\bar{T}_C}{\bar{T}_H} - \frac{\dot{Q}_{in}}{\dot{Q}_{out}}\right] = \bar{T}_C \sum \Delta \dot{S} \tag{3.33}$$

and

$$-\dot{Q}_{out}\frac{\dot{Q}_{in}}{\dot{Q}_{out}} = \bar{T}_C \sum \Delta \dot{S} - \dot{Q}_{out}\frac{\bar{T}_C}{\bar{T}_H} \tag{3.34}$$

Adding $\dot{Q}_{out}$ on both sides,

$$\dot{Q}_{out}\left[1 - \frac{\dot{Q}_{in}}{\dot{Q}_{out}}\right] = \bar{T}_C \sum \Delta \dot{S} + \dot{Q}_{out}\left[1 - \frac{\bar{T}_C}{\bar{T}_H}\right] \tag{3.35}$$

From Equation 3.30,

$$\dot{W}_{comp} = \dot{Q}_{out}\left[1 - \frac{\dot{Q}_{in}}{\dot{Q}_{out}}\right] \tag{3.36}$$

and similarly from Equation 3.8, with $\dot{W}_{comp\_Carnot}$ the compressor power required for a reversible cycle,

$$\dot{W}_{comp\_Carnot} = \dot{Q}_{out}\left[\frac{\bar{T}_H - \bar{T}_C}{\bar{T}_H}\right] \tag{3.37}$$

Equation 3.35 then becomes

$$\dot{W}_{comp} = \bar{T}_C \sum \Delta \dot{S} + \dot{W}_{comp\_Carnot} \tag{3.38}$$

or

$$\dot{W}_{comp} = \bar{T}_C \sum \Delta \dot{S} + \dot{Q}_{out}\left[\frac{\bar{T}_H - \bar{T}_C}{\bar{T}_H}\right] \tag{3.39}$$

Multiplying left and right parts of this equation with $\dfrac{\overline{T}_H}{T_H - \overline{T}_C}\dfrac{1}{\dot{W}_{comp}}$ leads to

$$COP_{hp\_Carnot} = \frac{\overline{T}_H}{\overline{T}_H - \overline{T}_C}\frac{1}{\dot{W}_{comp}}\overline{T}_C \sum \Delta \dot{S} + \frac{\dot{Q}_{out}}{\dot{W}_{comp}} \tag{3.40}$$

$$COP_{hp\_Carnot} = \left[\frac{\overline{T}_H}{\overline{T}_H - \overline{T}_C}\frac{1}{\dot{W}_{comp}}\overline{T}_C\right]\sum \Delta \dot{S} + COP_{hp} \tag{3.41}$$

This means that the Carnot COP is equal to the sum of the real COP and the sum of the entropy losses per component multiplied with a heat pump specific constant:

$$COP_{hp\_Carnot} = COP_{hp} + K_{hp}\sum \Delta \dot{S} \tag{3.42}$$

where

$$K_{hp} = \frac{\overline{T}_H}{\overline{T}_H - \overline{T}_C}\frac{1}{\dot{W}_{comp}}\overline{T}_C = \frac{COP_{hp}}{\dot{W}_{comp\_Carnot}}\overline{T}_C \tag{3.43}$$

Equation 3.42 in combination with the entropy production per component of the heat pump allows for the identification of the major sources of irreversibility in heat pump systems.

3.6.2 Entropy Production in the Components of the Heat Pump

The entropy production in a component indicates how irreversible the processes taking place in that component are. The entropy production per component of a heat pump is easy to quantify. As in previous sections, the condenser is discussed first. The assumption that the component is externally adiabatic applies also for the entropy production analysis. As indicated in Figure 3.6, the condenser is a heat exchanger in which the mass flow of refrigerant $\dot{m}_{ref}$ exchanges the useful heat $\dot{Q}_{out}$ with the process stream flow $\dot{m}_{sink}$. The refrigerant enters the condenser with entropy s_2 and leaves it with entropy s_3. The process fluid enters the heat exchanger with the entropy $s_{sin k\_in}$ and leaves it with $s_{sin k\_out}$. The entropy increase per time unit of the process stream $\dot{S}_{sink\_out} - \dot{S}_{sink\_in}$ can then be calculated from

$$\dot{S}_{sink\_out} - \dot{S}_{sink\_in} = \dot{m}_{sink}(s_{sin k\_out} - s_{sin k\_in}) \tag{3.44}$$

Taking the definition of thermodynamic averaged sink temperature given in Equation 3.13 into account, which assumes that the sink flow through the heat exchanger is reversible, the entropy increase can also be obtained from

$$\dot{S}_{sin k\_out} - \dot{S}_{sin k\_in} = \frac{\dot{Q}_{out}}{\overline{T}_H} \tag{3.45}$$

The entropy increase in the refrigerant side is

$$\dot{S}_3 - \dot{S}_2 = \dot{m}_{ref}(s_3 - s_2) \tag{3.46}$$

The entropy production in the condenser $\Delta\dot{S}_{cond}$ is obtained by summing the entropy increase terms of both fluids, Equations 3.45 and 3.46.

$$\Delta\dot{S}_{cond} = \dot{m}_{ref}(s_3 - s_2) + \frac{\dot{Q}_{out}}{\overline{T}_H} \tag{3.47}$$

Similarly, the entropy production in the evaporator $\Delta\dot{S}_{evap}$ is obtained by summing the entropy increase terms of refrigerant and source flow fluid. Refer to Figure 3.7 to view the relevant flows and fluid states:

$$\Delta\dot{S}_{evap} = \dot{m}_{ref}(s_1 - s_4) - \frac{\dot{Q}_{in}}{\overline{T}_C} \tag{3.48}$$

Because the heat flow leaves the source flow, it has a negative sign.

Both the compressor (Figure 3.8) and expansion valve are considered externally adiabatic. Also, through both compressor and expansion valve, only the refrigerant flow passes so that the entropy production results only from this flow:

$$\Delta\dot{S}_{comp} = \dot{m}_{ref}(s_2 - s_1) \tag{3.49}$$

and

$$\Delta\dot{S}_{exp} = \dot{m}_{ref}(s_4 - s_3) \tag{3.50}$$

The total entropy production in the heat pump cycle is obtained by adding the component contributions:

$$\Delta\dot{S}_{tot} = \Delta\dot{S}_{cond} + \Delta\dot{S}_{evap} + \Delta\dot{S}_{comp} + \Delta\dot{S}_{exp} = \Sigma\Delta\dot{S} = \frac{\dot{Q}_{out}}{\overline{T}_H} - \frac{\dot{Q}_{in}}{\overline{T}_C} \tag{3.51}$$

Notice that this follows from the second law of thermodynamics and has already been introduced in Equation 3.31.

Taking the same application case as discussed previously and the values given in Tables 3.1 and 3.2, the values listed in Table 3.6 can be obtained. For this specific heat pump, $K_{hp} = 0.885$ K/kW.

TABLE 3.6   Overview of Entropy Production in the Components of the Application Example

	Entropy Production (kW/K)	COP Contribution (−)	Percentage Loss (%)
$\Delta \dot{S}_{cond}$	0.207	0.18	2.8
$\Delta \dot{S}_{exp}$	1.087	0.96	14.7
$\Delta \dot{S}_{evap}$	1.708	1.51	23.1
$\Delta \dot{S}_{comp}$	1.742	1.54	23.5
COP_{hp}		2.35	35.9
COP_{hp_Carnot}		6.55	100

Comparison of Table 3.6 with Table 3.3 shows that the contribution to the irreversibilities of the system by the individual components is exactly the same when an exergy loss analysis or an entropy production analysis is applied.

3.7   REDUCING EXERGY LOSSES/ENTROPY PRODUCTION

3.7.1   Cycle Level

The previous sections illustrated that the choice of the operating medium has, to some extent, an effect on the level of the second law efficiency that can be attained. In Figure 3.3, only single-stage operation under a determined set of external conditions is shown, so that other specific conditions will give different results. Pentane and ammonia show the best performance. Butane and R245fa show slightly lower performance, but as indicated previously, butane has more favorable operating pressures. A group of refrigerants shows significantly lower performance and cannot operate under subcritical conditions when higher heat sink temperatures are required: R1234ze, R134a, R290 and R1234yf. Although it has not been discussed here, when the temperature lift between source and sink is large, multistage operation should be considered.

Depending on the fluid, application of an internal heat exchanger that subcools the liquid leaving the condenser by superheating the vapor before entering the compressor can be advantageous. This is illustrated in Figure 3.16 for the application case discussed in a previous

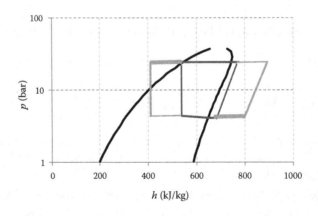

FIGURE 3.16   Effect of internal heat exchanger for the application example.

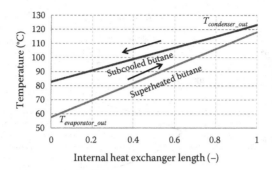

FIGURE 3.17 Temperature profile of liquid and vapor butane in the internal heat exchanger for the application example of Section 3.4.3 (conditions of Table 3.5).

section, with butane as the refrigerant. The figure shows that, in this case, it is significantly advantageous to apply an internal heat exchanger. The bold lines indicate the enthalpy change at the outlet of the condenser and at the outlet of the evaporator. The enthalpy change in the condenser is increased with a similar value, significantly increasing the *COP* of the heat pump: It increases from 2.83 (see Table 3.5) to 3.67. Figure 3.17 illustrates the temperature profiles of the liquid (hot side) and vapor (cold side) of the internal heat exchanger.

Generally, the advantage will not be this large, but for every heat pump system, the possibility of using such internal heat exchangers should be investigated. Notice that superheating implies a higher compressor discharge temperature, which may cause operational problems. In the present case, the compressor discharge temperature becomes 173.6°C, while it was only 125.1°C without an internal heat exchanger.

3.7.2 Condenser

The main cause of exergy loss in the condenser is the heat transfer at a finite temperature difference between refrigerant and sink medium. This is if the exergy loss associated with sink-fluid forced convection is not considered. These exergy losses may also be substantial –5% of the total exergy losses of the plant as compared to 14% due to heat transfer in the case of Figure 3.15. The temperature difference between condensing refrigerant and sink medium will always be a trade-off between energy consumption (and associated exergy losses) and investment costs.

Figure 3.18 illustrates, for specific conditions of a plant using an air-cooled condenser, that if the temperature difference is chosen too large (15 K between air inlet and condensation temperature, left) or too low (5 K between air inlet and condensation temperature, right), the total costs of energy plus investment are larger than for an intermediate case (10 K between air inlet and condensation temperature, center). In the first case, the energy costs are too large, while in the second case the investment costs are too large. It should be noted that the common practice in industry is designing condensers with 15 K between air inlet and condensation temperature.

Figure 3.19 illustrates that the number of operating hours of the refrigeration/heat pump plant plays an important role in the optimum temperature driving force for both condenser and evaporator.

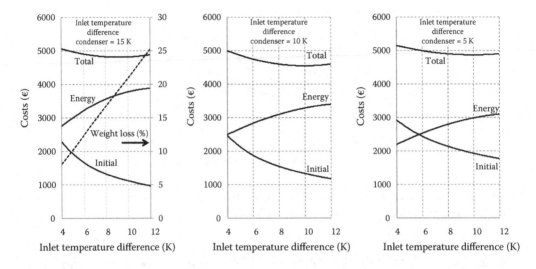

FIGURE 3.18 Temperature driving force and trade-off between energy and investment as a function of the temperature difference at the inlet of the evaporator.

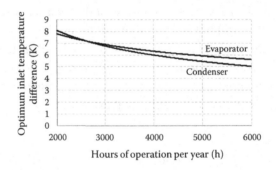

FIGURE 3.19 Optimum temperature driving force for condenser and evaporator for a specific application.

3.7.3 Evaporator

Similar considerations can be applied to the evaporator as for the condenser. Also, here the trade-off between energy and investment costs plays a role. Figures 3.18 and 3.19 illustrate the impact of evaporator driving force. However, because the temperature driving force also plays a role in the product quality, the impact of temperature driving force on product quality must be considered. The quantity of water vapor removed by the evaporator from a refrigerated/freezing room depends on the temperature driving force. With temperature differences of 5–6 K between air inlet and evaporating medium, there will be almost no water vapor removal, while for temperature driving forces of 10 K, about 20%–35% of the heat removal will be associated with latent heat effects. Products sensible to dry out may then lose weight and quality. The impact of evaporating temperature driving force on weight loss for long-term fruit storage applications is also shown in the left side of Figure 3.18.

In other cases, in freezing processes a large temperature driving force is used to guarantee high frozen product quality. For instance, this is the case for consumption ice manufacturing, where extremely large temperature driving forces are applied. The need for these extreme driving forces should be investigated: What is the impact of driving force on crystal size, and what is the impact of crystal size on ice taste?

3.7.4 Compressor

There are several causes for the irreversibilities in compressors: internal flow losses, internal heat transfer and heat conduction, mechanical friction, internal leakage, mixing and external convection and radiation. Depending on the compressor type and size, the contribution of the different effects may be quite different. The plant designer and owner are confronted with existing compressor designs so that the exergy losses of the compressor are associated with the compressor selection. Infante Ferreira and Touber (1999) gave an overview of the isentropic efficiency of the refrigerating compressors available in the Dutch market. In some application ranges, there is a large spread in compressor quality, as illustrated in Table 3.7.

Only the average values are listed. The minimum and maximum values may significantly deviate from the average values. The isentropic efficiency of semihermetic and hermetic compressors includes the electric motor losses. Contrary to the expectations, the larger compressors do not show higher isentropic efficiencies as compared to the smaller models.

3.7.5 Expansion Device

During recent years, two-phase expanders have been the topic of a number of research studies. In some applications, two-phase expanders appear to decrease significantly the exergy losses and so increase the performance of vapor compression refrigeration/heat pump cycles. Figure 3.20 shows for some refrigerants the effect of expander isentropic efficiency

TABLE 3.7   Average Isentropic Efficiency as a Function of Refrigerant Volume Flow (R22 Evaporating at −10°C)

Volume Flow (m³/h)	Reciprocating Open	Reciprocating Semihermetic	Reciprocating Hermetic	Screw Open	Screw Semihermetic	Scroll Hermetic
10–20	0.88	0.70	0.68			0.67
20–50	0.81	0.73	0.59			
50–100	0.85	0.77		0.71		
100–200	0.77	0.74		0.73	0.65	
200–300	0.75	0.74		0.75	0.66	
300–500	0.71			0.68		
>500	0.72			0.71		

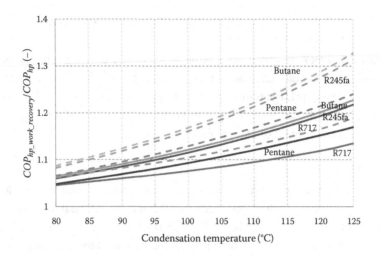

FIGURE 3.20 The COP gain for a heat pump when an expander is used to recover (part of) the expansion work for the same conditions as in Figure 3.2. Dashed lines apply for an expander efficiency of 80%; the full lines apply for an efficiency of 60%.

on the COP for a heat pump cycle. If isentropic efficiencies above 60% can be attained with practical expanders, then significant cycle performance gains can be obtained, especially for butane and R245fa.

3.8 CONCLUDING REMARKS

This chapter described the steps leading to the quantification of the thermodynamic losses in the different components of heat pump cycles, allowing for identification of the major sources of irreversibility. Furthermore, it showed how to determine the component specific exergy losses and how to compare these losses with the total exergy input into heat pump systems. Finally, it also showed how to determine the component specific entropy production and how to compare these values to the COP of reversible heat pumps.

In addition, the chapter analysed the major sources of irreversibility in the different components of heat pumps and how the associated losses can be maintained as small as possible.

LIST OF SYMBOLS

c_n	Velocity of flow at state n	ms^{-1}
c_p	Specific heat	$kJkg^{-1}K^{-1}$
ex	Exergy	$kJkg^{-1}$
Ex	Exergy flow	kW
g	Gravity constant	ms^{-2}
h	Enthalpy	$kJkg^{-1}$
K_{hp}	COP conversion factor	KkW^{-1}
$\dot{m}$	Mass flow	kgs^{-1}

n	Number of operating years	year
p	Pressure	kPa
$\dot{Q}$	Heat flow	kW
s	Entropy	kJkg^{-1}K^{-1}
$\dot{S}$	Entropy flow	kWK^{-1}
T	Temperature	K
$\bar{T}$	Equivalent temperature for gliding reservoirs	K
$\dot{W}$	Power	kW
z	Vertical coordinate	m

Greek Symbols

$\Delta\dot{S}$	Entropy production	kWK^{-1}
η	Efficiency	–

Subscripts

C	Source
$comp$	Compressor
$cond$	Condenser
$evap$	Evaporator
exp	Expansion valve
H	Sink
hp	Heat pump
in	At inlet
is	Isentropic
L	Saturated liquid
ln	Logarithmic
n	State number
0	Environment
out	At outlet
r	Refrigeration
ref	Refrigerant
s	Isentropic
t	Turbine
tot	Total
V	Vapor

Abbreviations

COP	Coefficient of performance
COSP	Coefficient of system performance
GWP	Global warming potential
TEWI	Total equivalent warming impact

REFERENCES

American Society of Heating, Refrigerating and Air-Conditioning Engineers (ASHRAE), *ASHRAE handbook of fundamentals*, 11th edition, ASHRAE, New York, 2009.

Chen J., Performance characteristics of a two-stage irreversible combined refrigeration system at maximum coefficient of performance, *Energy Conversion and Management*, 40 (1999), 1939–1948.

Infante Ferreira C. A., Touber S., The performance of refrigerating compressors for industrial applications, 20th International Congress of Refrigeration, IIR/IIF, Sydney, 1999, Volume 3, Paper 573.

Infante Ferreira C. A., van der Ree H., Touber S., The role of compressors in industrial refrigerating plants (an exergy analysis), 20th International Congress of Refrigeration, IIR/IIF, Sydney, 1999, Volume 3, Paper 578.

Lemmon E., Huber M., McLinden M., *NIST Standard reference database 23: Reference fluid thermodynamic and transport properties – REFPROP*, Version 9.1, National Institute of Standards and Technology, Gaithersburg, MD, 2013.

Moran M. J., Shapiro H. N., *Fundamentals of engineering thermodynamics*, 6th edition, Wiley, Chichester, UK, 2007.

Sahin B., Kodal A., Thermoeconomic optimisation of a two stage combined refrigeration system: A finite-time approach, *International Journal of Refrigeration*, 25 (2002), 872–877.

Sahin B., Kodal A., Koyun A., Optimal performance characteristics of a two-stage irreversible combined refrigeration system under maximum cooling load per unit total cost conditions, *Energy Conversion and Management*, 42 (2001), 451–465.

Touber S., *Thermische machines – Een compressie warmtepomp*, Lecture notes wb4200 (in Dutch), Werktuigbouwkunde en Maritieme Techniek, TU Delft, Delft, 1996.

CHAPTER 4

Pinch Analysis and Process Integration

4.1 INTRODUCTION

Energy efficiency is a primary concern for the chemical industry; hence, every process engineer has to find answers to questions related to the process energy patterns; for example:

- Are the existing chemical processes just as energy efficient as they should or could be?

- What changes can be made to increase the energy efficiency without significant costs?

- What acceptable investments can be made to improve the energy efficiency?

- How can new projects be evaluated with respect to their energy requirements and costs?

- What are the most appropriate types of utilities used for a particular chemical process?

- How can efficiency targets (e.g. minimum energy use, reduced emissions, increased plant capacity, improved product quality) be combined into a coherent strategic plan for the overall site?

Process integration is a holistic approach to process design that emphasizes the unity of the process and considers the interactions between different unit operations from the outset, rather than optimizing them separately. Process integration focuses on the efficient use of shared physical (heat) and chemical (e.g. solvents, reagents) resources in a process. The mass integration addresses the generation, separation and routing of streams and species throughout the process. This is being developed and applied to identify overall insights, synthesize strategies and address the root causes of mass processing and environmental problems at the heart of any chemical or biochemical process – as clearly illustrated in Figure 4.1 (Kiss et al., 2015).

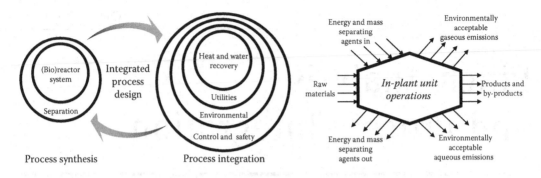

FIGURE 4.1   Sustainable process integration (left) and in-plant unit operations (right).

Mass integration is used as a systematic methodology that offers understanding of the overall flow of mass within the process, which is used in identifying the performance targets and optimizing the allocation, separation and generation of streams or species. *Energy integration* deals with the overall allocation, generation and exchange of energy throughout the process. The development of such methodologies for energy integration is strongly driven by the increasing cost of utilities within the chemical process industries. The aim is to identify cost-effective waste minimization systems, end-of-the-pipe separations, recycle systems and in-plant separation systems that allow an integrated solution (Dunn and El-Halwagi, 2003; Kiss et al., 2015).

Another meaning of process integration refers to heat integration, energy integration or pinch technology. Linnhoff and co-workers (Linnhoff et al., 1982, 1994), under the label of *pinch technology*, developed a systematic methodology optimizing energy use in process plants and introduced the concept of *process integration*, which evolved into a paradigm of process design that addresses the optimal use of resources. *Pinch analysis* is a technique for designing a process to minimize energy use and maximize heat recovery, and it designates the systematic research of innovative solutions in the area of energy savings. While pinch technology represents a new set of thermodynamically based methods that guarantee minimum energy use in heat exchanger networks (HENs), pinch analysis represents the application of the tools and algorithms of pinch technology for analyzing industrial processes.

This chapter aims to cover the pinch technology and pinch analysis topics only at an introductory level, sufficient for understanding the use of heat pumps in the broader context of energy-savings tools. A more detailed treatment of *process integration* and *pinch analysis* principles can be found in books by Smith (2005, 2015), Kemp (2007) and Dimian et al. (2014).

4.2 PINCH ANALYSIS

A simple methodology for the systematic analysis of chemical processes and the corresponding utility system can be used based on the laws of thermodynamics: The first law provides the energy equation for calculating the enthalpy changes in streams, while the second law determines the direction of the heat flow (from *hot* to *cold* only). Therefore, a hot stream can only be cooled practically to a temperature defined by the *temperature approach* of the heat exchanger, which is the minimum allowable temperature difference (ΔT_{min}) in the stream temperature profiles. The temperature level at which ΔT_{min} is observed is the *pinch point* that defines the minimum driving force allowed in the exchanger unit. Pinch analysis deals with the optimal structure of heat exchange between process streams and with the optimal use of utilities; hence, it is used to identify the pinch point, the energy costs and HEN capital cost targets for a process. The procedure provides, ahead of the design, the minimum requirements of external energy (utilities), exchange network area and the number of units needed for a given process at the pinch point. Afterwards, a HEN design that satisfies all these targets is synthesized, and the network is optimized by balancing the energy cost and capital cost such that the total annual cost is minimized. Therefore, the main objective of pinch analysis is to accomplish cost savings by improved process heat integration – in other words, maximizing process-to-process heat recovery and reducing the external utility loads. Pinch analysis is therefore a required procedure to be performed prior to considering the use of heat pumps in a chemical process. The main activities of pinch analysis include the following:

- Evaluate the reference basis of an energy-savings project:
  - *Minimum energy requirements* (MER), in terms of heating and cooling utility loads, for a minimum temperature approach (ΔT_{min}) assumed at the pinch point;
  - Maximum energy savings that are possible by process-process heat exchange;
  - Total investment cost and operating costs needed by the MER.
- Set optimal targets before the detailed design of the HEN:
  - Design targets for the HEN in terms of total heat exchange area required and the total number of equipment units needed to achieve the MER;
  - Nature and amount of utilities required to fulfill the optimal loads;
  - Integration of energy-saving options with power generation.
- Recommend changes in the process design with a significant impact on saving energy:
  - Optimal placement of heat pumps, heat engines and other types of units;
  - Optimization of the operating parameters for reactors and separation equipment;
  - Heat integration of distillation columns.

As a conceptual tool, pinch analysis uses intensively conceptual graphical tools that can be largely complemented by a different approach based on *mathematical programming* (MPR) and usable as an automatic design tool. The concept of pinch proved to have a generic value; thus, it was extended to the management of other valuable resources, such as water, solvents and hydrogen.

4.2.1 Basic Concepts

Composite curves (CCs) are fundamental concepts that allow visualizing heat flows between hot and cold process streams selected for heat integration, as illustrated in

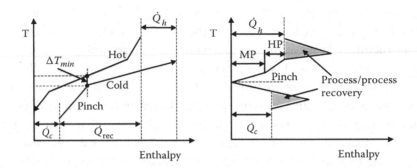

FIGURE 4.2 Composite curves (left) and grand composite curve (right). Identification of pinch point, minimum energy requirements ($\dot{Q}_h$ and $\dot{Q}_c$) and maximum heat recovery.

Figure 4.2, left (Dimian et al., 2014). A combined CC is obtained by plotting the cumulative enthalpy of streams (cold or hot) against temperature. The relative position of the (hot and cold) CCs depends on the minimum temperature difference ΔT_{min}) between the cold and hot streams. The place where the heat transfer is the most constrained is known as the *pinch point*. The CCs allow determining the MER for both hot and cold utility, the minimum network area and minimum number of heat exchanger units using only stream data and without designing any heat exchanger. The MER represent the minimum load of hot $\dot{Q}_h$ and cold $\dot{Q}_c$ utilities required for driving the HEN, with a minimum driving force ΔT_{min} at the pinch point.

Balanced CCs are similar to CCs, with the difference that the hot and cold utilities are also considered streams. As the hot and cold utilities cover any imbalance between the streams selected for integration, the enthalpy (heat) balance is closed. Moreover, the balanced grid diagram is used for the design of the HEN.

The *grand CC* (GCC) plots the difference between the enthalpy of the hot and cold streams against a conventional shifted temperature scale, as shown in Figure 4.2, right (Dimian et al., 2014). This graphical representation identifies the possibilities of heat recovery by internal process-process exchange and the optimal selection and placement of the cold and hot utilities (maximize cheaper utilities) to meet the overall energy requirements.

The *pinch principle* states that any process design where heat is transferred across the pinch requires more energy than MER. Thus, the heat recovery problem is practically divided into two subsystems, above and below the Pinch point.

Supertargeting consists of setting design targets for the whole process by an overall optimization procedure, without the need of detailed sizing of heat exchangers. While *energy and capital cost targeting* is used to calculate total annual cost of utilities and capital cost of HEN, *total cost targeting* is used to determine the optimum level of heat recovery or the optimum ΔT_{min} value by balancing energy and capital costs. Using this method, it is possible to obtain an accurate estimate (within 10%–15%) of the overall heat recovery system costs without having to design the system, the essence of the pinch approach being the quick economic evaluation. While the energy costs increase roughly proportionally to the driving force ΔT_{min}, the capital costs (e.g. heat exchangers) decrease sharply with

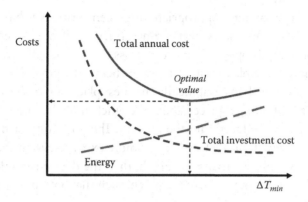

FIGURE 4.3   Targeting of energy and capital costs before HEN design.

the temperature difference ΔT_{min}, as illustrated in Figure 4.3 (Dimian et al., 2014). Several key observations can be made: (1) An increase of the ΔT_{min} value results in higher energy costs but lower capital costs; (2) a decrease of the ΔT_{min} value results in lower energy costs but higher capital costs and (3) an optimum ΔT_{min} exists where the total annual cost of both energy and capital costs is minimized. Therefore, by systematically changing the temperature approach, one can determine the optimum heat recovery level or the ΔT_{min} that is optimal for the overall process.

A *grid diagram* designates a working frame for developing the HEN. The *'bubble-headed'* stick linking two streams in Figure 4.4 symbolizes a heat exchanger (Dimian et al., 2014). The development of the HEN is based on feasibility rules belonging to the *pinch design* method. A HEN developed for MER is optimal when both the energy and the capital costs are considered. Additional reduction in capital costs is possible by removing small units. However, this operation might transfer heat across the pinch and thus increase the total requirements of utilities.

Appropriate placement principles provide insights for proper integration of key equipment units, such as distillation columns, evaporators, furnaces, heat engines and heat pumps, to reduce the utility requirements of the combined system. The *appropriate placement* rules must be respected when positioning the unit operations with respect to the

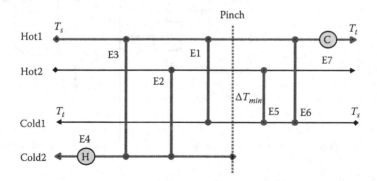

FIGURE 4.4   Grid diagram used for the development of the heat exchanger network (HEN).

pinch to achieve energy savings. Appropriate placement principles have been developed for distillation columns, evaporators, heat engines, furnaces, heat engines and heat pumps. For example, a heat pump is optimally placed across the pinch, while heat engines or distillation columns have to be placed either above or below the pinch (not across it). Figure 4.5 (Dimian et al., 2014) explains graphically that best placement of heat pumps is indeed across the pinch. If a heat pump is placed above the pinch and takes off the amount $\dot{Q}_e$, then an amount $\dot{Q}_e + \dot{W}$ is reintroduced into the process. The hot utility load decreases by $\dot{W}$, but this additional energy (typical electricity) is more expensive than the thermal energy. But, if the heat pump is placed across the pinch, then the device can take an amount $\dot{Q}_e$ from below the pinch and bring it above the pinch, such that both utility loads are reduced by the same amount $\dot{Q}_e$.

The *plus/minus principle* states that a reduction in the cold and hot utility requirements can be obtained if the following actions are taken:

- Increase (+) the total cold/hot stream load below/above the pinch, respectively.

- Decrease (−) the total cold/hot stream heat duty above/below the pinch, respectively.

By combining CCs with the rules of appropriate placement, changes in process design with significant effects on energy saving can be recommended. If a cold stream is removed from the region above the pinch and placed below the pinch, the shape of the cold composite changes accordingly. By removing a hot stream, the hot utility is reduced by exactly the same amount as the cold utility load. Similarly, moving a hot stream from the region below

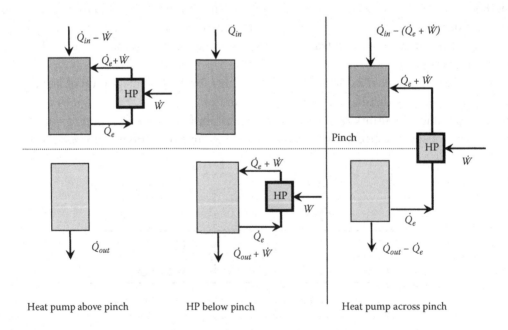

FIGURE 4.5   Appropriate placement of heat pumps (HP).

to above the pinch reduces the cold utility load and as a result the hot utility requirements. These observations can be extended to a generic heuristic in energy integration:

- Shift hot streams from below to above the pinch.

- Shift cold stream from above to below the pinch.

The process changes that can help achieve certain stream shifts involve changes in some operating parameters, such as reactor pressure/temperature; distillation column settings (temperatures, reflux ratio, feed and pump around conditions); storage vessel temperature; evaporator pressure and so on. For instance, reducing the pressure of a feed vaporizer may shift the vaporization duty from above to below the pinch, thus leading to a reduction in both hot and cold utilities.

Figure 4.6 presents an example (Dimian et al., 2014) of a distillation column that is originally placed across the pinch – the reboiler above and the condenser below the pinch (see dotted lines). Note that the reboiler is considered a cold stream, which has to be heated. By lowering the operating pressure of the column, the reboiler moves from above to below the pinch. Therefore, a cold stream is removed from above the pinch (–) and placed below (+). Hence, a reduction in both hot and cold utilities is obtained. Accordingly, from the viewpoint of the whole process, the distillation column can run at zero net utility consumption, as compared with the initial situation.

4.2.2 Pinch Technology Approach

Pinch analysis consists of a systematic screening of maximum energy savings that can be obtained in a plant by internal process-process exchange, as well as the optimal use of the available utilities. The method is capable of assessing optimal design targets for a HEN well ahead of the detailed design of the equipment. Furthermore, pinch analysis may suggest improvements in the original design that could enhance significantly the energetic performance of the process, from changes in the parameters of the operational units to structural modifications in the flowsheet.

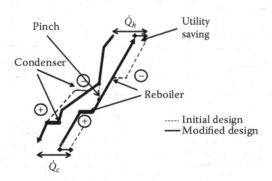

FIGURE 4.6   Plus/minus principle.

The overall approach in pinch analysis is presented in Figure 4.7 (Dimian et al., 2014). The initial step consists of identifying the hot, cold and utility streams in the process. *Hot streams* are those that must be cooled or are available to be cooled (e.g. product cooling before storage), while *cold streams* are those that must be heated (e.g. feed preheat before a reactor). *Utility streams* are used to heat or cool the process streams when heat exchange between process streams is not practical or economic. A number of hot utilities (e.g. steam, hot water, flue gas) and cold utilities (e.g. cooling water [CW], air, refrigerants) are used in industry and discussed in a further section. The next step is the extraction of *stream data* from a process flowsheet simulation that describes the material and energy balances. An important factor for efficient heat integration is the proper selection and treatment of streams by segmentation, as well as the selection of utilities. Based on initial ΔT_{min} value, the CCs and the GCC are constructed. Additional information is needed about the partial heat transfer coefficients of the different (segments of) streams and utilities, as well as the cost of utilities and the cost laws for heat exchangers. The *targeting stage* that follows consists of finding the optimal ΔT_{min} as a trade-off between energy and capital costs. On this basis, targets for MER can be determined, along with the overall heat exchange area and the number of units. If the economic data are not reliable, selecting a practical ΔT_{min} is recommended.

Then, the appropriate placement of unit operations is checked as this may suggest some design modifications by applying the plus/minus principle. Utility options are tested again and revised if deemed necessary. As capital cost is a trade-off against energy cost, the procedure may imply iteration between targeting and process revision. Major changes could require reviewing the flowsheet simulation. The iterative procedure is ended when no further improvement can be achieved. Although information about the heat transfer coefficients is required, the individual heat exchangers are never sized during the different steps of the procedure. The detailed sizing of the HEN takes place only after completing the overall design targets. Optimization can be employed to refine the design, but then the final solution is checked by rigorous process simulations.

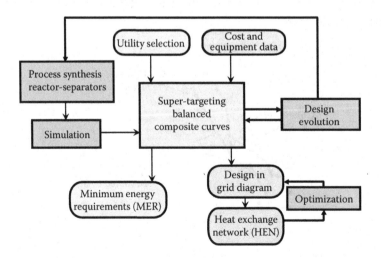

FIGURE 4.7   Overall approach in designing a heat exchanger network by pinch analysis.

The practical use of pinch analysis is facilitated by several software tools available on the market, usable for both pinch analysis and HEN design:

Academic packages:

- Centre for Process Integration of the University of Manchester, United Kingdom (www.cpi.umist.ac.uk):

  - SPRINT: for energy targets, HEN design, automatic grassroots design and retrofit
  - STAR: for optimization of site utility and cogeneration systems
  - WORK: for the design of low-temperature processes with mixed refrigerants

- Online Pinch Analysis Tool: an interactive Web-based application hosted by the College of Chemical Engineering at the University of Illinois at Chicago in the United States (www.uic-che.org/pinch/)

Commercial software:

- SuperTarget: pinch analysis, HEN optimization (KBC/Linnhoff-March, www.kbcat .com)

- HEXTRAN: all-purpose package for HEN design and rigorous heat exchanger design, retrofit and operation studies (Simulation Science, USA, iom.invensys.com /EN/pages/home.aspx)

- PINCH-Int: targets process energy and capital utilization base on pinch technology (Process Integration Ltd. UK, www.processint.com)

- Aspen Energy Analyzer: Optimal HEN design (AspenTech, USA, www.aspentech .com)

- Simulis Pinch: a tool that can be used directly in Excel, dedicated to the diagnosis and the energy integration of the processes (ProSim SA)

Freeware packages:

- Pinchleni: freeware implementation of the pinch method (Laboratoire d'Energétique Industrielle de l'Ecole Polytechnique Fédérale de Lausanne, Switzerland)

4.3 MINIMUM ENERGY AND CAPITAL COST TARGETS

For any given value of heat transfer load $\dot{Q}$, the heat recovery and the exchange area requirements increase when smaller values of ΔT_{min} are chosen. Conversely, if a higher value of ΔT_{min} is selected, then the heat recovery and exchange areas decrease, while the demand for external utilities increases. Therefore, the selection of the ΔT_{min} value has crucial implications for both capital and energy costs. The typical values for ΔT_{min} vary per industrial sector, from 20–40 K (oil refining) and 10–20 K (petrochemical and chemical industry) to 3–5 K (low-temperature processes, cryogenics).

4.3.1 Composite Curves

4.3.1.1 Stream Data

The stream data chosen to illustrate the construction of CCs is presented in Table 4.1 (Dimian et al., 2014). The following minimum elements are necessary:

- Stream or segment temperatures: supply temperature T_s and target temperature T_t.

- Heat capacity of each stream or segment, defined as $CP = \Delta H/\Delta T$ (kW/K), where ΔH is the enthalpy change over temperature interval ΔT. The enthalpy change of a stream segment is

$$\Delta H = CP \times (T_t - T_s) \tag{4.1}$$

The energy equation states that $\Delta H = \dot{Q} \pm \dot{W}$, but in a heat exchanger no work is being performed, so $\dot{W} = 0$ and $\Delta H = \dot{Q}$. Also, note that CP is actually defined as $\dot{m} \times c_p$ (mass flow rate by the mass heat capacity), while enthalpy is actually defined as an enthalpy flux or heat flux. The hypothesis of constant CP is fundamental in pinch analysis. Therefore, if the enthalpy-temperature relation is not linear, then the stream must be 'segmented'.

The heat balance of the streams in Table 4.1 (Dimian et al., 2014) shows an excess of 1000 kW, but simply adding 1000 kW of cold utility is not sufficient as the second law of thermodynamics requires a minimum temperature difference between the hot and cold streams, so the real use of energy is much higher in practice.

Composite curves plot the temperature T versus the cumulated enthalpy of all streams H, hot or cold, available in a temperature interval between the extreme supply and target temperatures. Any stream with a constant heat capacity value is plotted on the T-H diagram as a straight line going from the stream supply temperature to the target temperature. The construction of hot and cold CCs simply involves the addition of enthalpy changes of the streams in the respective temperature intervals. A complete hot or cold CC consists of a series of connected straight lines: Each change in slope represents a change in overall hot stream heat capacity flow rate CP. ΔT_{min} can be measured directly from the T-H profiles as the minimum vertical difference between the hot and cold curves, this point being a bottleneck in heat recovery and thus referred to as the *pinch point*. Increasing the ΔT_{min} value leads to shifting the hot and cold curves horizontally apart, resulting in reduced process-to-process heat exchange and higher utility requirements. At a particular ΔT_{min} value, the overlap of the hot and cold curves shows the maximum possible heat recovery

TABLE 4.1 Stream Data for Composite Curves

Stream	Name	T_s (°C)	T_t (°C)	CP (kW/K)	ΔH (kW)
1	hot1	220	60	100	−16,000
2	hot2	180	90	200	−18,000
3	cold1	50	150	150	15,000
4	cold2	130	180	400	20,000
	Total				1000

within the process. The hot and cold end overshoots indicate the minimum hot and cold utility requirements for the chosen ΔT_{min}. The relation enthalpy-temperature can be calculated using the following formula:

$$H = H_0 + \sum_i \left[\sum_j (CP_j) \right]_{i-1}^i \Delta T_i \tag{4.2}$$

Note that the value H_0 can shift the position of the CC. CP_j values are heat capacities of the active streams in the temperature interval ΔT_i. The partition in temperature intervals is based on the analysis of the stream population. For the streams given in Table 4.1, there are three intervals for the hot streams and three for the cold streams, as shown in Figure 4.8 (Dimian et al., 2014).

Figure 4.9 (top) explains the graphical construction of the *hot CC* (Dimian et al., 2014). The two streams, **hot1** and **hot2**, are represented by the segments *ab* and *cd* with $CP_1 = 100$ and $CP_2 = 200$ kW/K, respectively. The total enthalpy change is $\Delta H_h = \Delta H_1 + \Delta H_2 = 16,000 + 18,000 = 34,000$ kW. The interval between the target and supply temperatures is divided into three subintervals: 60°C–90°C, 90°C–180°C and 180°C–220°C. In each interval, the overall CP can be obtained simply by adding the CP values of the active streams. For example, in the first and third interval there is only **hot1**, so $CP = 100$. In the second interval, both **hot1** and **hot2** are active; therefore, $CP = CP_1 + CP_2 = 300$. Thus, each change in the slope of the CC corresponds to the entry or to the exit of a stream. A slope close to zero (horizontal position) means very high CP, as typically is the case for phase transitions.

The same method can be applied to draw the *cold CC*, as shown in Figure 4.9, bottom (Dimian et al., 2014). There are three temperature intervals, 50°C–130°C, 130°C–150°C and 150°C–180°C, where $CP_3 = 150$, $CP_3 + CP_4 = 550$ and $CP_4 = 400$. The enthalpy is then calculated as follows: $\Delta H_c = \Delta H_3 + \Delta H_4 = 15,000 + 20,000 = 35,000$ kW. Both CCs can be plotted on the same diagram, as shown in Figure 4.10 (Dimian et al., 2014). The hot CC may remain in the original position, while the cold CC shifts to the right by adding an amount

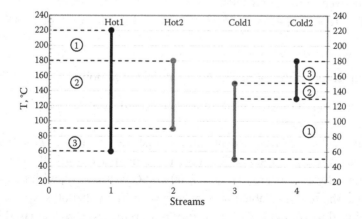

FIGURE 4.8　Temperature intervals of the hot and cold streams.

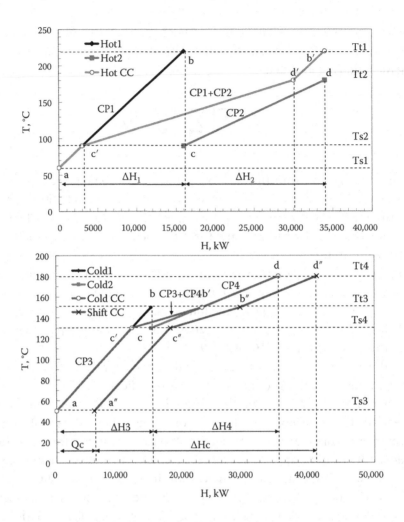

FIGURE 4.9 Construction of a hot (top) and cold (bottom) composite curve.

of heat such as to achieve ΔT_{min}. For $\Delta T_{min} = 10$ K, $\dot{Q}_c = 6000$ kW and $\dot{Q}_h = 7000$ kW, with pinch located between 130°C and 140°C.

The graphical representation has identified two key elements of a heat integration problem:

- Minimum temperature approach (ΔT_{min}) at the pinch point

- Minimum energy requirements as utility targets for heat recovery

The MER and ΔT_{min} are interdependent, as illustrated in Figure 4.11 (Dimian et al., 2014). If ΔT_{min} is set to 15°C, the hot CC keeps the same place, while the cold CC shifts to the right. Graphically, results are that $\dot{Q}_c = 7500$ and $\dot{Q}_h = 8500$ kW, with the pinch located at 130°C–145°C. The heat available for recovery was initially 28,000 kW, while in the new case, it has reduced to 26,500 kW. Both utility requirements have increased by 1500 kW. Thus, increasing ΔT_{min} requires more utilities and reduces the energy savings.

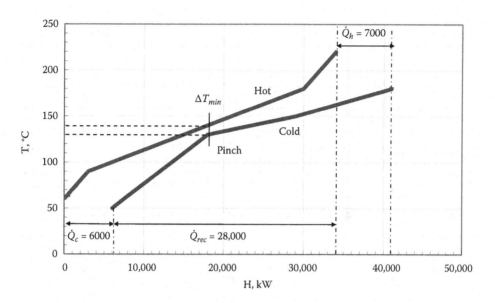

FIGURE 4.10   Composite curves.

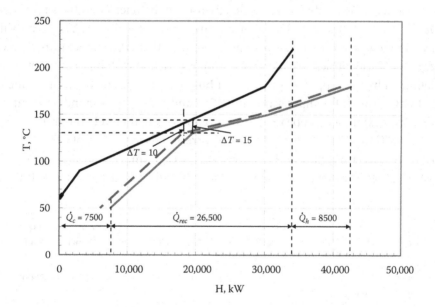

FIGURE 4.11   Shifting the composite curves modifies ΔT_{min} and utility targets.

The *problem table algorithm* (PTA) is a means of determining the utility requirements and the location of the process pinch in a process. When investigating the heat integration between hot and cold streams available in a temperature interval, one must ensure that the driving force is at least equal to ΔT_{min}, with this problem elegantly solved by PTA (Linnhoff and Flower, 1978). The temperature scale is modified to accommodate a minimum driving force ΔT_{min}, as shown in Figure 4.12 (Dimian et al., 2014). Hot streams are represented on the left scale. Cold streams are plotted on the right scale, where the temperature is shifted

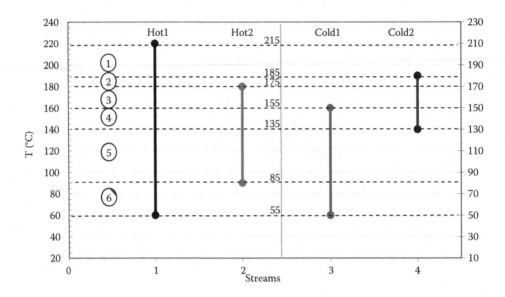

FIGURE 4.12 Shifted temperature scale.

by ΔT_{min} (assumed 10 K). Both hot and cold streams can be referred to the *shifted temperature scale*, where the hot stream temperatures are moved down by $\frac{1}{2}\Delta T_{min}$, and similarly the cold stream temperatures are shifted up by $\frac{1}{2}\Delta T_{min}$. Then, the temperature intervals for heat integration are identified. In the same example, there are six temperature intervals. The first is delimited by the supply temperature of **hot1** (220°C) and by the target temperature of **cold2** (180°C). The shift temperatures are 215°C and 185°C, corresponding to an interval $\Delta T = 30$°C. The next interval appears due to the entry of stream **hot2**. The shift temperature varies from 185°C to 175°C. The third interval corresponds to the exit of **cold1** and so on.

The next step is setting up the PTA. Table 4.2 gives a sample from an Excel workbook (Dimian et al., 2014). The first three columns contain hot and cold stream temperatures, as well

TABLE 4.2 Problem Table Algorithm

T_{hot} (°C)	Shift T (°C)	T_{cold} (°C)	Interval	ΔT_{int} (K)	CP_1	CP_2	CP_3	CP_4	CP_{int} (kW/K)	ΔH (kW)	Heat
220	215	210									
			1	30	−1	0	0	0	−100	−3000	Surplus
190	185	180									
			2	10	−1	0	0	1	300	3000	Deficit
180	175	170									
			3	20	−1	−1	0	1	100	2000	Deficit
160	155	150									
			4	20	−1	−1	1	1	250	5000	Deficit
140	135	130									
			5	50	−1	−1	1	0	−150	−7500	Surplus
90	85	80									
			6	30	−1	0	1	0	50	1500	Deficit
60	55	50									

as shifted temperatures. The next columns give the temperature intervals and the corresponding ΔT_{int}, as well as an indication of the active streams necessary for CP calculation (zero for inactive streams, –1 for hot active streams and 1 for cold active streams). This information can be related to Table 4.1 by computing the CPs of each interval, as given in the column CP_{int}. Then, the enthalpy variation of each interval can be calculated by multiplying each CP by ΔT_{int}.

The rightmost column gives a qualitative message: A negative value signifies heat surplus, while a positive value means a deficit. Excess heat can be removed by cold utility, while deficit heat can be covered by hot utility. Obviously, this approach would mean poor use of energy, so one should try to combine the heat content of different temperature intervals. Thus, the deficit of cold streams could be covered by the surplus of the hot process streams, and only the unbalanced amount should be covered by hot utilities. This strategy allows the identification of the maximum recoverable energy by process-process heat exchange and of the minimum amount of hot and cold utilities needed to cover the heat balance. The condition of a minimum temperature approach between coupled hot and cold streams must be fulfilled, but only in a restricted zone, namely, that of the pinch. Note that in regions far from the pinch, the temperature difference could be larger than ΔT_{min}.

The coupling of intervals can be found by organizing the flow of heat in a cascade manner. In a first attempt (Table 4.3, left), let us assume that no heat is transferred from the hot utility (Dimian et al., 2014). The first interval has an excess of 3000 kW that can be transferred to the second one, resulting in a net enthalpy flow of 0 – (–3000) = 3000 kW. The second interval has a deficit of 3000 kW, so that the net heat flow after this interval becomes zero; hence, the first and second intervals perfectly match each other. The third interval has a deficit of 2000 kW, and a net heat flow deficit of 2000 appears at its exit, which would further increase at –7000 kW on the fourth interval. Clearly, a negative heat flow cannot be cascaded further, so this solution is not feasible. In a second attempt, one may consider a hot utility load of 7000 kW that could compensate the deficit noted previously. The result of cascading heat flow can be seen in Table 4.3, right (Dimian et al., 2014). After the first interval,

TABLE 4.3   Cascade Diagram

T_{shift}	ΔH	H_{flow}	T_{shift}	ΔH	H_{flow}
215	↓	0	215	↓	7000
	–3000			–3000	
185	↓	3000	185	↓	10,000
	3000			3000	
175	↓	0	175	↓	7000
	2000			2000	
155	↓	–2000	155	↓	5000
	5000			5000	
135	↓	–7000	135	↓	0
	–7500			–7500	
85	↓	500	85	↓	7500
	1500			1500	
55	↓	–1000	55	↓	6000

the net heat flow is 7000 − (−3000) = 10,000 kW. The second interval delivers 10,000 − 3000 = 7000 kW. The cascade of heat flow goes on until the lowest interval is reached, and now all the net flows (except one) are positive. The location where the heat flow is zero is the *pinch point* (at shifted temperature 135°C or 130°–140°C expressed in real stream temperatures).

The *GCC* was introduced by Itoh et al. (1982) as an auxiliary tool that shows the variation of heat supply and demand within the process. Using this diagram, the designer can find which utilities are to be used and aim to maximize the use of cheaper utility levels and minimize the use of expensive utility levels. Low-pressure (LP) steam and CW are preferred instead of high-pressure (HP) steam and refrigeration, respectively. Note that the GCC can be drawn using information that comes directly from the PTA. A GCC is obtained by plotting the heat content of each temperature interval (*x* axis) against the shifted temperature scale (*y* axis). The method involves shifting down the hot CC by $\frac{1}{2}\Delta T_{min}$ and shifting up the cold CC up by $\frac{1}{2}\Delta T_{min}$ along the vertical axis. Basically, the curves are shifted by subtracting part of the allowable temperature approach from the hot stream temperatures and adding the remaining part to the cold stream temperatures. The end result is a scale based on process temperature having an allowance for temperature approach (ΔT_{min}). The GCC is then constructed from the enthalpy differences (horizontal axis) between the shifted CCs at different temperatures. Note that while the CCs provide overall energy targets, they do not clearly indicate the amount of energy that must be supplied by different utility levels. The utility mix is actually determined by the GCC. Figure 4.13 presents the GCC corresponding to the streams given in Table 4.1 for a value of $\Delta T_{min} = 10$ K (Dimian et al., 2014). The following observations can be noted:

- The pinch point divides the diagram in two regions (above and below the pinch), in which the heat recovery problem can be analyzed separately.

- Above the pinch, there is a need only for hot utilities.

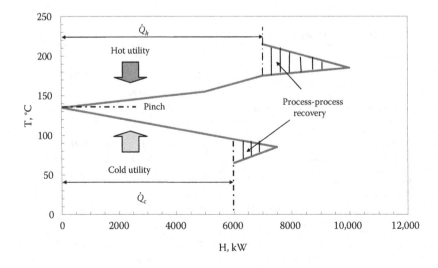

FIGURE 4.13 Grand composite curve (GCC).

- Below the pinch, there is a need only for cold utilities.

- The so-called pockets of the GCC designate the heat recovery possible by process-process exchange.

4.3.2 Pinch Point Principle

The pinch separates the heat recovery space into distinct regions, as shown in Figure 4.14 (Dimian et al., 2014), so the original problem can be decomposed in two subproblems, above and below the pinch. *Pinch point principles* were formulated as follows (Linnhoff and Hindmarsh, 1983):

- Do not transfer heat across the pinch.

- Do not use cold utility above the pinch.

- Do not use hot utility below the pinch.

If there is a heat flow across the pinch, then the energy usage is higher than the minimum necessary, as both hot and cold utility consumption will increase by the same amount XP above the minimum targets:

$$\textit{Actual Energy (A) = Target (T) + Cross-Pinch Energy Flow (XP)} \tag{4.3}$$

Therefore, transferring heat across the pinch is a double loss in energy. However, during the effective design of HENs, the initial targets must be revised to accommodate constraints, such as a smaller number of units or some imposed loads. The actual energy use could increase above the minimum targets, but the designer should try to keep the pinch violation as small as possible. The temperature approach (ΔT_{min}) is a key variable in pinch analysis. The recommended values are 20–40 K in oil refining, 10–20 K in petro-/chemical industry and 3–5 K in cryogenics. More rigorously, ΔT_{min} can be determined by a *supertargeting* procedure that performs an optimization between the costs

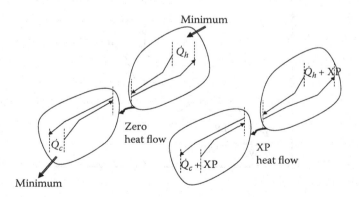

FIGURE 4.14   Pinch point principle.

of utilities and of equipment. Beside stream data, some supplementary information is needed:

- Type, temperature and costs of utilities

- Partial heat transfer coefficients of streams (or of segments) and of utilities

- Maximum heat transfer area of the heat exchangers

As the energy targets increase linearly with ΔT_{min}, the cost of utilities follows the same trend, but the capital cost actually decreases nonlinearly with ΔT_{min}. Moreover, the cost function exhibits a jump when the number of units is changed, so reducing the number of units is actually more critical for the overall cost optimization than the incremental reduction of the heat transfer area.

4.3.3 Balanced Composite Curves

In case of balanced CCs, shown in Figure 4.15 (Dimian et al., 2014), both utility and process streams are considered as usual streams. When more than one hot or cold utility is used, each supplementary utility introduces a new pinch. Therefore, it is possible to have a main process pinch and several utility pinches. The selection of utilities may be guided by some heuristics:

- Add heat at the lowest temperature level relative to the process pinch.

- Remove heat at the highest temperature level relative to the process pinch.

4.3.4 Stream Segmentation

The hypothesis of constant *CP* is the fundamental assumption in pinch analysis. In fact, the stream enthalpy is not strictly a linear function of temperature. This is particularly true for mixtures undergoing phase change. In this case, we should decompose a *T-H* curve in segments of constant *CPs* by an operation called *stream segmentation*. Let us examine the

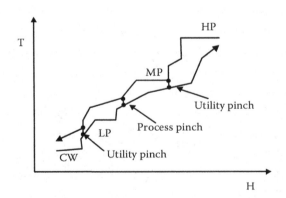

FIGURE 4.15   Balanced composite curves.

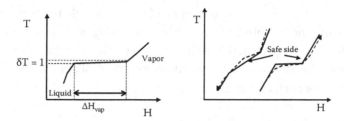

FIGURE 4.16 Phase transition (left) and stream segmentation (right).

vaporization/condensation of a pure component. We may define a virtual *CP* by considering a small temperature change, say 1 K, as shown in Figure 4.16, left (Dimian et al., 2014). Thus, the phase transition of a pure component stream may be represented by a (large) horizontal segment. In the case of a mixture, we can divide the stream in segments on which the *CP*s are constant. Note that the segmentation always has to be done on the 'safe side', meaning that the segments should be placed below a hot stream or above a cold stream, as shown in Figure 4.16, right (Dimian et al., 2014).

4.3.5 Thermal Data Extraction

The selection of streams is crucial for efficient heat integration. As most studies are based on results obtained from a process simulator, the automatic transfer of data from the simulator to the pinch analysis software is a key feature of many simulation platforms (commonly called *data extraction*), which is assisted sometimes by an expert system. However, one should be aware of potential problems in heat integration caused by inappropriate stream extraction, even if it is computer assisted. For more accurate data extraction, several recommendations can be followed:

- Check the accuracy of the property model used for enthalpy calculation. For example, models based on equations of state sometime underestimate the enthalpy of phase transition.

- Consider the plant decomposition in smaller sections. The proximity of integrated streams is an important practical constraint, and short energy paths avoid time delays in process control.

- List carefully the process constraints, including controlled temperatures (e.g. before reactors, separation flashes and distillation columns), for which heaters (furnaces) or coolers (refrigerant) are needed. Some imposed heat exchangers may be eliminated from the analysis.

- Implicit duties that are unimportant in the process simulation could become significant for heat integration, such as those included in flashes.

- Evaluate carefully the integration of chemical reactors, especially for exothermic reactions. Avoid the feedback of energy that might lead to multiple steady states and unstable behavior.

- Do not consider initially the condensers and reboilers of distillation columns, except if there is a specific incentive to save utilities by using process streams. These duties have a high energetic potential, but the heat integration of separators with other process streams is limited by serious constraints on controllability. Use the plant or site utility system for such exchange.

4.3.6 Targets for Energy and Capital Cost

The pinch point principle allows the determination of the load and cost of utilities for a given ΔT_{min}, and these minimum requirements become targets in energy-saving projects. The capital cost is the other key part of the targeting procedure, being controlled by the total heat exchange area, equipment type, equipment layout, number of units, number of shells, material of construction, pressure, pressure drop and piping layout.

Total energy cost ($/h) can be calculated using the following energy equation:

$$Total\, Energy\, Cost = \sum_{U=1}^{U} \dot{Q}_U C_U \tag{4.4}$$

where $\dot{Q}_U$ (kW) is the duty of utility U, C_U ($/kWh) is the cost of utility U and U is the total number of utilities used in the process. A plant operating time of 8000 h per year is typically used.

4.3.6.1 Number of Heat Exchange Units

The number of the heat exchangers N_{HX} involving N_S streams, including utilities, is given by a simple relationship usable either above or below the pinch:

$$N_{HX} = N_S - 1 \tag{4.5}$$

Note that the minimum number of heat exchanger units N_{min} required for MER can be evaluated prior to the design by using a simplified form of Euler's graph theorem. In designing for MER, no heat transfer is allowed across the pinch; hence, a realistic target for the minimum number of units is the sum of the targets evaluated both above and below the pinch, separately:

$$N_{min,MER} = (N_H + N_C + N_U - 1)_{AP} + (N_H + N_C + N_U - 1)_{BP} \tag{4.6}$$

where N_H is the number of hot streams, N_C is the number of cold streams, N_U is the number of utility streams, all counted above pinch (AP) and below pinch (BP).

4.3.6.2 Heat Exchange Area

The heat exchange area can be estimated from the balanced CCs. The simplest way is considering countercurrent operation and vertical heat transfer driving force. Note that ΔT_{min} for process-utility heat exchangers may be different from that for process-process heat exchangers. The total area A_{1-1} is obtained by summing the differential heat exchange area in different temperature intervals, as expressed by the relation

$$A_{1-1} = \sum_{k}^{intervals} \frac{1}{\Delta T_{LMk}} \left(\sum_{i}^{hot} \frac{\dot{Q}_i}{\alpha_i} + \sum_{j}^{cold} \frac{\dot{Q}_j}{\alpha_j} \right) \tag{4.7}$$

where $\dot{Q}_i$ and $\dot{Q}_j$ are the enthalpy variations over a temperature interval with a mean-logarithmic temperature difference (ΔT_{LMk}), and α_i and α_j are partial heat transfer coefficients, including fouling.

4.3.6.3 Number of Shells

The number of shells can be calculated by dividing the target of total area by the area of the selected type of heat exchanger. Shell-and-tube heat exchangers are still the most used in process industries, the simplest being the 1-1 type (1 shell pass, 1 tube pass). However, there are situations when multipass heat exchangers are more beneficial, the most common type being 1-2 (1 shell pass, 2 tube passes). The deviation from purely countercurrent flow can be accounted for by means of the F_T factor (with a practical value of $F_T = 0.9$, values lower than 0.75 should be avoided). The overall heat exchange area can be calculated with the relation

$$A_{1-2} = \sum_{k}^{intervals} \frac{1}{\Delta T_{LMk} F_{Tk}} \left(\sum_{i}^{hot} \frac{Q_i}{\alpha_i} + \sum_{j}^{cold} \frac{Q_j}{\alpha_j} \right) \tag{4.8}$$

In this case, the calculated area is increased by an oversizing factor of 10%–20%. The total number of shells can be obtained by knowing the area of a single shell. Note that the manufacturing of shell-and-tube heat exchangers is regulated by standards, the most well known being the Tubular Exchanger Manufacturers Association (TEMA) standards.

Capital cost laws can be used to estimate the capital of a heat exchanger, such as the one given by

$$Capital = A + B \times (Area)^n \tag{4.9}$$

where the constants A, B and n are functions of the type of exchanger, pressure and material.

The targets for the minimum area A_{min} and the number of units N_{min} can be combined with the heat exchanger cost law to determine the targets for HEN capital cost C_{HEN}:

$$C_{HEN} = N_{min} \left[A + B \left(\frac{A_{min}}{N_{min}} \right)^n \right]_{AP} + N_{min} \left[A + B \left(\frac{A_{min}}{N_{min}} \right)^n \right]_{BP} \tag{4.10}$$

Note that the installed costs of a heat exchanger include insulation, piping and instrumentation. A typical assumption is that the total heat transfer area is divided equally into a number of units, constrained by the maximum shell area. The capital costs must be aligned with the energy costs on an annual basis. A payback time of usually 3 years is assumed. This method gives good results (within 10% errors) when the variation of the heat transfer coefficients is in the same order of magnitude (Smith, 2005). The search of preliminary targets can be summarized as follows:

- Screen design options in conceptual design are reflected in alternative mass and energy balances.

- Try different utility options with implications in both energy and capital costs.

- Start preliminary process optimization.

The targeting procedure allows rapid estimation of energy and capital targets, and this can be extended for designing a complete HEN by decomposing the plant in subsystems, where local integration may be applied (Polley and Heggs, 1999). The reduced number of streams leads finally to simple but reliable practical solutions, even if these might be considered suboptimal.

4.4 PLACEMENT OF UTILITIES

This section emphasizes the advantage of selecting utilities well ahead of the detailed design of heat exchangers. Again, a number of key design decisions with great impact on the process economics can be taken at the targeting level.

4.4.1 Threshold Problems

Not all heat recovery problems have a pinch, and sometimes only hot or cold utility is required. Figure 4.17 (left) illustrates such a typical situation (Dimian et al., 2014). The initial analysis indicates a pinched problem with both hot and cold utilities, but by lowering

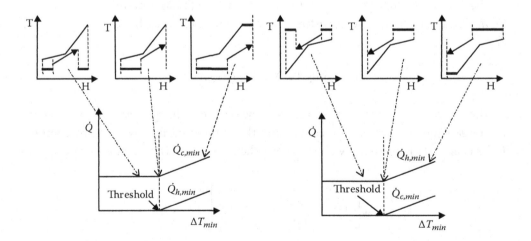

FIGURE 4.17 Threshold problems: only cold utility (left); only hot utility (right).

the ΔT_{min} the cold CC shifts to the left up to a position where there is no need for hot utility. The value of ΔT_{min} when this situation occurs is called *threshold*. Lowering ΔT_{min} below the threshold leads to the need for a second cold utility, this time at the hot end. Similarly, a problem with both hot and cold utility can turn into a problem that needs only hot utility below a threshold value of ΔT_{min} (Figure 4.17, right). Hence, a problem apparently with both cold and hot utilities might hide a threshold problem. Below the threshold ΔT_{min}, the cost of energy remains constant, so the trade-off between capital and energy can be considered only at ΔT_{min} values equal to or above the threshold. Because the occurrence of utility at the opposite end increases the number of units, the probability of an optimum near ΔT_{min} at threshold is more likely.

A threshold problem may turn into a pinch problem when multiple utilities are used. Figure 4.18 gives a graphical description (Dimian et al., 2014). In Figure 4.18 (left), the cold utility is covered by both CW and steam generation, while a new utility pinch occurs at the steam generation side. In Figure 4.18 (right), only hot utility is required, and if the utility is covered by HP and LP steam, then a new pinch occurs at the LP process side. The only utility above this pinch is HP steam.

4.4.2 Multiple Utilities

Several hot or cold utilities can be used in a heat recovery project, and GCC is the appropriate conceptual tool. Reviewing the heat integration of the streams presented in Table 4.1, the same hot load as in Figure 4.13 can be shared between two steam levels: HP and LP steam. In the GCC, the solution is represented by horizontal segments placed at temperatures corresponding to their pressures, as shown in Figure 4.19 for three steam levels (Dimian et al., 2014). This feature makes it possible to specify exactly the amount required by each utility. Simple targeting can be applied to optimize their amount if the prices are significantly different. Similarly, the cold utility can be split into cold water and other cold utilities (e.g. refrigerant or brine). Increasing the pressure of LP steam is another option of cooling at temperatures above 100°C. In this case, boiling feed water (BFW) is used to recover a part of energy by preheating. The LP steam can be upgraded at higher temperature by thermal compression. Each additional utility above the highest and coldest utilities introduces a new utility pinch, and a minimum temperature approach is required for each.

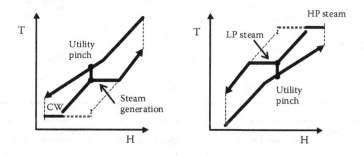

FIGURE 4.18 Threshold problem turned into a problem with utility pinch.

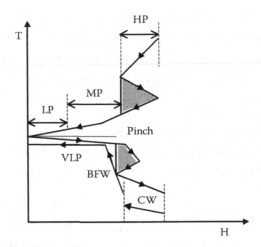

FIGURE 4.19   Multiple utilities. VLP, very low pressure.

4.4.3 Variable Temperature Utilities

This category considers flue gases, hot oil or other thermal fluids. The main problem is that the use of such utilities is more constrained with respect to the temperature level. Figure 4.20 illustrates two situations encountered when using flue gases for heating (Dimian et al., 2014). Firstly, larger ΔT_{min} below the utility process must be assumed (e.g. 50°C) due to the poor heat transfer on the gas side. Figure 4.20 (left) presents the case when the constraint is the exit gas temperature. The graphical representation allows an easy visualization of the stack losses. Increasing the theoretical flame temperature by air preheating or by low air excess can reduce the heat losses. Other factors than the pinch can also limit the flue gas temperature. This can be a process away from the pinch, such

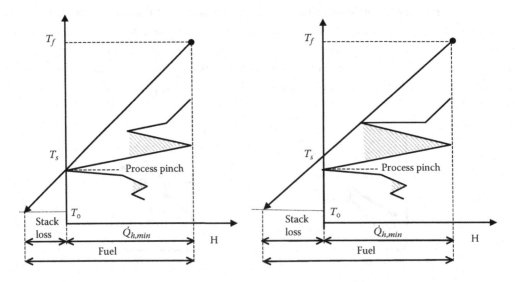

FIGURE 4.20   Flue gas as a variable temperature utility: constraint by process Pinch (left); limitations by a process away from the pinch (right).

as the one shown in Figure 4.20 (right) or the dew point of the acid components in the gas stack.

4.5 HEAT EXCHANGER NETWORK DESIGN

The systematic procedure for designing the HEN based on the pinch principle is known as the *pinch design method* (Linnhoff and Flower, 1978; Linnhoff and Hindmarsh, 1983). The systematic application of the pinch design method allows the design of a HEN that achieves the energy targets, within practical limits. The method has two key features: (1) It recognizes that the pinch region is the most constrained part of the problem, so it starts the design at the pinch and develops by moving away, and (2) it allows the designer to choose between match options. The design of HEN examines which *hot streams* can be matched to *cold streams* via heat recovery by employing *tick-off heuristics* to identify the heat load on the pinch heat exchanger. Every match brings one stream to its target temperature. The pinch divides the heat exchange system into two thermally independent regions, so the HENs for above and below pinch regions are designed separately. When the heat recovery is maximized, the remaining thermal needs must be supplied by hot and cold utilities. The graphical method of representing flow streams and heat recovery matches is called a *grid diagram*.

Basically, the pinch design method consists of the following steps:

1. Select streams for integration. Only streams useful for effective energy integration should be considered. Some simple rules should be followed:

   a. Examine the heat integration of reactors separately.

   b. Do not consider reboilers and condensers in the first approximation.

   c. Decompose large problems in subsystems, taking into account the proximity of streams.

2. Select the type of utilities.

3. Estimate process-process saving and process/utility loads.

4. Construct the CC and GCC. Identify process and utility pinches.

5. Determine design targets: ΔT_{min}, MER, number of units, capital and utility costs.

6. Construct the grid diagram.

7. Find the matches of heat exchangers by applying the rules of the pinch design method.

8. Improve the design by eliminating the small heat exchangers.

9. Optimize the HEN.

Note that the first five points belong to the targeting procedure, while the remaining points form the core of the pinch design method that is explained next.

4.5.1 Topological Analysis

4.5.1.1 Number of Matches

It can be demonstrated that the minimum number of units (matches) N_E necessary to recover the energy between N_S process streams using N_U utilities is (Linnhoff, 1993)

$$N_E = N_S + N_U - 1 \qquad (4.11)$$

If loops are present in the network, then the number of heat exchangers is given by the formula:

$$N_E = N_S + N_U - 1 + (Loops) \qquad (4.12)$$

4.5.1.2 Paths and Loops

The concepts of paths and loops are useful for optimizing a HEN. The application of the pinch design method generates redundancy in the number of units, and reducing the number of heat exchangers can contribute more significantly in the cost saving than the incremental optimization of the exchange area. However, merging some units could violate the pinch principle in the sense of increasing both hot and cold energy loads. In a number of cases, an important reduction in capital may be obtained only with a marginal increase of energy above minimum requirements. This problem can be solved efficiently by the approach explained hereafter.

A *path* means a physical connection through streams and heat exchangers for shifting energy between the hot and cold utilities. Consequently, a path allows the modification of the temperature difference between the hot and cold streams, as illustrated in Figure 4.21,

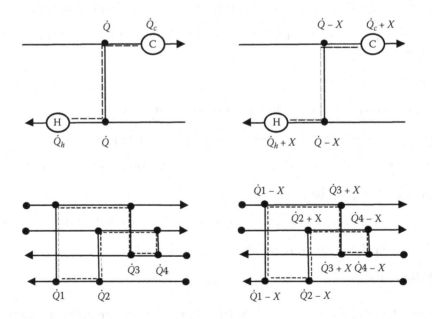

FIGURE 4.21   Concept of paths (top) and loops (bottom).

top (Dimian et al., 2014). Initially, the utility loads are $\dot{Q}_c$ and $\dot{Q}_h$, and in between there is a heat exchanger of duty $\dot{Q}$ (Figure 4.21, top-left). By rising $\dot{Q}_h$ with X units, the cold utility duty $\dot{Q}_c$ must go up exactly with the same amount. As a result, the heat exchanger duty becomes $Q-X$), as shown in Figure 4.21 (top right). The outlet temperature of the hot side will decrease, and the same happens with the outlet temperature of the cold side. Therefore, the temperature differences at both ends are enlarged. Note that the heat exchangers outside the path are not affected, but the utility heat exchangers are affected.

A *loop* is a closed trajectory passing through several heat exchangers (Figure 4.21, bottom), and it can link several heat exchangers with the same utility. Changing a duty with the amount X modifies the duties of the other exchangers with exactly the same amount because of energy conservation. Setting the duty to zero eliminates the heat exchanger. The operation will change the temperatures of the units involved in the loop and might lead to infeasible temperature profiles. Feasible ΔT can be restored if the loop is connected with both hot and cold utilities through a path.

4.5.2 HEN Design in the Grid Diagram

The design of the HEN takes place in the balanced grid diagram. The utilities have to be placed on the GCC: The hot streams run from left to right at the top, and cold streams run countercurrently at the bottom. The pinch splits the diagram in two regions: above (left side) and below (right side) the pinch. A heat exchanger is represented by a vertical 'bubble-headed' stick linking two streams, while temperature, duty and area are also normally displayed. The design starts at the pinch, and the match procedure has to respect some feasibility rules. For a match above the pinch, the following heuristic has to be respected:

$$CP_{hot} \le CP_{cold} \tag{4.13}$$

The rule can be justified intuitively by the fact that above the pinch there is need only for hot utility, such that the CP of cold streams must be greater than the CP of hot streams; keep in mind that in this context, CP means (Flow rate) × (Specific heat capacity). A more accurate reason can be found in the fact that the temperature difference $(T_h - T_c)$ at the end looking to the pinch is always smaller, at most equal to the temperature difference on the other side. It is also clear why the rule presented is not so strict for a match away from the pinch. Similarly, below the pinch, the following heuristic holds:

$$CP_{hot} \ge CP_{cold} \tag{4.14}$$

Both heuristics can be conveniently captured by the following relation:

$$CP_{in} \le CP_{out} \tag{4.15}$$

Thus, for a match close to the pinch, the CP of the incoming stream must be smaller than the CP of the outgoing stream.

4.5.3 Stream Splitting

By stream splitting, practical solutions can be found for cases for which the matches at pinch are not feasible apparently because the rules for *CP* do not hold. Some common situations are shown in Figure 4.22 (Dimian et al., 2014). The following explanations address the part above the pinch:

1. The number of cold streams is smaller than the number of hot streams. In Figure 4.22 (top left), there are two hot streams against one cold stream. Regardless of *CP*, there are not sufficient cold streams because cold utility cannot be used above the pinch. Thus, in addition to *CP* rule a 'count rule' regarding the pairing of streams at pinch must be satisfied. Above the pinch, this rule is

$$N_{hot} \leq N_{cold} \tag{4.16}$$

   Two matches become possible by splitting the cold stream in two segments. Moreover, the split must be done such to respect the *CP* rule, as in Figure 4.22 (top right).

2. The count rule is satisfied, but not the *CP* rule. Figure 4.22 (bottom left) illustrates this situation by one hot stream and two cold streams. The hot stream must be split in two parts, such as the *CP* of hot streams becomes smaller than the *CP* of the corresponding cold streams, as in Figure 4.22 (bottom right). It may appear also that the count rule is satisfied, but the *CP* rule is fulfilled only partially. In this case, the largest cold stream should be split. The analysis can be extended below the pinch:

$$N_{cold} \leq N_{hot} \tag{4.17}$$

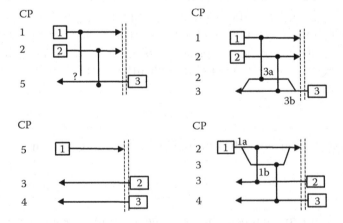

FIGURE 4.22 Principle of stream splitting at the pinch.

Considering that the hot and cold streams are *in* and *out*, respectively, the count rule becomes

$$N_{out} \leq N_{in} \tag{4.18}$$

The rules described previously can be organized into a general design procedure at pinch, as illustrated by Figure 4.23 (Dimian et al., 2014). Firstly, the stream count rule is checked, and if not fulfilled, a first stream split is performed to balance streams (cold stream above or hot stream below pinch). Then, the *CP* rule is checked for matches close to the pinch, and if not fulfilled, stream splitting is executed but this time opposite to the first. These rules might be not respected away from the pinch.

4.5.4 Reducing the Heat Exchanger Network

The synthesis of the HEN for MER ensures the best energy recovery from the thermodynamic viewpoint. However, the cost of heat exchangers could substantially change the analysis. The decomposition of the system in subsystems increases the number of units above those required by the first law analysis. The difference is given by the loops crossing the pinch. Further reduction is paid by increasing both heating and cooling requirements. The trade-off depends on the costs of energy and hardware in a specific situation. The following heuristics are recommended:

- Firstly, break the loop including the smallest heat exchanger.

- Remove the unit with the smallest load in a loop.

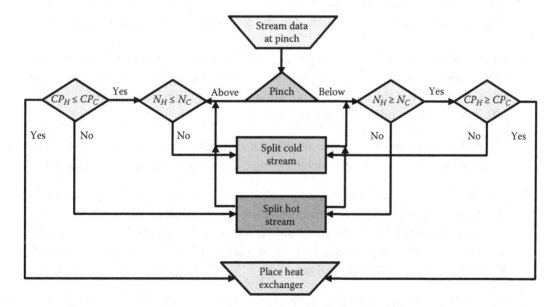

FIGURE 4.23   General HEN design procedure at the pinch.

The explanation is that for the same incremental area, the capital reduction for small units is by far more important than for large units. For small heat exchangers, the contribution of the fixed capital term, which accounts for installation, instrumentation, control, piping, civil and maintenance costs, can largely overcome the cost of area.

4.5.5 Network Optimization

The pinch design method generates a network with a certain degree of redundancy. The reduction of units can contribute to major saving in capital costs, but more energy has to be introduced into the network to restore infeasible driving forces. This task can be treated elegantly by optimization. Another application is the split of streams at the pinch to fulfill the feasibility criteria of matches. Although some insights can be obtained from the grid diagram, the exact split fraction is a typical optimization problem. The same holds for splitting streams far from the pinch. Notably, the bypasses of streams around heat exchangers can be used for temperature control.

Not surprisingly, optimization is a powerful method in handling the design of a HEN (Duran and Grossmann, 1986; Floudas et al., 1986; Floudas, 1995). As the whole approach is based on a linear relation enthalpy-temperature, the problem can be conveniently solved by *mixed-integer linear programming* (MILP). The optimization is a parametric one because the network structure is not affected. If the streams involved in matches are no longer constrained by the *CP* rule – as is the case for matches away from the pinch – then the optimization becomes structural, and the problem is of another type: *mixed integer nonlinear programming* (MINLP). Optimization is also a powerful manner for designing networks subjected to constraints. Usually, these can be duties, inlet/outlet temperatures, area, heat transfer coefficients, split and bypass fractions of streams. The approach is particularly powerful for revamping existing HENs, where the number of old exchangers is by far larger than the new units to be inserted.

To summarize, the pinch analysis method can be described as follows (Dimian et al., 2014):

- Pinch analysis is a systematic design methodology using a number of concepts and techniques that ensure optimal energy use. The pinch designates the location where the heat recovery is the most constrained, with a minimum temperature difference ΔT_{min} between hot and cold streams.

- The PTA is the primary computational tool used in pinch analysis, as it allows finding the pinch location, as well as determining energy targets for hot and cold utilities.

- The net heat flow across the pinch is zero; hence, the system can be split into two stand-alone subsystems: above and below the pinch. Above and below the pinch, there is only a need for hot and cold utility, respectively. The MER are given by the hot and cold utility requirements for a given ΔT_{min}. Clearly, no design can achieve MER if there is a cross-pinch heat transfer.

- Using subsystems may introduce redundancy in the number of heat exchangers, so when the capital cost is high, it might be necessary to remove the pinch constraint to reduce the number of units. This is paid by supplementary energy use, which has to be optimized against the reduction in capital costs. The result is that a heat recovery problem becomes an optimization of energy and capital costs, constrained by a minimum temperature approach in designing the heat exchangers.

- Stream selection and data extraction are essential in pinch analysis for effective heat integration.

- The key computational assumption in pinch analysis is the constant *CP* on the interval where the streams are matched. If that is not the case, then stream segmentation is necessary.

- The countercurrent heat flow of the streams selected for integration may be represented by means of CCs. The GCC allows the visualization of the excess heat between hot and cold streams against temperature intervals. This feature helps the selection and placement of utilities, as well as the identification of the potential process-process matches.

- The synthesis of a HEN consists of three main activities:

  - *Targeting* sets a reference basis for energy integration in terms of MER, utility selection and placement, number of units, heat exchange area, cost of energy and hardware at MER.

  - *Synthesis of heat exchanger network* (HEN) for minimum energy requirements and maximum heat recovery, determination of matches in subsystems and generation of alternatives.

  - *Network optimization* reduces redundant elements (e.g. small heat exchangers or small split streams) and finds the trade-off between utility consumption, heat exchange area and number of units while considering constraints.

- The design improvement can be realized by *appropriate placement* and using the *plus/minus principle*. Appropriate placement defines the optimal location of individual units against the pinch, and it applies to heat pumps, heat engines, distillation columns, evaporators, furnaces and any other unit operation that can be represented in terms of heat sources and sinks.

- The *plus/minus principle* helps to identify major flowsheet modifications that can significantly improve energy recovery. Alternating between appropriate placement, plus/minus principle and targeting allows one to formulate near-optimum targets for HEN without sizing the units.

Note that the pinch is important in directing process changes rather than controlling the design. Although it is possible to construct composite lines for every heat sink or source stream involved in a process, heat integration often results in complex integrated systems that may be difficult to control in operation. The systematic addition or subtraction of streams to or from the analysis helps to keep it simple, leading to near-optimal but controllable systems (Polley and Heggs, 1999).

4.6 TOTAL SITE INTEGRATION

The concept of *total site analysis* (TSA) enables the analysis of overall energy usage for an entire site that consists of several processes served by a central utility system. Production processes operate typically as parts of large sites that serve some processes using a centralized utility system, while having both consumption and recovery of process steam via steam mains, as well as imports and exports of power to balance the on-site power generation. The sitewide fuel demand via the utility system is dictated by the process stream heating and cooling demands and cogeneration potential.

The individual production processes and central services are controlled by different departments, which usually operate independently; thus, the site infrastructure suffers from inadequate integration. The integration can be improved by a simultaneous approach that considers the individual process issues alongside sitewide utility planning. TSA can be used to calculate energy targets for the entire site (e.g. steam requirements, amount of power that can be generated). Key process changes that will reduce the overall site utility consumption can also be identified.

4.6.1 Process Utilities

Utilities refer to the assembly of means necessary for enabling the operation of a process plant, in particular with respect to the energy needs. These include electricity; steam and hot water; thermal agents (e.g. oil, fluids, salts); fluids for cooling (CW, chilled water, brines); process water (washing water, demineralized water); compressed air and inert means (usually nitrogen). In most cases, the utilities are provided by specialized site suppliers (e.g. steam, electricity, CW, compressed air and inert means), while some utilities could be produced locally (on site).

Figure 4.24 displays the schematic of a typical utility system (Klemes et al., 2011; Dimian et al., 2014). Very high-pressure (VHP) steam, produced by a boiler house at 60–100 bar

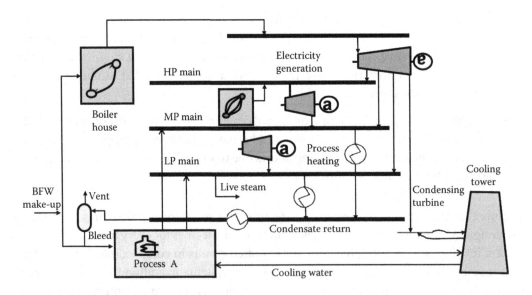

FIGURE 4.24   Power, heat and cooling utility system.

and 480°C–600°C, is used mainly for generating electricity and heat by cogeneration. The steam extracted by back-pressure turbines is distributed over the site at several levels (mains), such as the following:

- HP steam: around 40 bar with saturation temperature at 250°C

- Medium-pressure (MP) steam: around 20 bar with saturation pressure at 212°C

- LP steam: around 6 bar (159°C) or 3 bar (133°C)

The pressure in the mains is kept constant by means of let-down stations. Steam only for heating can be supplied by boilers, furnaces or different processes involving exothermal reactions. The steam produced on site can be integrated with the site network. Steam turbines could be used for generating power, while waste heat from exhaust gases or steam can be recovered by heat pumps. The steam pressure can be adjusted by means of expansion valves, and live steam can be drawn out for direct use. The return condensate from different levels is collected for recycling to the boiler house. The enthalpy may be used for heating at lower temperatures, usually 60°C–120°C. After conditioning operations (e.g. venting of dissolved gases and bleeding of salt and impurities), the water is recycled to the boiler house. The inventory of the BFW circulating through the steam system is kept constant by adding makeup water, after chemical treatment.

Another feature highlighted in Figure 4.24 is the circuit of the CW. CW is necessary when using condensing turbines or refrigeration engines, as well as for efficient cooling in heat exchange operations. Hot water is sent to a *cooling tower*, where the temperature drops by adiabatic evaporation. The effect depends on the hot water temperature and local air condition (temperature, humidity). The difference between the hot and cold temperature is designated by the *cooling range*, usually 3–11 K. When selecting the cooling range, it should be kept in mind that the hottest water temperature should be kept below 40°C to avoid the formation of deposit by salt (hydrocarbonate) precipitation. For this reason, the temperature difference between inlet and outlet flows in CW-driven operations should be kept below 10 K, in most cases between 5 and 6 K. The assessment of a cooling tower can be done by means of hygrometric charts, which can be found in classical handbooks (Green and Perry, 2008). The wet bulb temperature is about 19°C for air at 25°C and 60% humidity. In practice, the cooling towers operate with a temperature approach of at least 2.8 K (Towler and Sinnott, 2013), so instead of 19°C, the cold CW would have an inlet temperature of about 22°C. Using air and finned heat exchangers is much cheaper but is limited by the seasonal conditions and less-efficient heat transfer. The temperature of the hot air during the summer period could be reduced by evaporative cooling by spraying water.

4.6.2 Site Integration

The concepts of pinch analysis can be extended from a single plant to an *industrial site*. The principle of export-import of utilities via the site system is a simple way for integration, with key economic advantages. A further expansion in the energy supply can be done via the central utility plant and distributed via the existing network. Another example is steam produced by exothermal reactions, which can be exported at the maximum allowed level of pressure. The optimization of steam distribution in individual plants is typically more profitable when the total site is considered.

The analysis of a site integration problem can be executed in diagrams known as *total site profiles* (Linnhoff et al., 1994). This representation can give a central view of the targets for fuel, turbine loads and cooling. The problem of emissions can be better handled as a site integration problem because the reduction in waste and effluents could increase the fuel consumption and thus also the gaseous emissions. Larger steam consumption is needed for waste separations, which in turn are produced in fired boilers. There is a trade-off between process emissions and fuel-related emissions. Site integration topics are treated in detail in the books of Smith (2005) and Klemes et al. (2011).

4.7 MATHEMATICAL PROGRAMMING

Mathematical programming (MP) represents a class of optimization tools that can be employed to solve energy-saving problems. A good introduction in this area can be found in the book of Biegler et al. (1997), where a clear difference is made between *sequential* and *simultaneous* approaches.

The *sequential method* makes possible the development of partial optimized networks that satisfy one of the criteria among minimum utility, minimum investment or the minimum number of units. The solution of the minimum utilities may be seen as analogous to the *transshipment problem* in linear programming. Hot streams can be regarded as sources and cold streams as destinations. The heat can be transferred from sources to destinations through some intermediates exchanging places, where a ΔT_{min} driving force must be ensured. This is equivalent to temperature intervals in the PTA. The energy starts to flow from the hot utility source. The excess energy that cannot be transferred to cold streams can be cascaded at lower-temperature intervals and finally to the cold utility. The result is a full targeting of the recovery problem, where a pinch can be clearly defined. The problem can be extended to take into account physical matches by means of an MILP algorithm. It is worth noting that the solution may not be unique, and there might be several networks for the same cost of utility and hardware. Conversely, the same network may have different distributions of heat loads if loops are present. Remarkably, the method can directly treat forbidden or imposed matches.

In the *simultaneous method*, an existing superstructure is optimized rigorously by an MINLP algorithm. The modeling is based on a stagewise buildup of the superstructure for each temperature interval. Within each stage potential, heat exchange between any pair of hot and cold streams may take place. Accordingly, each hot stream is split and directed to match all cold streams and reciprocally. In a first approximation, the outlets of the exchangers are well mixed, and these are also optimization variables. The number of stages can be set equal to the temperature intervals. Finding the number of matches is similar to the extended transshipment model. However, the constraint of the pinch can be removed more easily, consequently leading to networks that minimize simultaneously units, capital and energy. Other structural features, such as steam splitting and bypassing, can also be considered.

Pinch analysis and MP are highly complementary, but pinch analysis remains an invaluable systematic investigation method, solidly anchored in thermodynamics and based on simple intuitive graphical tools. Moreover, it offers a stepwise global picture of the problem to be solved, suggests intermediate solutions and thus stimulates creativity greatly. However, when the computations become tedious or excessive, the use of an automatic algorithmic tool is highly desirable. Not surprisingly, MP is embedded today into packages based on pinch analysis and allows a reliable solution of most energy-saving problems. Generic optimization software tools are also available, as for example GAMS (General Algebraic Modeling System, GAMS Development Corp., USA); LINDO (nonlinear and stochastic programming, LINDO Systems, USA); MIPSYN (Carnegie Mellon, USA, and University of Maribor, Slovenia); AIMMS (Advanced Interactive Multidimensional Modeling System); AMPL (A Mathematical Programming Language, AMPL Optimization Inc.) and Gurobi Optimizer.

4.8 CONCLUDING REMARKS

The most efficient energy saving can be obtained by examining the integration of different units in the context of the whole process rather than by improving individual performance. Pinch analysis is able to set energy and capital cost targets for an individual process or for an entire production site ahead of design, such that the scope for energy savings and investment requirements is known prior to identifying any projects. An important challenge is to properly integrate the pinch tools into the conceptual process design, as decisions made in the conceptual process design phase affect the entire life cycle of a chemical process. The role of pinch technology is to identify *what it might be*, while the input from other engineering disciplines (e.g. process specialists) ultimately determines *what it can be*.

A key concept is the appropriate placement of units with respect to the pinch. The correct placement for energy sources is above the pinch, and for sinks it is below the pinch. Heat pumps should be placed across the pinch, but unit operations involving both sink and sources (e.g. distillation columns, heat engines) must be placed on the same side of the pinch, either above or below.

Total site integration allows the identification of major opportunities for saving both energy and capital investment. Significant savings can be obtained by structural flowsheet modifications or by reconsidering the site policy regarding energy and utilities. However, saving energy by very tight integration might conflict with the process control and operability. In addition to energy recovery, pinch analysis enables process engineers to achieve additional process improvements, such as the following:

- *Update or modify process flow diagrams*: The pinch quantifies the savings available by changing the process itself, showing where process changes reduce the overall energy target.

- *Conduct process simulation studies*: Pinch analysis replaces old energy studies with information that can be easily updated by simulation studies that help avoid unnecessary capital costs by identifying energy savings with a smaller investment before the projects are implemented.

- *Set practical targets*: Theoretical targets are modified by taking into account the practical constraints (e.g. layout, safety, difficult fluids), such that they can be realistically achieved.

- *Debottlenecking*: Compared to a conventional revamp, pinch analysis can lead to important benefits, such as reduction in capital costs and decrease in specific energy demand.

- *Determine opportunities for combined heat and power (CHP) generation*: Pinch analysis shows the best type of CHP system that matches the inherent thermodynamic opportunities on the site, but heat recovery should be optimized by pinch analysis before specifying CHP systems.

- *Decide the use of low-grade waste heat*: Pinch analysis shows what waste heat streams could be recovered and provides insight into the most effective means of recovery.

By analogy with heat saving, the pinch concept has been extended to operations involving mass exchange. The analogy consists of identifying the key bottleneck in the use of the driving force, as well as in the internal recycle. The pinch concept proved to be useful for the treatment of other valuable resources in process industries, such as process water, hydrogen, oxygen, waste and emissions (Staine and Favrat, 1996; Foo, 2012; Klemes, 2013). Saving water can be treated systematically by water pinch methodology, while the inventory of hydrogen in refineries can be efficiently handled by hydrogen pinch methods. Other applications of industrial interest have been developed in the field of waste and emissions minimization. For a deeper study of various and more subtle aspects of energy integration in process design, the reader is directed to the expert books of Smith (2005), El-Halwagi (2006), Klemes et al. (2011), Klemes (2013) and Dimian et al. (2014), while for the fundamentals of engineering thermodynamics, refer to the book of Moran and Shapiro (2007).

LIST OF SYMBOLS

A	Area	m^2
C	Cost	$\$kW^{-1}yr^{-1}$
c_p	Specific heat capacity	$kJkg^{-1}K^{-1}$
CP	Flow heat capacity	kW/K
F_T	Temperature correction factor	–
H	Enthalpy or energy (heat flow)	kW
$\dot{m}$	Mass flow	kg/s
N	Number of units	–
N_C	Number of cold streams	–
N_{HX}	Number of heat exchangers	–
N_H	Number of hot streams	–
N_S	Number of streams	–
N_U	Number of utility streams	–
$\dot{Q}$	Enthalpy variation	kW

T	Temperature	K
T_s	Supply temperature	K
T_t	Target temperature	K
$\dot{W}$	Power	kW

Greek Symbols

α	Partial heat transfer coefficient	W/m²K
ΔT	Temperature difference	K
ΔT_{LM}	Log mean temperature difference	K
ΔT_{min}	Minimum temperature approach	K

Subscripts

C, c	Cold stream
e	External
H, h	Hot stream
in	Inlet
int	Interval
min	Minimum
rec	Recovery
out	Outlet
S	Stream
U	Utility

Abbreviations

AP	Above pinch
BFW	Boiler feed water
BP	Below pinch
CC	Composite curve
CHP	Combined heat and power
CW	Cooling water
GCC	Grand composite curve
HEN	Heat exchanger network
HP	High pressure (steam)/Heat pump
LP	Low pressure (steam)/Linear programming
MER	Minimum energy requirements
MILP	Mixed integer linear programming
MINLP	Mixed integer nonlinear programming
MP	Medium pressure (steam)/Mathematical programming
PTA	Problem table algorithm
VHP	Very high pressure
VLP	Very low pressure
TEMA	Tubular Exchanger Manufacturers Association
TSA	Total site integration

REFERENCES

Biegler L., Grossmann I. E., Westerberg A., *Systematic methods of chemical process design*, Prentice Hall, Englewood Cliffs, NJ, 1997.

Dimian A. C., Bildea C. S., Kiss A. A., *Integrated design and simulation of chemical processes*, 2nd edition, Elsevier, New York, 2014.

Dunn R. F., El-Halwagi M. M., Process integration technology review: background and applications in the chemical process industry, *Journal of Chemical Technology and Biotechnology*, 78 (2003), 1011–1021.

Duran M. A., Grossmann I. E., Simultaneous optimization and heat integration of chemical processes, *AIChE Journal*, 32 (1986), 123–138.

El-Halwagi M. M., *Process integration*, Elsevier, New York, 2006.

Floudas C. A., *Nonlinear and mixed-integer optimization. Fundamentals and applications*, Oxford University Press, New York, 1995.

Floudas C. A., Ciric A. R., Grossmann I.E., Automatic synthesis of optimum heat exchanger network configuration, *AIChE Journal*, 32 (1986), 276–290.

Foo D. C. Y., *Process integration for resource conservation*, CRC Press, Boca Raton, FL, 2012.

Green D. W., Perry R. H., *Chemical engineer's handbook*, 8th edition, McGraw-Hill, New York, 2008.

Itoh J., Shiroko K., Umeda T., Extensive applications of the T-Q Diagram to heat integrated system synthesis, International Symposium on Process Systems Engineering, Kyoto, Japan, August 23–27, 1982, Technical Report 92–99.

Kemp I., *Pinch analysis and process integration*, Butterworth-Heinemann, Elsevier, New York, 2007.

Kiss A. A., Grievink J., Rito-Palomares M., A systems engineering perspective on process integration in industrial biotechnology, *Journal of Chemical Technology and Biotechnology*, 90 (2015), 349–355.

Klemes J., *Handbook of process integration (PI): minimisation of energy and water use, waste and emissions*, Woodhead, Cambridge, UK, 2013.

Klemes J., Friedler F., Bulatov I., Varbanov P., *Sustainability in the process industry: Integration and optimization*, McGraw-Hill, New York, 2011.

Linnhoff B., Pinch analysis. A-state-of-the-art overview, *Chemical Engineering Research and Design*, 71 (1993), 503–522.

Linnhoff B., Flower J. R., Synthesis of heat exchanger networks, *AIChE Journal*, 24 (1978), 633–654.

Linnhoff B., Hindmarsh B., The pinch design method of heat exchanger networks, *Chemical Engineering Science*, 38 (1983), 745–763.

Linnhoff B., Townsend D. W., Boland D., Hewitt G. F., Thomas B., Guy A. R., Marsland R. H., *User guide on process integration for the efficient use of energy*, Institution of Chemical Engineers, Rugby, UK, 1982.

Linnhoff B., Townsend D. W., Boland D., Hewitt G. F., Thomas B., Guy A. R., Marsland R. H., *User guide on process integration for the efficient use of energy*, 2nd edition, Institution of Chemical Engineers, Rugby, UK, 1994.

Moran M. J., Shapiro H. N., *Fundamentals of engineering thermodynamics*, 6th edition, Wiley, New York, 2007.

Polley G. T., Heggs P. J., Don't let the Pinch pinch you, *Chemical Engineering Progress*, 95 (1999), 27–36.

Smith R., *Chemical process design and integration*, Wiley, New York, 2005.

Smith R., *Chemical process design and integration*, 2nd edition, Wiley, New York, 2015.

Staine F., Favrat D., Energy integration of industrial processes based on the pinch analysis method extended to include exergy factors, *Applied Thermal Engineering*, 16 (1996), 497–507.

Towler G., Sinnott R., *Chemical engineering design: Principles, practice and economics of plant and process design*, Elsevier, New York, 2013.

CHAPTER 5

Selection of Heat Pumps

5.1 INTRODUCTION

Before deciding what type of heat pump is the most appropriate for a specific application, the feasibility of using a heat pump in the first place must be evaluated. In many cases, a quick check is possible simply by answering some questions:

Is the heat demand of the process optimized already?

Is there any heat source (e.g. waste heat stream) available with sufficient heat capacity?

Is the waste heat available at the same time as the heat demand?

Is the temperature of the heat source lower than the required heating temperature?

Is the temperature difference between the heat source and the heat sink lower than 80 K?

Is the heat demand constant and for long periods?

A positive answer to all questions indicates that it is feasible to apply a heat pump for a specific process.

However, when selecting a heat pump for a specific application, several aspects should be considered, such as position of the heat pumps in relation to the pinch point, heat pump duty, temperature levels of the sink and source, temperature lift, sources of irreversibilities, energy savings and type of heat pump (open and closed heat pump cycles). This chapter discusses all these aspects and proposes a method to estimate the investment level of different heat pump concepts, as well as a procedure for the selection of the most appropriate heat pump type.

5.2 THERMAL AND HYDRAULIC REQUIREMENTS

Heat pumps must meet the specifications that follow from a heat exchanger network (HEN) analysis, taking the pinch temperature of the relevant process flows into account. Ranade (1988) investigated the optimal thermodynamic integration of heat pumps in industrial sites, making use of the grand composite curves of the sites. Taking into account the additional capital expenditure for increased heat transfer area, the maximum temperature lift can be identified such that the payback time of the investment remains within 2 years. Typically, heat exchangers account for 40%–50% of the capital cost of semiopen cycles, while they account for over 60% for the closed-type cycles (Rossiter et al., 1988). These numbers did not change over the years, and presently heat exchangers have a similar impact on the investment level. Heat pumps should preferably make use of the cold utility requirement as the heat source and deliver heat to fulfill (part of) the hot utility requirement. In this case, the heat pump specifications follow directly from the process network optimization, as clearly illustrated in Figure 5.1. The heat flow determined by the cold utility requirement $\dot{Q}_{cold\_utility}$ gives the duty available as a source for the heat pump. The in- and outlet temperatures of the hot streams in this region determine the operating temperatures of the source flow, while its heat capacity follows from duty and temperature change.

Preferably the heat pump delivers heat to fulfill (part of) the hot utility requirement of the same network of heat exchangers. In this case, the heat delivered is the cold utility duty plus the external power input $\dot{W}_{hp}$. The in- and outlet temperatures of the cold streams in this region determine then the operating temperatures of the heat sink flow. Table 5.1 gives an overview of possible heat pump applications in relation to a network of heat exchangers and indicates how the thermal requirements for the heat pump can be quantified. If the duty required by the hot utility is smaller than the duty required by the cold utility, the sink determines the maximum heat delivered by the heat pump and only part of the cold utility can be served by the heat pump.

As discussed in the previous chapter about pinch analysis, it is also possible to integrate heat pumps below or above the pinch, but this will imply that electricity is consumed instead of fuel, thus resulting in a rather expensive solution. In specific cases, heat pumps can make use of the cold utility requirement of a process network and deliver the higher-temperature heat to a nonrelated process. Also, it is possible that the heat pump delivers the

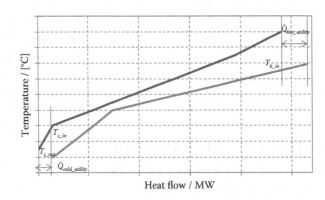

Heat flow / MW

FIGURE 5.1 Grand composite lines for hot streams (top) and cold streams (bottom) showing the minimum cold and hot utility requirement for the selected minimum approach temperature.

TABLE 5.1 Determination of the Thermal Heat Pump Requirements in Relation to Heat Exchanger Networks

Use in	$\dot{Q}_{duty}$	T_{source_in}	T_{source_out}	T_{sink_in}	T_{sink_out}
Same heat exchanger network	$\dot{Q}_{cold\_utility} + \dot{W}_{hp}$ if $< \dot{Q}_{hot\_utility}$	T_{c_in}	T_{c_out}	T_{h_in}	$T_{h\_in} + \dfrac{\dot{Q}_{cold\_utility} + \dot{W}_{hp}}{(\dot{m}c_p)_c}$
Same heat exchanger network	$\dot{Q}_{hot\_utility}$ if $< \dot{Q}_{cold\_utility} + \dot{W}_{hp}$	T_{c_in}	$T_{c\_in} - \dfrac{\dot{Q}_{hot\_utility}}{(\dot{m}c_p)_h}$	T_{h_in}	T_{h_out}
Application not related to heat exchanger network	$\dot{Q}_{cold\_utility} + \dot{W}_{hp}$	T_{c_in}	T_{c_out}	$T_{external\_sink\_in}$	$T_{external\_sink\_out}$
Source from unrelated heat exchanger network	$\dot{Q}_{hot\_utility}$	$T_{external\_source\_in}$	$T_{external\_source\_out}$	T_{h_in}	T_{h_out}

hot utility duty required by making use of an external heat source. In these cases, generally the specifications follow from the external process, but sometimes the cold/hot utility requirement determines the heat pump specifications.

The addition of heat exchangers to the HEN implies an increase in total pressure drop for the streams of the network. This leads to an increase in the pumping power requirements. Finally, the specifications should include the following:

- Amount of heat available at the cold side or alternatively required at the hot side

- Source and sink fluid inlet and outlet temperatures

- Allowable pressure drop for these fluids (or allowable pumping power)

5.3 TEMPERATURE LIFT AND SYSTEM IRREVERSIBILITIES

5.3.1 Temperature Lift and System Performance

The required temperature lift for the heat pump can be derived from its thermal specifications. The heat pump must convert the heat extracted from the source into heat to be delivered to the heat sink, making use of an external source of power, as conveniently illustrated in Figure 5.2.

Generally, as indicated in Figures 5.1 and 5.2, the temperature of the source and sink streams is not constant. In these cases, the equivalent heat sink temperature is given by

$$\overline{T}_{sink} = \frac{T_{h\_out} - T_{h\_in}}{\ln \dfrac{T_{h\_out}}{T_{h\_in}}} \tag{5.1}$$

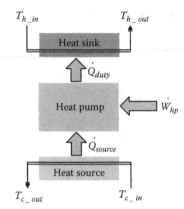

FIGURE 5.2 Temperature levels of the hot and cold streams associated with the heat pump.

and by

$$\bar{T}_{source} = \frac{T_{c\_in} - T_{c\_out}}{\ln\dfrac{T_{c\_in}}{T_{c\_out}}} \tag{5.2}$$

The equivalent heat sink and heat source temperatures are the temperatures that lead to the same entropy change as the gliding temperatures. The temperature lift is therefore obtained from the difference between these temperatures:

$$\Delta T_{lift\_glide} = \bar{T}_{sink} - \bar{T}_{source} \tag{5.3}$$

This illustrates that the reversible efficiency of systems with temperature glides in the sink $(T_{h\_out} - T_{h\_in})$ or source $(T_{c\_in} - T_{c\_out})$ will be larger compared to systems for which a constant temperature applies. For heat pump cycles that cannot cope with the temperature glide of the streams, a different definition should be used:

$$\Delta T_{lift} = T_{h\_out} - T_{c\_out} \tag{5.4}$$

This definition of temperature lift is adopted in further discussions. The energetic performance of the reversible cycle COP_{Carnot} can be obtained as follows:

$$COP_{Carnot} = \frac{\dot{Q}_{duty}}{\dot{W}_{hp\_reversible}} = \frac{T_{h\_out}}{\Delta T_{lift}} = \frac{T_{h\_out}}{T_{h\_out} - T_{c\_out}} \tag{5.5}$$

For a system, as shown in Figure 5.2, applying the first law of thermodynamics leads to

$$COP_{Carnot} = 1 + \frac{\dot{Q}_{source}}{\dot{W}_{hp\_reversible}} = 1 + \frac{T_{c\_out}}{\Delta T_{lift}} = 1 + \frac{T_{c\_out}}{T_{h\_out} - T_{c\_out}} \tag{5.6}$$

Example 5.1

A reversible heat pump that cannot take advantage of temperature glides is to be applied to upgrade heat that needs to be removed by the cold utility to the temperature level of the hot utility. $\dot{Q}_{cold\_utility} = 1.0$ MW, and the hot stream(s) need to be cooled from $T_{c\_in} = 60°C$ to $T_{c\_out} = 40°C$; the hot utility needs to heat the cold stream(s) from $T_{h\_in} = 110°C$ to $T_{h\_out} = 120°C$. If the heat pump is to deliver exactly the hot utility requirement, what is the value of $\dot{Q}_{hot\_utility}$? Please consult Figures 5.1 and 5.2 for the symbols used.

$$COP_{Carnot} = \frac{T_{h\_out}}{\Delta T_{lift}} = \frac{T_{h\_out}}{T_{h\_out} - T_{c\_out}} = \frac{393.15}{120 - 40} = 4.91$$

From Equation 5.6, it follows

$$\dot{W}_{hp\_reversible} = \frac{\dot{Q}_{source}}{COP_{Carnot} - 1} = \frac{1.0}{4.91 - 1} = 0.256 \text{ MW}$$

$$\dot{Q}_{hot\_utility} = \dot{Q}_{source} + \dot{W}_{hp\_reversible} = 1.0 + 0.256 = 1.256 \text{ MW}$$

Example 5.2

A reversible heat pump that can take advantage of temperature glides is to be applied to upgrade heat that needs to be removed by the cold utility to the temperature level of the hot utility. $\dot{Q}_{cold\_utility} = 1.0$ MW, and the hot stream(s) need to be cooled from $T_{c\_in} = 60°C$ to $T_{c\_out} = 40°C$; the hot utility needs to heat the cold stream(s) from $T_{h\_in} = 110°C$ to $T_{h\_out} = 120°C$. If the heat pump is to deliver exactly the hot utility requirement, what is the value of $\dot{Q}_{hot\_utility}$? Please consult Figures 5.1 and 5.2 for the symbols used. Now, the equivalent sink and source temperatures should be used because the heat pump can profit from the temperature glides.

$$\overline{T}_{sink} = \frac{T_{h\_out} - T_{h\_in}}{\ln \dfrac{T_{h\_out}}{T_{h\_in}}} = \frac{120 - 110}{\ln \dfrac{120 + 273.15}{110 + 273.15}} = 388.13 \, K = 114.98°C$$

$$\overline{T}_{source} = \frac{T_{c\_in} - T_{c\_out}}{\ln \dfrac{T_{c\_in}}{T_{c\_out}}} = \frac{60 - 40}{\ln \dfrac{60 + 273.15}{40 + 273.15}} = 323.05 \, K = 49.90°C$$

$$COP_{Carnot\_glide} = \frac{\overline{T}_{sink}}{\Delta T_{lift\_glide}} = \frac{\overline{T}_{sink}}{\overline{T}_{sink} - \overline{T}_{source}} = \frac{388.13}{388.13 - 323.05} = 5.96$$

From Equation 5.6, it follows

$$\dot{W}_{hp\_reversible\_glide} = \frac{\dot{Q}_{source}}{COP_{Carnot} - 1} = \frac{1.0}{5.96 - 1} = 0.20 \text{ MW}$$

$$\dot{Q}_{hot\_utility} = \dot{Q}_{source} + \dot{W}_{hp\_reversible\_glide} = 1.0 + 0.2 = 1.2 \text{ MW}$$

These two examples illustrate that if the process streams to be heated or cooled have glides, then heat pumps that can profit from these glides lead to better solutions, at least from the energy point of view. Remarkably, in these examples the solution with glide requires 28% less external power.

Generally, the duties for cold and hot utilities will not exactly match such that part of the hot or cold utility has to be realized using conventional technologies. In this case, T_{h_out} or T_{c_out} is not known and follows from the heat capacity of the stream with the largest heat capacity. For instance, if $\dot{Q}_{hot\_utility} > \dot{Q}_{cold\_utility} + \dot{W}_{hp}$, then it follows

$$T_{h\_out} = T_{h\_in} + \frac{\dot{Q}_{source} + \dot{W}_{hp}}{(\dot{m}c_p)_{cold\_stream}} \tag{5.7}$$

Because $\dot{W}_{hp}$ depends on T_{h_out} and T_{h_out} is needed to calculate it, solving Equation 5.7 requires an iteration loop. This type of calculation is illustrated in Example 5.3.

Example 5.3

A reversible heat pump is to be applied to upgrade heat that needs to be removed by the cold utility to the temperature level of the hot utility. $\dot{Q}_{cold\_utility} = 1.0$ MW, and the hot streams need to be cooled from $T_{c\_in} = 60°C$ to $T_{c\_out} = 40°C$; the hot utility needs to heat cold streams from $T_{h\_in} = 110°C$ to $T_{h\_out} = 120°C$. If the heat capacity $\dot{m}c_p$ of the cold stream needing heating is 0.2 MW/K, determine the temperature of this stream after it has been heated. What is the remaining $\dot{Q}_{hot\_utility}$?

Assume first that $T_{h\_out} = 120°C$. Then, doing the same steps as in the previous example, it follows that $\dot{Q}_{hot\_utility} = \dot{Q}_{source} + \dot{W}_{hp\_reversible} = 1.0 + 0.256 = 1.256$ MW such that from Equation 5.7

$$T_{h\_out} = T_{h\_in} + \frac{\dot{Q}_{source} + \dot{W}_{hp}}{(\dot{m}c_p)_{cold\_stream}} = 110 + \frac{1.256}{0.2} = 116.28°C$$

$$COP_{Carnot} = \frac{T_{h\_out}}{\Delta T_{lift}} = \frac{T_{h\_out}}{T_{h\_out} - T_{c\_out}} = \frac{389.43}{116.28 - 40} = 5.10$$

$$\dot{W}_{hp\_reversible} = \frac{\dot{Q}_{source}}{COP_{Carnot}-1} = \frac{1.0}{5.10-1} = 0.243\,\text{MW}$$

$$T_{h\_out} = T_{h\_in} + \frac{\dot{Q}_{source}+\dot{W}_{hp}}{(\dot{m}c_p)_{cold\_stream}} = 110 + \frac{1.0+0.243}{0.2} = 116.22°\text{C}$$

and

$$\dot{Q}_{extra\_utility} = (\dot{m}c_p)_{cold\_stream}(T_{utility\_out} - T_{h\_out}) = 0.2 \times (120-116.2) = 0.76\,\text{MW}$$

5.3.2 Temperature-Driving Forces in the Heat Exchangers of Heat Pumps

Figure 5.3 illustrates that the temperature of the working fluid of the heat pump needs to be slightly higher (sink side) or slightly lower (source side) than the temperature of the streams requiring heating and cooling, respectively. In fact, these temperature differences cause a deviation from the Carnot cycle for which the temperature differences between working fluid and source or sink is zero.

The figure shows that there is a minimum temperature approach required to make the design of the heat exchangers economically feasible. In practice, the minimum approach temperature is in the range of 5–15 K, and its value follows from economic optimization. As indicated in Figure 5.3, pure fluids with a constant temperature in the two-phase liquid-vapor region are generally used in vapor compression (VC) heat pumps. This results in an increasing temperature difference from the minimum approach temperature to the other

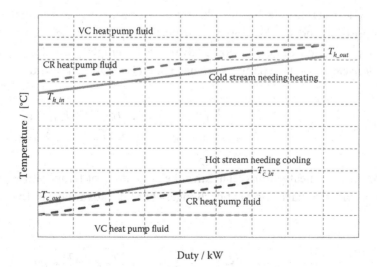

FIGURE 5.3 Temperature-driving forces in the heat exchangers of heat pumps.

side of the heat exchanger, such that the average temperature difference is significantly larger than the minimum value:

$$\Delta T_{\ln} = \frac{\Delta T_{\max} - \Delta T_{\min}}{\ln \dfrac{\Delta T_{\max}}{\Delta T_{\min}}} \tag{5.8}$$

with

$$\Delta T_{\min} = T_{c\_out} - T_{hp\_low} \quad \text{or} \quad \Delta T_{\min} = T_{hp\_high} - T_{h\_out} \tag{5.9}$$

$$\Delta T_{\max} = T_{c\_in} - T_{hp\_low} \quad \text{or} \quad \Delta T_{\max} = T_{hp\_high} - T_{h\_in} \tag{5.10}$$

Figure 5.3 also illustrates that compression-resorption (CR) heat pumps make use of fluids with a temperature glide during the phase change. Ideally, this glide should be identical to the temperature glide of the process streams requiring heating or cooling, respectively. In this case,

$$\Delta T_{\min} = \Delta T_{\max} = \Delta T_{approach}$$

The coefficient of performance (COP) for the heat pumps with real heat exchangers becomes, for (semi)open heat pump cycles,

$$COP_{Real\_heat\_exchangers\_open} = \frac{\dot{Q}_{duty}}{\dot{W}_{hp\_heat\_exchangers}} = \frac{T_{h\_out} + \Delta T_{\min\_\sin k}}{\Delta T_{lift} + \Delta T_{\min\_\sin k}} \tag{5.11}$$

because open-cycle heat pumps will generally only have a heat exchanger on the heat sink side.

Vapor compression and similar closed-cycle heat pumps make use of condensation and evaporation of pure fluids and cannot take advantage of temperature glides of the process streams. In these cases, heat exchangers are needed on both sink and source sides:

$$COP_{Real\_heat\_exchangers\_VC} = \frac{\dot{Q}_{duty}}{\dot{W}_{hp\_heat\_exchangers}} = \frac{T_{h\_out} + \Delta T_{\min\_\sin k}}{\Delta T_{lift} + \Delta T_{\min\_\sin k} + \Delta T_{\min\_source}} \tag{5.12}$$

Finally, closed-cycle heat pumps that can take advantage of temperature glides as the CR heat pumps will also have two heat exchangers but take advantage of the smaller temperature lift:

$$COP_{Real\_heat\_exchangers\_CR} = \frac{\dot{Q}_{duty}}{\dot{W}_{hp\_heat\_exchangers}} = \frac{\overline{T}_{\sin k} + \Delta T_{\min\_\sin k}}{\Delta T_{lift\_glide} + \Delta T_{\min\_\sin k} + \Delta T_{\min\_source}} \tag{5.13}$$

Example 5.4

A reversible heat pump is to be applied to upgrade heat that needs to be removed by the cold utility to the temperature level of the hot utility. $\dot{Q}_{cold\_utility} = 1.0$ MW, and the hot stream(s) need to be cooled from $T_{c\_in} = 60°C$ to $T_{c\_out} = 40°C$; the hot utility needs to heat the cold stream(s) from $T_{h\_in} = 110°C$ to $T_{h\_out} = 120°C$. Compare the COP of the open cycle and closed cycles that do not or do take advantage of the temperature glides. Assume that the minimum temperature approach for the heat exchangers is 10 K ($\Delta T_{min\_sin k} = \Delta T_{min\_source} = 10$ K).

$$COP_{Real\_heat\_exchangers\_open} = \frac{T_{h\_out} + \Delta T_{min\_sink}}{\Delta T_{lift} + \Delta T_{min\_sink}} = \frac{120 + 10 + 273.15}{(120 - 40) + 10} = 4.48$$

$$COP_{Real\_heat\_exchangers\_VC} = \frac{T_{h\_out} + \Delta T_{min\_sink}}{\Delta T_{lift} + \Delta T_{min\_sink} + \Delta T_{min\_source}} = \frac{120 + 10 + 273.15}{(120 - 40) + 10 + 10} = 4.03$$

and making use of the results of Example 5.2:

$$COP_{Real\_heat\_exchangers\_CR} = \frac{\overline{T}_{sink} + \Delta T_{min\_sink}}{\Delta T_{lift\_glide} + \Delta T_{min\_sink} + \Delta T_{min\_source}} = \frac{388.13 + 10}{(388.13 - 323.05) + 10 + 10} = 4.68$$

In this example, closed heat pump cycles that take advantage of the process stream glides lead to the best energetic performance, followed by the open cycles and finally by the closed cycles that cannot take advantage of the glides. It should be noted that this is a specific example with a large temperature lift. Generally, for small temperature lifts, the open cycle will lead to higher COP values. This is clearly illustrated in Example 5.5.

Example 5.5

A reversible heat pump is to be applied to upgrade heat that needs to be removed by the cold utility to the temperature level of the hot utility. The hot stream(s) need to be cooled from $T_{c\_in} = 110°C$ to $T_{c\_out} = 105°C$; the hot utility needs to heat the cold stream(s) from $T_{h\_in} = 110°C$ to $T_{h\_out} = 115°C$. Compare the COP of the open cycle and closed cycles that do not or do take advantage of the temperature glides. Assume that the minimum temperature approach for the heat exchangers is 5 K ($\Delta T_{min\_sin k} = \Delta T_{min\_source} = 5$ K).

$$COP_{Real\_heat\_exchangers\_open} = \frac{T_{h\_out} + \Delta T_{min\_sink}}{\Delta T_{lift} + \Delta T_{min\_sink}} = \frac{115 + 5 + 273.15}{(115 - 105) + 5} = 26.21$$

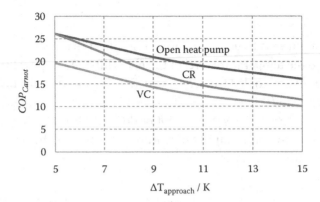

FIGURE 5.4 Effect of temperature-driving forces in the heat exchangers for the three heat pump types.

$$COP_{Real\_heat\_exchangers\_VC} = \frac{T_{h\_out} + \Delta T_{min\_sink}}{\Delta T_{lift} + \Delta T_{min\_sink} + \Delta T_{min\_source}} = \frac{115 + 5 + 273.15}{(115 - 105) + 5 + 5} = 19.66$$

$$COP_{Real\_heat\_exchangers\_CR} = \frac{\overline{T}_{sink} + \Delta T_{min\_sink}}{\Delta T_{lift\_glide} + \Delta T_{min\_sink} + \Delta T_{min\_source}} = \frac{385.64 + 5}{(385.64 - 380.64) + 5 + 5} = 26.04$$

This example makes clear that if the temperature lift and the minimum approach temperature of heat exchangers are small, then the open pump cycles will be more beneficial. In this case, $\Delta T_{lift} = 10\ K$ and $\Delta T_{min} = 5\ K$, such that the open heat pump cycles start showing an advantage in comparison to the alternatives. For the conditions of this example, Figure 5.4 illustrates how the COP decreases as the temperature-driving forces in heat exchangers increase. As long as the process streams show a glide, heat pump cycles that cannot take advantage of the glide of the process streams will always show lower performance.

5.3.3 Nonisentropic Compression Processes

It can be demonstrated that the efficiency of heat pump systems is linearly dependent on the isentropic efficiency of the compressor (van de Bor and Infante Ferreira, 2013), such that

$$COP_{irreversible\_open} = \frac{\dot{Q}_{duty}}{\dot{W}_{hp\_heat\_exchangers}} = \eta_{is} \times \frac{T_{h\_out} + \Delta T_{min\_sink}}{\Delta T_{lift} + \Delta T_{min\_sink}} \tag{5.14}$$

$$COP_{irreversible\_VC} = \frac{\dot{Q}_{duty}}{\dot{W}_{hp\_heat\_exchangers}} = \eta_{is} \times \frac{T_{h\_out} + \Delta T_{min\_sink}}{\Delta T_{lift} + \Delta T_{min\_sink} + \Delta T_{min\_source}} \tag{5.15}$$

$$COP_{irreversible\_CR} = \frac{\dot{Q}_{duty}}{\dot{W}_{hp\_heat\_exchangers}} = \eta_{is} \times \frac{\overline{T}_{sink} + \Delta T_{min\_sink}}{\Delta T_{lift\_glide} + \Delta T_{min\_sink} + \Delta T_{min\_source}} \quad (5.16)$$

The isentropic efficiency of compressors is defined as

$$\eta_{is} = \frac{\Delta h_{is}}{\Delta h_{real}} \quad (5.17)$$

with Δh_{is} the isentropic compression work and Δh_{real} the real compressor work. Large centrifugal compressors can reach efficiencies of 0.85, while the efficiency of positive-displacement compressors is generally lower, around 0.70. In positive-displacement compressors, the pressure ratio between discharge and inlet pressure has a significant impact on the isentropic efficiency of the compressor. Equation 5.18 and Figure 5.5 give, for a specific compressor, an example of how the isentropic efficiency changes with the pressure ratio between the outlet and the inlet of the compressor.

$$\eta_{is} = -0.0014 \times \left(\frac{p_{discharge}}{p_{suction}}\right)^4 + 0.0297 \times \left(\frac{p_{discharge}}{p_{suction}}\right)^3 - 0.236 \times \left(\frac{p_{discharge}}{p_{suction}}\right)^2$$

$$+ 0.8477 \times \left(\frac{p_{discharge}}{p_{suction}}\right) - 0.4579 \quad (5.18)$$

This equation (quartic function) applies for pressure ratios between 1 and 9.

The isentropic efficiency of centrifugal compressors is strongly dependent on the volume flow displaced by the compressor, as illustrated in Figure 5.6. For preliminary calculations, $\eta_{is} = 0.70$ gives a good approximation of real compressor efficiency. This value generally also includes mechanical losses.

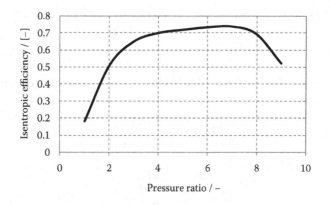

FIGURE 5.5 Isentropic efficiency of a specific positive-displacement compressor as a function of the ratio between discharge and suction pressures.

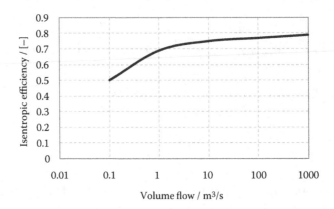

FIGURE 5.6 Isentropic efficiency of centrifugal compressors as a function of the suction volume flow they displace.

5.3.4 Absorption Heat Pumps

Absorption cycles are driven by heat instead of work. These cycles combine a power cycle that operates between the driving temperature level T_{gen} and the heat application temperature level T_{sink}, with a heat pump that operates between source T_{source} and heat application level T_{sink}. Industrial heat pumps are frequently driven by boilers, but also rest (waste) heat at a sufficiently elevated temperature can be used to drive the systems. Its reversible performance is given by

$$COP_{Carnot\_absorption\_hp} = \left(\frac{T_{gen} - T_{h\_out}}{T_{gen}} \right) \left(\frac{T_{h\_out}}{T_{h\_out} - T_{c\_out}} \right) \tag{5.19}$$

Performance of the absorption heat pump when temperature-driving forces are considered is given by

$$COP_{ahp} = \frac{(T_{gen} - \Delta T_{gen}) - (T_{h\_out} + \Delta T_{min\_sink})}{(T_{gen} - \Delta T_{gen})} \times \frac{T_{h\_out} + \Delta T_{min\_sink}}{(\Delta T_{lift} + \Delta T_{min\_sink} + \Delta T_{min\_source})} \tag{5.20}$$

Including the conversion from primary fuel to heat at generator temperature η_h,

$$COP_{ahp\_irreversible} = \left(1 - \frac{T_{gen}}{T_{adiabatic}} \right) \frac{(T_{gen} - \Delta T_{gen}) - (T_{h\_out} + \Delta T_{min\_sink})}{(T_{gen} - \Delta T_{gen})}$$

$$\times \frac{T_{h\_out} + \Delta T_{min\_sink}}{(\Delta T_{lift} + \Delta T_{min\_sink} + \Delta T_{min\_source})} \tag{5.21}$$

with $T_{adiabatic}$ the adiabatic flame temperature ($T_{adiabatic}$ = 2223 K).

Example 5.6

An irreversible absorption heat pump that cannot take advantage of temperature glides is to be applied to upgrade heat that needs to be removed by the cold utility to the temperature level of the hot utility. $\dot{Q}_{cold\_utility} = 1.0\,\text{MW}$, and the hot stream(s) need to be cooled from $T_{c\_in} = 60°C$ to $T_{c\_out} = 40°C$; the hot utility needs to heat the cold stream(s) from $T_{h\_in} = 110°C$ to $T_{h\_out} = 120°C$. If the heat pump is to deliver exactly the hot utility requirement, what is the value of $\dot{Q}_{hot\_utility}$? Condensing steam at 250°C is used to deliver heat to the generator of the absorption heat pump, $T_{gen} = 250°C = 523.15\,K$. The minimum approach temperature for the heat exchangers is 5 K. Also, the average temperature-driving force in the generator ΔT_{gen} is 5 K. Equation 5.20 gives

$$COP_{ahp} = \frac{(523.15-5)-(120+273.15+5)}{(523.15-5)} \times \frac{120+273.15+5}{(120-40+5+5)} = 0.232 \times 4.42 = 1.03$$

and taking the boiler efficiency into account, the efficiency related to the gas consumption becomes

$$COP_{ahp\_irreversible} = \left(1-\frac{T_{gen}}{T_{adiabatic}}\right) \times 1.03 = \left(1-\frac{523.15}{2223}\right) \times 1.03 = 0.765 \times 1.03 = 0.79$$

This example shows that absorption heat pumps cannot economically cope with large temperature lifts. Its primary energy consumption is practically larger than required by a directly fired boiler. For convenience, Example 5.7 illustrates a more attractive application for absorption heat pumps.

Example 5.7

A reversible heat pump is to be applied to upgrade heat that needs to be removed by the cold utility to the temperature level of the hot utility. The hot stream(s) need to be cooled from $T_{c\_in} = 110°C$ to $T_{c\_out} = 105°C$; the hot utility needs to heat the cold stream(s) from $T_{h\_in} = 110°C$ to $T_{h\_out} = 115°C$. Calculate the COP of an absorption heat pump that can fulfill these requirements. Condensing steam at 212°C is used to deliver heat to the generator of the absorption heat pump, $T_{gen} = 212°C = 485.15\,K$. The minimum approach temperature for the heat exchangers is 5 K. Also, the average temperature-driving force in the generator ΔT_{gen} is 5 K.

$$COP_{ahp} = \frac{(485.15-5)-(115+273.15+5)}{(485.15-5)} \times \frac{115+273.15+5}{(115-105+5+5)} = 0.181 \times 19.66 = 3.56$$

$$COP_{ahp\_irreversible} = \left(1-\frac{T_{gen}}{T_{adiabatic}}\right) \times 3.56 = \left(1-\frac{485.15}{2223}\right) \times 3.56 = 0.917 \times 3.56 = 3.26$$

5.3.5 Other Loss Mechanisms

In addition to the effect of temperature-driving forces in heat exchangers and isentropic efficiency of the compressor, there are several other sources of efficiency losses in heat pumps, such as pressure drop, superheating, throttling losses and a mismatch between temperature glide of process and heat pump fluids.

Pressure drop is a design parameter. Modification of the flow passage area of the fluids through the heat exchangers and lines of the heat pump allows for an adjustment, but this has little impact in the area requirements for the heat exchangers. Depending on the fluids involved, the pressure drop in the two-phase region corresponds to a temperature drop, and it has a negative effect on the cycle performance. Pressure drop has a huge impact for low-pressure systems, but only a limited impact for high-pressure systems.

Superheating can have both positive and negative effects, and this is strongly dependent on the thermodynamic properties of the fluids used in the heat pump cycle. For instance, superheating in heat pumps that use ammonia as working fluid leads always to a decrease in energetic performance of the cycle. On the other side, some fluids, such as R134a, can benefit from superheating. This effect can only be quantified when the heat pump fluids have been defined.

Throttling losses depend again on the thermodynamic properties of the fluids involved. The higher the pressure ratio in the cycle, the higher these losses are, but the absolute value depends on the selected fluids. From the first law of thermodynamics,

$$dh = Tds + vdp \tag{5.22}$$

It follows that, for a throttling process ($dh = 0$),

$$ds = -\frac{v}{T}dp \tag{5.23}$$

The larger the entropy production (ds) is, the larger the irreversibilities and the lower the energetic performance of the cycle will be.

In Equations 5.13 and 5.16, it was assumed that there is a perfect match between the glide of the sink and source fluids and the heat pump fluid. In practice, there will be a *mismatch between process and heat pump fluids*, leading to an increase in the temperature-driving force.

5.3.6 Example of the Impact of Irreversibilities

Assume that a heat pump is to be installed between the condenser of a distillation column and the reboiler of the same column. The process fluid has a significant temperature glide during condensation: $T_{c\_in} = 60°C$; $T_{c\_out} = 40°C$. Also in the reboiler, the process shows a large temperature glide: $T_{h\_in} = 80°C$; $T_{h\_out} = 100°C$. From Equation 5.1, $\overline{T}_{sink} = 363.1$ K = 89.9°C, and from Equation 5.2, $\overline{T}_{source} = 323.0$ K = 49.9°C, such that $\Delta T_{lift\_glide} = 40$ K and $\Delta T_{lift} = 60$ K. The COP for a Carnot heat pump cycle is then $COP_{Carnot\_glide} = 9.08$ and

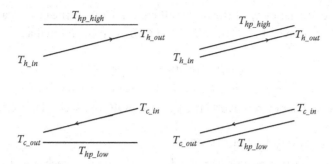

FIGURE 5.7 Temperature-driving forces for a heat pump fluid with no glide (left) and with glide (right).

COP_{Carnot} = 6.22. The temperature-driving forces, as given in Equations 5.9 and 5.10, are illustrated in Figure 5.7.

Assuming a minimum temperature approach of 5 K between process fluids and heat pump working fluid, Equations 5.8 through 5.10 give the sink (hot) side ΔT_{min} = 5 K and ΔT_{max} = 25 K for heat pump fluids with no glide, and ΔT_{max} = 5 K for heat pump fluids with a matching glide. The average temperature difference for heat pumps without glide becomes ΔT_{ln_sink} = 12.43 K. Similarly, in the source (cold) side, ΔT_{min} = 5 K and ΔT_{max} = 25 K for heat pump fluids with no glide and ΔT_{max} = 5 K for heat pump fluids with a matching glide. The average temperature difference for heat pumps without glide becomes in this case also ΔT_{ln_source} = 12.43 K. The isentropic efficiency of the compressor is assumed to be η_{is} = 0.70. Table 5.2 summarizes the results for the cycle efficiencies.

The efficiency of the heat pump cycles strongly depends on both temperature differences (compression and absorption cycles) and compressor efficiency (compression cycles). Also, the grid efficiency and the conversion from primary energy to electricity (η_e = 0.42) or heat (η_h ≈ 0.78–0.86) play a significant role. Heat-driven heat pumps can become effective when the COP of a compressor-driven heat pump becomes small.

$$COP_h > COP_e \frac{\eta_e}{\eta_h} \tag{5.24}$$

TABLE 5.2 Effect of Irreversibilities on Heat Pump Cycle Performance

	COP for Open Cycle	COP for Cycle with No Glide	COP for Cycle with Glide	COP for Absorption Cycle (T_{gen} = 250°C)
COP_{Carnot}	6.22	6.22	9.08	1.78
Effect of temperature-driving forces	5.82	5.40	7.36	1.46
Effect of nonisentropic compressor (η_{is} = 0.70)	4.07	3.78	5.15	n/a
$\eta_{Carnot} = COP/COP_{Carnot}$	0.65	0.61	0.57	0.82
Effect of primary fuel conversion in power plant (η_e = 0.42)	1.71	1.59	2.16	n/a
Effect of primary fuel conversion (burner efficiency, η_h = 0.765)	n/a	n/a	n/a	1.17

where η_e is the grid efficiency, η_h is the boiler efficiency, COP_e is the electrically driven COP and COP_h is the heat-driven COP. The electrical COP_e can be defined as

$$COP_e = COP_{Carnot} \times \eta_{Carnot} \tag{5.25}$$

where η_{Carnot} is the efficiency relative to the Carnot COP (the second law efficiency).

5.3.7 Savings from Energetic Performance

The energy cost savings per megawatt hour delivered heat for mechanical and heat-driven heat pumps are

$$Savings_{Mechanical\_heat\_pump} = -\frac{C_{gas}}{\eta_h} + \frac{C_e}{COP_{hp}} \tag{5.26}$$

$$Savings_{Heat\_driven\_heat\_pump} = -\frac{C_{gas}}{\eta_h} + \frac{C_{gas}}{COP_{ahp}} \tag{5.27}$$

with C_{gas} the cost of gas (considered here as €32/MWh) and C_e the cost of electricity (here taken as €65/MWh). Industrial systems are expected to operate continuously throughout the year, such that it is expected that the system will operate 8000 h/yr (92% availability).

5.4 SELECTION OF HEAT PUMPS

Just as for any other equipment, the final selection of heat pumps is based on economics. Capital, installation and operating costs (costs of maintenance and driving energy) should be considered. In fact, governments stimulate the use of heat pumps to reduce primary energy use and simultaneously reduce the emission of carbon dioxide. However, investments will always be based on economic advantages. The economic benefits result from the energy savings and associated emission cost reduction, as well as from the reduction in costs of cold utility. Kew (1982) was one of the first researchers investigating the technical and economic criteria that must be met before heat pumps used to recover heat from industrial waste streams can provide adequate payback times. Omideyi et al. (1984) also studied the factors affecting the overall economics of heat pumps specifically designed for integration in distillation systems. These authors included the global design of the major components based on overall heat transfer performance obtaining a more accurate estimation of the capital costs. These authors presented a flow chart for the selection algorithm. Wallin et al. (1990) related the selection of heat pumps for industrial processes with the location of the pinch point and the shape of the grand composite curves. Also, Benstead and Sharman (1990) made use of pinch technology and the grand composite curves to identify possible applications for heat pumps. Fonyo and Mizsey (1994) emphasized that pinch technology provides insight into the appropriate placement of heat pumps. They proposed measuring the attractiveness of heat pump processes by the payback time of excess

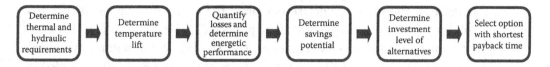

FIGURE 5.8 Steps to be undertaken in the selection of heat pumps.

capital. Fonyo and Benko (1996) extended the previous study, including equations for the calculation of capital and utility costs as functions of the minimum approach temperature of the involved heat exchangers. A flow chart for the design strategy of process integration, including eventually a heat pump, was also presented, but the selection of heat pump alternatives takes place on the basis of payback time.

Bagajewicz and Barbaro (2003) presented a targeting procedure that gives optimal cost savings and temperature levels for heat pump applications. This procedure does not require performing interval partitions. The authors described optimal solutions also with heat pumps that do not cross the pinch.

The method proposed here for the selection of a heat pump type is based on the same principles: capital, installation and operating costs versus economic benefits of the heat pump. The approach is schematically given in Figure 5.8.

5.5 HEAT PUMP TYPES

Most researchers (Rossiter et al., 1988; Wallin et al., 1990; Boot et al., 1998) distinguish two main heat pump types with very distinct properties: (semi)open cycles and closed cycles. This is illustrated in Figure 5.9, which includes only the heat pumps available in the market. The CR heat pumps are not really available (although Hybrid Energy started their commercialization), but because these cycles use temperature glides, they are included to indicate their potential.

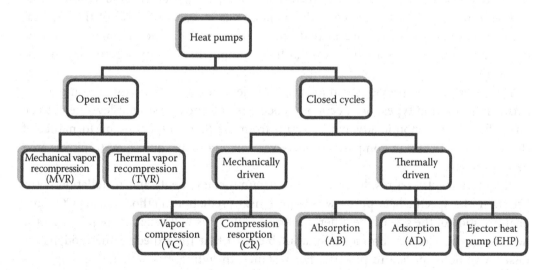

FIGURE 5.9 Classification of heat pumps in relation to driving method and degree of being hermetic.

This classification allows for the determination of the required components such that the irreversibilities and investment costs can be predicted. Table 5.3 gives an overview of the required components and specific requirements of the different heat pump types. This table allows the determination of the number and size of the heat exchangers needed for each of the alternative types. It is clear that absorption heat pumps require a larger number of heat exchangers, and that the open cycles (mechanical vapor recompression [MVR] and thermal vapor recompression [TVR]) require only one heat exchanger.

There are more methods to create a heat-pumping effect in addition to the ones mentioned in Figure 5.9. However, these methods are often in a very early research stage or not yet accepted by the chemical process industry.

Wallin et al. (1990) considered the compression heat pump, the open-cycle heat pump and the heat transformer. These authors also considered compression heat pumps working with a nonazeotropic mixture and CR heat pumps. Wall (1986) restricted his thermoeconomic optimization of heat pumps to VC heat pumps. Rossiter et al. (1988) considered two different types of heat pumps: closed cycle and semiopen cycle. The closed-cycle system contained a working fluid that undergoes repeated evaporation and condensation through heat exchange with process fluids in the heat pump evaporator and condenser; the authors mainly considered VC cycles. The type of semiopen cycle considered in this study, on the other hand, used a vapor-phase process fluid as its heat source and had no heat pump evaporator (vapor recompression heat pump). Both types of systems were powered by electrically driven compressors. Newbert and Martin (1983) indicated that most interest in heat pump developments lies in the VC cycle and in the absorption cycle. The VC cycle is a fairly well-known technology and has been adapted to heat pumps. The VC system is advanced, having benefited from many years of development for refrigeration, but there is still scope for further improvement. The main options are different drives, improved efficiency and higher temperatures.

Omideyi et al. (1984) focused only on VC heat pumps and their specific requirements. Although the absorption cycle could be looked at as one sort of drive of the VC heat pump, it has many special features that open a whole range of developments into new fluids and systems, such as the chemical heat pump, which in turn gives rise to new applications.

This overview of previously published studies indicates that, for a comparison between heat pump types, both economic benefits and the investment level need to be quantified. The economic advantages result from the thermodynamic performance of the heat pump (COP) in comparison to conventional heating equipment (topic covered in Section 5.3).

The investment level results from the heat pump components and installation costs. Selection of a specific heat pump will require next prediction methods for its COP and quantification of the size of the different components. This will generally require that the heat pumps be already commercially available such that the expected thermodynamic losses have been verified in practice. For instance, including magnetic heat pumps in a selection procedure is practically impossible, as experimental COPs are not available and design procedures for the system components are not well defined.

TABLE 5.3 Relation between Heat Pump Type and Its Characteristics

	MVR	TVR	VC	CR	AB
Driving mechanism	Shaft power	Steam ejector	Shaft power	Shaft power	Heat
Heat exchanger source side	No	No	Yes, evaporator (no glide)	Yes, desorber (with glide)	Yes, evaporator (no glide)
Duty source side (kW)	–	–	$\dot{Q}_{duty} \times \left(\dfrac{COP-1}{COP}\right)$	$\dot{Q}_{duty} \times \left(\dfrac{COP-1}{COP}\right)$	$\dot{Q}_{duty} \times \left(\dfrac{COP-1}{COP}\right)$
U-value source side (kW/m²K)	–	–	1.2	1.0	1.2
Heat exchanger sink side	Yes, condenser (no glide)	Yes, condenser (no glide)	Yes, condenser (no glide)	Yes, resorber (with glide)	Yes, absorber (with glide) + condenser (no glide) + recuperator
Duty sink side (kW)	$\dot{Q}_{duty}$	$\dot{Q}_{duty}$	$\dot{Q}_{duty}$	$\dot{Q}_{duty}$	Absorber $\approx 0.6 \times \dot{Q}_{duty}$ Condenser $\approx 0.4 \times \dot{Q}_{duty}$ Recuperator $\approx 0.15 \times \dot{Q}_{duty}$
U-value sink side (kW/m²K)	1.5	1.5	1.5	1.0	Absorber 1.0 Condenser 1.5 Recuperator 3.0

5.6 INVESTMENT COSTS

If reliable cost data are not available for the economic calculation, cost equations can be used as a first approximation (van de Bor and Infante Ferreira, 2013). The correlations concern installed equipment costs and have the following form:

$$C_{equipment} = a + b \times S^n \tag{5.28}$$

with $C_{equipment}$ the installed equipment cost expressed in euros (base year 2012), a and b are constants given in Table 5.4, S is the size parameters (units also listed) and n is the exponent for that specific type of equipment. The application range of the equations is also given in terms of size parameter. Sinnott and Towler (2009) discussed in more detail the estimation of the purchased and installed equipment costs, including a more extensive range of equipment.

5.6.1 Size of Heat Exchangers

The size of the required heat exchangers can be derived from Table 5.3.

$$A = \frac{\dot{Q}_{hex}}{U \times F_T \times \Delta T_{ln}} \tag{5.29}$$

where $\dot{Q}_{hex}$ is the amount of heat exchanged, U is the overall heat transfer coefficient for the specific heat exchanger type (see Table 5.3) and ΔT_{ln} is the logarithmic temperature difference that applies for the heat exchanger and is obtained with Equation 5.8. The factor F_T accounts for the deviation from purely countercurrent flow and is equal to 1.0 for heat exchangers with one of the fluids condensing/evaporating at a constant temperature (closed cycles with no glide). In other cases, for U-tube shell-and-tube heat exchangers, F_T can be obtained from (Sinnott and Towler, 2009)

$$F_T = \frac{\sqrt{(R^2+1)}\ln\left[\dfrac{1-S}{1-RS}\right]}{(R-1)\ln\left[\dfrac{2-S\left[R+1-\sqrt{(R^2+1)}\right]}{2-S\left[R+1+\sqrt{(R^2+1)}\right]}\right]} \tag{5.30}$$

TABLE 5.4 Installed Heat Pump Equipment Costs for Heat Exchangers and Compressor

Equipment	Units for Size, S	S_{lower}	S_{upper}	a	b	n
U-tube shell-and-tube heat exchanger, carbon steel	Area, m²	20	500	0	7151.8	0.65
U-tube shell-and-tube heat exchanger, stainless steel	Area, m²	20	500	0	10,947.4	0.65
Plate and frame heat exchanger, stainless steel	Area, m²	1.0	500	1350	860.9	0.95
Screw compressor	Driver power, kW	150	3000	0	7827.9	0.7849
Centrifugal compressor	Driver power, kW	150	3000	0	751,687	0.2416
Direct fired boiler	Duty, kW	600	18,000	0	515.4	0.85

with

$$R = \frac{T_{h\_in} - T_{h\_out}}{T_{c\_out} - T_{c\_in}}$$ (5.31)

and

$$S = \frac{T_{c\_out} - T_{c\_in}}{T_{h\_in} - T_{c\_in}}$$ (5.32)

In Table 5.5, the example of Section 5.3.6 is extended to illustrate the difference between the several heat pump options. For the open-cycle TVR, it has been assumed that the

TABLE 5.5   Comparison between Heat Pump Types for a 10-MW Heat Pump

	MVR	TVR	VC	CR	AB
COP_{real}	4.07	2.00	3.78	5.15	1.46
Q_{hex} (kW) source side	0	0	7354	8058	3151
U-value (kW/m²K) source side			1.2	1.0	1.2
ΔT_{ln} (K)	0	0	12.43	5	12.43
A (m²) required source side	0	0	493	1612	211
Q_{hex} (kW) sink side	10,000	10,000	10,000	10,000	Absorber: 6000 Condenser: 4000 Recuperator: 1500 Desorber: 6849
U-value (kW/m²K) sink side	1.5	1.5	1.5	1.0	Absorber: 1.0 Condenser: 1.5 Recuperator: 3.0 Desorber: 1.5
ΔT_{ln} (K)	12.43	12.43	12.43	5	Absorber: 5 Condenser: 12.43 Recuperator: 5
A (m²) required sink side	536	536	536	2000	Absorber: 1200 Condenser: 215 Recuperator: 100 Desorber: 913
Costs installed heat exchangers (source side) (k€)	0	0	402.5	869.4	231.8
Costs installed heat exchangers (sink side) (k€)	425.0	425.0	425.0	1000.2	Absorber: 717.6 Condenser: 234.7 Recuperator: 142.7 Desorber: 600.8 Total: 1695.8
W_c/Q_{boiler} (kW)	2457	5814	2646	1942	6849
Cost installed compressor/ boiler (k€)	4956	1016	5046	4683	938
Cost of heat pump installed (k€)	5381	1441	5874	6553	2866
Yearly savings (k€)	1699	1488	1601	1967	1223
Simple payback time (years)	3.2	1.0	3.7	3.3	2.3
CO_2 emission ($\times 10^6$ kg CO_2)	23.4	27.0	25.2	18.5	31.9

entrainment ratio (between suction nozzle flow and motive nozzle flow) is 0.5, such that the steam boiler delivers 5 MW heat. The investment in the ejector has been taken as €200,000, which probably is too optimistic.

From Table 5.5, it becomes clear that open cycles require no heat exchangers on the source side, and as such they require much smaller heat exchanger areas. On the other hand, absorption heat pumps require a large amount of heat exchangers with large areas. Considering the payback time of the different alternatives for this specific application, the shortest payback is attained with TVR (1.0 year), absorption heat pump (2.5 years), MVR (3.2 years), CR heat pump (3.3 years) and finally VC heat pump (3.7 years). Considering the related CO_2 emissions, the lowest emissions (and also largest primary energy savings) are obtained with CR heat pumps, followed by MVR, VC heat pumps, TVR and absorption heat pumps. This illustrates that solutions with the largest energy-saving potential are not necessarily economically the best solutions.

5.7 COMPARISON OF THE PERFORMANCE OF DIFFERENT CYCLES

All heat pumps perform best when working with low-temperature lifts. The simplest and most developed heat pump types are VC and vapor recompression heat pumps. Although they are the most developed, their temperature lifts are limited. Like VC and vapor recompression heat pumps, CR heat pumps reach their highest efficiency at low-temperature lifts. However, their relative advantages compared to the more conventional VC and vapor recompression heat pumps increase.

Infante Ferreira et al. (2009) ranked a number of heat pump types for a specific industrial application, based only on energy savings. The ranking for the types discussed here was MVR, CR, VC, AB. Note that TVR was not included in that overview. Depending on the temperature lift that the heat pump needs to cover, MVR and CR will show the largest energy savings, but from the example in Table 5.5, it will become clear that economics can significantly change the ranking. Unless governments compensate for the energy-saving advantage, economy is generally the main reason for a selection, so investors may select less-attractive solutions from the point of view of energy-saving potential only.

5.8 OTHER RELEVANT CRITERIA

In addition to the economic criteria related with investments for the selection of heat pumps, other criteria can play an even bigger role. These criteria include aspects related to exploitation, such as maintenance costs; sensibility to internal corrosion and susceptibility to internal fouling, sound-level production and reliability.

Other aspects include, but are not limited to, process control schemes and procedures, part load behavior, suitability for operation at high temperatures (Table 5.6 gives an

TABLE 5.6   Relation between Heat Pump Type and Its Characteristics

	MVR	TVR	VC	CR	AB
Maximum temperature lift with energy savings	30 K	20 K	70 K	100 K	50 K
Maximum temperature delivered heat	200°C	180°C	115°C	180°C	115°C

indication of the applicability of the different heat pump types for what concerns temperature range), safety (fire and toxicity), environmental impact, startup/shutdown behavior, disturbances the heat pump can cause on associated process streams and acceptability of rotating equipment by operators.

5.9 CONCLUDING REMARKS

This chapter described the steps leading to the selection of the best heat pump for specific operating conditions. These steps can be summarized as follows:

- *Thermal and hydraulic requirements.* Heat pumps must meet the specifications indicated by the process integration analysis. This includes

  - Amount of heat to be delivered by the heat pump

  - Sink and sources fluids and their respective inlet and outlet temperatures

  - Allowable pressure drop of the process streams

- *Compatibility with fluids and operating conditions*:

  - Suitability of materials of construction of heat exchangers, compressors, ejectors or other heat pump components for the process fluids/heat pump fluids without excessive corrosion

  - Ability for operation under fouling conditions of the process streams; ability of cleaning the equipment

  - Ability to cope with fluid pressure stresses and stresses due to temperature differences; tolerances in relation to differences in thermal expansion of materials

- *Maintenance*:

  - Requirements for maintenance of rotating equipment/cleaning of heat exchangers

  - Requirements for replacement of parts of unit

  - Floor space requirements for maintenance

  - Ease of modifications

- *Availability*:

  - Delivery time

  - Limitations of design methods in relation to size. Equipment set up in parallel leads to additional piping and installation costs and may lead to flow distribution problems

- *Economic factors*:

  - Final choice should be based on economics

  - Consideration of the economic advantages resulting from energy savings, emission cost reduction and eventually from reduction of the cold utility costs

- Consider capital, installation and costs of driving energy

- Consider the economic advantages that can follow from governmental subsidies for implementation of heat pumps with reduced energy use

- *Initial selection*:

  - If one heat pump type shows much lower payback time than the others (say a difference in payback time of more than 1.5), then this type should be selected and detailed

  - If several heat pump types show similar payback times, then estimate the three best types in more detail

LIST OF SYMBOLS

A	Area	m^2
C	Cost	€
c_p	Specific heat	$kJkg^{-1}K^{-1}$
F_T	Temperature correction factor	–
h	Enthalpy	$kJkg^{-1}$
$\dot{m}$	Mass flow	kgs^{-1}
p	Pressure	kPa
$\dot{Q}$	Heat flow	kW
s	Entropy	$kJkg^{-1}K^{-1}$
S	Size parameter	Several
T	Temperature	K
$\bar{T}$	Equivalent temperature for gliding reservoirs	K
U	Overall heat transfer coefficient	$kWm^{-2}K^{-1}$
v	Specific volume	m^3kg^{-1}
$\dot{W}$	Power	kW

Greek Symbols

Δh	Enthalpy change	$kJkg^{-1}$
ΔT	Temperature difference	K
ΔT_{lift}	Temperature lift of heat pump	K
ΔT_{max}	Largest temperature difference	K
ΔT_{min}	Minimum approach temperature	K
η	Efficiency	–

Subscripts

ahp	Absorption heat pump
c	Cold stream
e	Grid efficiency, electrically driven
gen	Generator
h	Hot stream, gas driven

hex	Heat exchanger
high	High temperature side
hp	Heat pump
in	At inlet
is	Isentropic
ln	Logarithmic
low	Low temperature side
out	At outlet

Abbreviations

AB	Absorption
AD	Adsorption
COP	Coefficient of performance
CR	Compression-resorption
EHP	Ejector heat pump
HEN	Heat exchanger network
HP	Heat pump
MVR	Mechanical vapor recompression
TVR	Thermal vapor recompression
VC	Vapor compression

REFERENCES

Bagajewicz M. J., Barbaro A. F., On the use of heat pumps in total site heat integration, *Computers and Chemical Engineering*, 27 (2003), 1707–1719.

Benstead R., Sharman F. W., Heat pumps and pinch technology, *Heat Recovery Systems*, 10 (1990), 387–398.

Boot H., Nies J., Verschoor M. J. E., de Wit J. B., *Handboek industriële warmtepompen*, Kluwer Bedrijfsinformatie, Deventer, the Netherlands (in Dutch), 1998.

Fonyo Z., Benko N., Enhancement of process integration by heat pumping, *Computers and Chemical Engineering*, 20 (1996), S85–S90.

Fonyo Z., Mizsey P., Economic applications of heat pumps in integrated distillation systems, *Heat Recovery Systems & CHP*, 14 (1994), 249–263.

Infante Ferreira C. A., Spoelstra S., Hamoen E., How successful are heat pumps in Dutch process industry applications? *RCC Koude & Luchtbehandeling*, 102 (2009), 14–20 (in Dutch).

Kew P. A., Heat pumps for industrial waste heat recovery – A summary of required technical and economic criteria, *Heat Recovery Systems*, 2 (1982), 283–296.

Newbert G. J., Martin D. J., Trends and developments in non-domestic heat pump applications, *Heat Recovery Systems*, 3 (1983) 69–79.

Omideyi T. O., Kasprzycki J., Watson F. A., The economics of heat pump assisted distillation systems – I. A design and economic model, *Heat Recovery Systems*, 4 (1984), 187–200.

Ranade S. M., New insights on optimal integration of heat pumps in industrial sites, *Heat Recovery Systems*, 8 (1988), 255–263.

Rossiter A. P., Seetharam R. V., Ranade S. M., Scope for industrial heat pump applications in the United States, *Heat Recovery Systems & CHP*, 8 (1988), 279–287.

Sinnott R. K., Towler G. P., *Chemical engineering design*, 5th edition, Elsevier, New York, 2009.

van de Bor D. M., Infante Ferreira C. A., Quick selection of industrial heat pump types including the impact of thermodynamic losses, *Energy*, 53 (2013), 312–322.

Wall G., Thermoeconomic optimization of a heat pump system, *Energy*, 11 (1986), 957–967.

Wallin E., Frank P. A., Berntsson T., Heat pumps in industrial processes – An optimization methodology, *Heat Recovery Systems*, 10 (1990), 437–446.

CHAPTER 6

Mechanically Driven Heat Pumps

6.1 INTRODUCTION

This chapter discusses the characteristics of mechanically driven heat pumps. In 2015, Science Direct from Elsevier identified 837 papers related to vapor compression (VC) heat pumps, 85 papers related with vapor recompression (VRC) heat pumps, 66 papers related to compression-resorption heat pumps (CRHPs), 69 papers related to transcritical heat pumps (TCHPs) (mainly CO_2 systems) and 99 papers related to Stirling heat pumps. Vapor compression is the most applied and developed technology, and it is not surprising that a large number of papers are dedicated to these systems. Maybe surprising is that Stirling heat pumps have received more attention than other systems presently commercialized.

In this chapter, first the vapor compression heat pumps are discussed. These heat pumps are not only the most studied type but also the most frequently implemented type, and there is a large body of experience in designing and applying these heat pumps for several applications in industry and in the built environment. Heat pumps and refrigeration cycles are the same except for what concerns which reservoir is at the application temperature. For this reason, the relevant topics for heat pumps are similar to the topics discussed in the Refrigeration Fundamentals course at the Delft University of Technology by Stolk (1990), with the main difference that the focus is on applications of heat pumps in the chemical process industry. It identifies the ideal (reference) cycle, the Carnot cycle, and discusses the different sources for a departure from the ideal cycle. The chapter further discusses the vapor recompression cycles, which are generally the preferred solutions for industrial applications. Both working principle and specific industrial applications are presented. Compression-resorption heat pumps combine the advantages of absorption cycles with the advantages of vapor compression cycles and are promising for applications with significant temperature glides both on the process stream application side and in the heat source side. The differences in comparison to the other mechanically driven heat pumps are discussed as well as the calculation procedure to predict their performance. Because CO_2 as a refrigerant has a limited environmental impact, transcritical CO_2 heat pumps are popular these days and are also discussed in a dedicated section, which also describes when these cycles could be advantageous. These cycles show a large temperature glide in the application side, while the heat is absorbed from the heat source at constant temperature, making the cycle specifically interesting when such conditions exist in a chemical plant. Finally, the Stirling heat pump and its calculation method are discussed. As mentioned, these cycles have received a lot of attention because they are expected to perform significantly better than vapor compression cycles. Nevertheless, application up to now has been limited to niche markets for very low-temperature refrigeration cycles. The cycle is discussed with enough detail to allow simple modeling and performance prediction.

6.2 VAPOR COMPRESSION HEAT PUMPS

6.2.1 Reversible Cycle and Efficiency

This section introduces and discusses vapor compression heat pump cycles, which are the cycles with the largest number of implemented heat pumps. This cycle is, in its simplest form, composed of four components (see Figure 6.1). The components are compressor (1), condenser (2), throttling device (3) and evaporator (4). The components are interconnected by tubing so that the system composed of these components forms a closed cycle through which the working fluid (generally called the refrigerant) flows. The components and connecting lines form a system enclosed by the borderline indicated by the dotted rectangle in the figure. In the evaporator, the heat flow $\dot{Q}_{in}$ enters the system, in the compressor

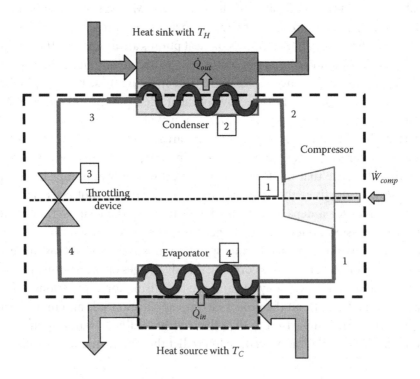

FIGURE 6.1 Vapor compression heat pump.

the shaft power $\dot{W}_{comp}$ is added to the system and finally in the condenser the heat flow $\dot{Q}_{out}$ is delivered to the heat pump application. The energy balance for the system is

$$\dot{Q}_{in} + \dot{W}_{comp} - \dot{Q}_{out} = 0 \qquad (6.1)$$

Note that in this equation, energy storage in the refrigerant and construction materials is neglected, as are energy losses. Most of these systems operate under steady-state conditions for long periods so that liquid levels are maintained and changes in stored energy can be neglected. Generally, insulating materials will be used to prevent energy exchange with the environment so that neglecting energy losses is an acceptable assumption.

Heat flow $\dot{Q}_{out}$ is delivered at a higher temperature than the temperature of heat flow $\dot{Q}_{in}$ that enters the system. The compression process takes place as follows: The vapor formed in the evaporator (4) is compressed by compressor (1) so that its pressure increases. Due to the added power $\dot{W}_{comp}$, the temperature of the vapor rises also. In the condenser (2), the vapor cools due to the removal of heat flow $\dot{Q}_{out}$ and is finally condensed into the liquid phase. The pressure for which this takes place is the saturation pressure of the refrigerant that circulates through the heat pump system at the condensation temperature imposed by the (external) heat pump application. This high-pressure liquid is then throttled to a lower pressure in the throttling device (3) and finally enters the evaporator. Generally, flashing takes place so that the refrigerant enters the evaporator always as a mixture of liquid and vapor. The liquid is evaporated by adding heat flow $\dot{Q}_{in}$ from a heat source at environmental conditions. Evaporation takes place at a specific pressure, which corresponds to the saturation temperature imposed by the (external) heat source. The refrigerant leaves the evaporator in vapor form and reenters the compressor, closing the cycle.

Note that the useful effect takes place in the condenser, where heat $\dot{Q}_{out}$ is delivered to the external heat pump application stream. The system can be subdivided in two parts: low-pressure and high-pressure parts. In Figure 6.1, the central horizontal dotted line indicates the separation between high- and low-pressure levels in the system. This line goes through the compressor (1) and the throttling device (3). All processes taking place within the system border correspond to changes of state of the refrigerant along a cycle. This cycle can be represented in state diagrams of the working fluid as commonly used in thermodynamics. The refrigerant is the fluid that facilitates the transport of heat from the evaporator (low temperature) to the condenser (high temperature). The compressor allows for the energy conversion from lower to higher temperature level.

The Carnot cycle, which comprises two isothermal and two adiabatic reversible processes, can be seen as a heat pump if the cycle runs as indicated in Figure 6.1. Figure 6.2 shows the Carnot process in a T-s diagram. In the case of a heat pump, the process is followed anticlockwise, while for a power cycle the direction is clockwise.

The entropy change for a reversible process is given by

$$ds = \frac{d\dot{Q}}{T} \qquad (6.2)$$

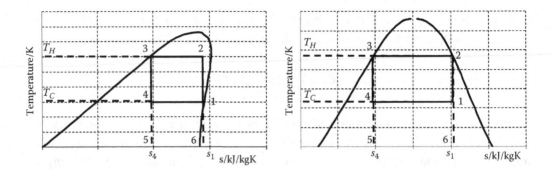

FIGURE 6.2 Carnot heat pump cycle in *T-s* diagram. Left: A fluid with decreasing saturated vapor entropy as temperature decreases (reentrant saturated vapor line). Right: A fluid with increasing saturated vapor entropy as temperature decreases (bell-shaped saturated line).

Integration gives

$$\int_{2}^{3} d\dot{Q} = \int_{2}^{3} T ds \qquad (6.3)$$

The numbers 2 and 3 correspond to the states 2 and 3 in the diagram of Figure 6.2. In this case, it applies that

$$\int_{2}^{3} d\dot{Q} = -\dot{Q}_{out} \quad \text{and} \quad -\int_{2}^{3} T ds = -T_H(s_3 - s_2) \quad \text{so that} \quad \dot{Q}_{out} = T_H(s_2 - s_3) \qquad (6.4)$$

Rectangle 2-3-5-6 represents the heat flow $\dot{Q}_{out}$ delivered at temperature T_H to the heat pump application. Similarly, the following applies for the evaporating process between states 4 and 1 in Figure 6.2:

$$\int_{4}^{1} d\dot{Q} = \dot{Q}_{in} \quad \text{and} \quad \int_{4}^{1} T ds = T_C(s_1 - s_4) \quad \text{so that} \quad \dot{Q}_{in} = T_C(s_1 - s_4) \qquad (6.5)$$

which is represented by rectangle 1-4-5-6 in Figure 6.2. From Equation 6.1, it can be derived that

$$\dot{W}_{comp} = \dot{Q}_{out} - \dot{Q}_{in} \qquad (6.6)$$

so that rectangle 1-2-3-4 represents the work input to the cycle. The efficiency of the heat pump cycle is the ratio between the useful heat delivered to the application $\dot{Q}_{out}$ and energy required to drive the compressor $\dot{W}_{comp}$.

$$COP = \frac{\dot{Q}_{out}}{\dot{W}_{comp}} \tag{6.7}$$

For a Carnot cycle, this becomes

$$COP_{Carnot} = \frac{\dot{Q}_{out}}{\dot{W}_{comp}} = \frac{\dot{Q}_{out}}{\dot{Q}_{out} - \dot{Q}_{in}} = \frac{T_H(s_2 - s_3)}{T_H(s_2 - s_3) - T_C(s_1 - s_4)} \tag{6.8}$$

And, because $s_2 = s_1$ and $s_3 = s_4$,

$$COP_{Carnot} = \frac{T_H(s_2 - s_3)}{T_H(s_2 - s_3) - T_C(s_1 - s_4)} = \frac{T_H(s_1 - s_4)}{T_H(s_1 - s_4) - T_C(s_1 - s_4)} = \frac{T_H}{T_H - T_C} \tag{6.9}$$

The temperature of heat uptake T_C is determined by the environment that delivers the heat source of the heat pump, and it is also more or less limited. The heat release has to take place at temperature T_H, the heat pump application temperature. From Equation 6.9, it follows that, for a fixed T_C, the higher the required heat delivery temperature T_H, the lower the coefficient of performance (COP) of a heat pump. Because it consists of only reversible processes, the Carnot cycle gives a reference for the maximum feasible efficiency COP_{Carnot} of a heat pump cycle. The COP of a heat pump is always larger than 1.

6.2.2 Theoretical Cycle

The Carnot cycle is technically not realizable because real processes always show some degree of irreversibility. For this reason, often a cycle in which all processes are feasible is considered a reference cycle. Figure 6.3 illustrates such a cycle in a T-s diagram. The figure shows the saturated liquid line (left side) and the saturated vapor line (right side) of the refrigerant with the liquid-vapor region in between. The two lines meet at the critical point of the fluid, CP in the diagram. For pure fluids, in the liquid-vapor region isobars and isotherms coincide and are thus horizontal lines. In the liquid region, the isobars follow the saturation line as long as the

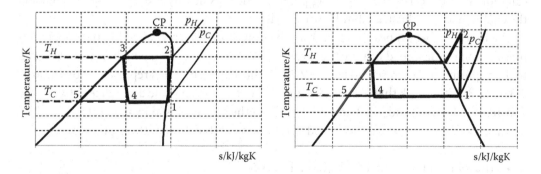

FIGURE 6.3 Theoretical heat pump cycle in T-s diagram. Left: A fluid with decreasing saturated vapor entropy as temperature decreases (reentrant saturated vapor line). Right: A fluid with increasing saturated vapor entropy as temperature decreases (bell-shaped saturated line).

pressure is significantly lower than the critical pressure. In the vapor region, the isobars show an exponential character rising to the right side. An adiabatic reversible process is isentropic and as such a vertical line in the *T-s* diagram. This is also evident in Figure 6.3. In vapor compression heat pumps, in the vapor-liquid region, isenthalpic lines move obliquely downwards.

The theoretical cycle can be described making use of the *T-s* diagram in Figure 6.3. In state 1, at the outlet of the evaporator, all liquid has been vaporized. This state has saturated vapor conditions and lies in the border of the vapor-liquid region and the vapor region. The vapor is compressed in an adiabatic reversible way, thus vertically along an isentropic line. The compression process is represented by line 1-2. As work is added during compression, the vapor enthalpy rises during the process. The pressure rises from isobar p_C (evaporator pressure level) to isobar p_H (condenser pressure level). Depending on the fluid properties, state 2 lies in the liquid-vapor region (left) or vapor region (right, indicating that the vapor is superheated when it leaves the compressor). In the condenser, the refrigerant is first cooled along isobar p_H until state 3a and then condensed from state 3a until state 3. For the fluids represented by the left part of Figure 6.3, there is no superheating zone, and condensation starts directly. In state 3, the liquid line has been reached, and all refrigerant has just become liquid. This liquid flows until the throttling device, in which it is flashed to state 4. In this throttling process, there is no heat exchange and no work exchange with the environment, so that the enthalpy of the refrigerant remains unchanged. States 3 and 4 lie in the same isenthalpic line. The throttling process itself is difficult to describe, but the initial and final states can easily be defined. The process is drawn along a line of constant enthalpy. The throttling is not reversible, so that the entropy of the refrigerant rises as the refrigerant passes the throttling device. It is evident that part of the liquid becomes vapor because the isenthalpic line enters the liquid-vapor region and state 4 lies in this region. State 4 is also the inlet of the evaporator, so that a mixture of liquid and vapor enters the evaporator. The quality (vapor fraction) at the evaporator inlet can be obtained from the ratio of lines 5-4 and 5-1. The liquid evaporates along the length of the evaporator until it totally has become vapor in state 1, where it reenters the compressor, closing the cycle.

In the *T-s* diagram, the work required to drive the cycle is no longer given by the enclosed area as discussed for the reversible cycle. This should be applied carefully specifically for what concerns the throttling process. The reason why the work extends outside the enclosed area is illustrated in Figure 6.4 (left).

The process in which heat is taken from the heat source is represented by the evaporation line 4-1 and the absorbed heat by the area 4-1-8-7-4. The process in which heat is delivered to the application is represented by line 2-3a-3. The delivered heat is represented by the area 2-3a-3-6-8-2. The work that needs to be supplied to the system is the difference between the two areas 2-3a-3-6-7 and 4-1-2 and is larger than the area enclosed by the cycle.

The throttling process can be considered to be a combination of three processes:

- The temperature drop of the liquid, line 3-5.

- The pressure reduction in state 5 in which isobars p_H and p_C are superposed.

- The evaporation of part of the liquid, line 5-4.

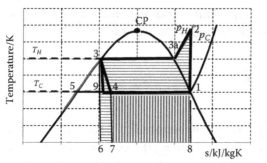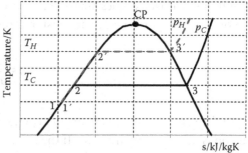

FIGURE 6.4 Left: Effect of throttling process on the work input required by the heat pump cycle. Right: Isobars in a *T-s* diagram.

If we consider the process to take place as proposed, then the added work is given by area 2-3a-3-5-1-2. This area must be equal to the area enclosed by 2-3a-3-6-7-4-1-2. This implies that area 9-4-7-6-9 is equal to the area of the triangle 3-5-9.

The work released during the throttling process is not used and is totally lost in internal losses. These losses consist of friction losses and evaporation of part of the liquid.

6.2.3 Isobars in a *T-s* Diagram

Making use of the right-hand side of Figure 6.4, the following can be noted about the *T-s* diagram. Assume that at some distance from the critical point, liquid is heated. The heat needed is

$$q = \int c_v \, dT = \int T \, ds \tag{6.10}$$

Not too close to the critical point, liquid expansion is limited, and there is little difference between c_p and c_v. In the liquid, c_v is practically independent of temperature, so that it can be stated that c_v = constant.

$$dq = c_v dT = T ds \tag{6.11}$$

so that

$$ds = c_v \frac{dT}{T} \tag{6.12}$$

and

$$s = c_v \ln T + A \tag{6.13}$$

with *A* a constant. This equation applies at constant pressure and thus for an isobar. The isobar is a logarithmic line as line 1-2 in the right-hand side of Figure 6.4. State 2 lies in the liquid line. The isobar goes horizontally through the liquid-vapor region until state 3 in the vapor line. If vapor is heated at constant pressure, the following applies:

$$q = \int c_p \, dT \tag{6.14}$$

In the vapor phase, c_p is strongly dependent on the temperature. The isobar in the vapor region is not a logarithmic line. If the pressure is increased to p_H, then states 2′ and 3′ come closer to each other. The isobar in the liquid region 1′-2′ coincides practically with the low-pressure isobar and the saturated liquid line. Because $c_v \approx c_p$ and is pressure independent, the entropy is practically only a function of the temperature. This is the reason why isobars in the liquid region practically coincide.

Close to the critical point, c_v is no longer constant. In this region, the isobars do not coincide anymore with the saturation line. At the critical point, states 2 and 3 join together because the saturated liquid line joins the saturated vapor line. Above the critical point, only a dense gas phase exists.

6.2.4 Processes in Practical Cycle

The practical cycle is discussed step by step, and the theoretical processes are compared with the practical processes.

6.2.4.1 Compression

In the theoretical process, the compression process takes place in an adiabatic reversible way, thus along an isentropic line. In the left-hand side of Figure 6.5, this is indicated by the vertical line that connects the isobars p_C and p_H, line 1-2. In the real cycle, the evaporator outlet conditions are given by state 1″. In the line from evaporator to compressor and in the suction chamber of the compressor itself, the vapor flow undergoes further superheating due to heat flow from outside, resulting in state 1‴. There is also some pressure loss. The suction valve of the compressor creates a significant pressure drop as the vapor flows into the compression cavity. Taking the shape of suction valves into account, the process taking place is similar to the throttling process across a nozzle. The process takes place

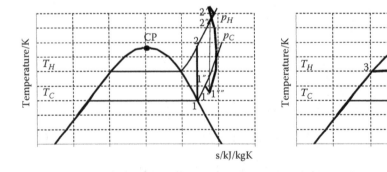

FIGURE 6.5 Left: Steps during compression process. Right: Steps during condensation process.

along an isenthalpic line from 1‴ to 1‴′. For a piston compressor, as the piston switches direction, the suction valve closes, and the actual compression process starts. During compression, the vapor is first heated by the piston and cylinder wall and later releases heat to the piston and cylinder. The entropy increases first and later it reduces again, as indicated in the left-hand side of Figure 6.5. The real compression process takes place from state 1‴′ to state 2′. State 2′ has a higher pressure than p_H because also the discharge valve creates a pressure drop, and the pressure at the end of the compression process needs to be higher to overcome this pressure drop. Losses take place due to internal friction in the vapor flow and heat release due to friction of the piston along the cylinder wall. The entropy increases during compression: State 2′ is located to the right of the isentropic line that goes through state 1‴′. If the cylinder wall is cooled, then it is possible that the entropy reduces and state 2′ is located to the left of this isentropic line. In the discharge valve, a throttling process takes place, after which the vapor flow is slightly cooled in the discharge chamber of the compressor. Also, in the line from compressor to condenser, a further pressure drop and heat release to the environment take place so that the vapor reaches the condenser with state 2″. The position of this state depends on the amount of cooling and can eventually be positioned to the left of the isentropic line that goes through state 1‴′. Common numbers are 4 to 20 K superheating in the evaporator and 0.2 to 1.0 bar pressure drop in lines and valves. The compression process itself takes place practically along an isentropic process line. Note that the process steps have been exaggerated in Figure 6.5 so that they are clearly visualized.

6.2.4.2 Condensation

Theoretically, the heat rejection starts at 2 and moves along isobar p_H through 3a to 3. In the real cycle, the process starts at 2″. There is some pressure loss in the condenser, so that the pressure of state 2″ is slightly higher than pressure p_H, which in the right-hand side of Figure 6.5 is the pressure at the outlet of the condenser. From state 2″ to state 3a′, the vapor is cooled to the saturation temperature. From state 3a′ to state 3, condensation takes place. Generally, the condenser is designed in such a way that the pressure drop is negligibly small so that the condensed liquid does not experience problems in leaving the large number of parallel flows normally used in condensers.

6.2.4.3 Expansion Process

The expansion valve is located in the generally long line between the outlet of the condenser and the inlet of the evaporator. The line, expansion device and possibly a liquid distributor constitute the system in which the expansion process takes place. Ideally, the process takes place without heat exchange with the environment so that the process is externally adiabatic. Nevertheless, this process is irreversible, so it does not follow an isentropic line. Because there is no work or heat exchanged with the environment, the first law of thermodynamics tells that the process is isenthalpic. The energy that is released in the process is totally consumed by friction losses and by the evaporation of part of the refrigerant liquid: The entropy increases. In principle, only the beginning and the end of the process can be indicated in the state diagram. The real process does not exactly follow an isenthalpic line

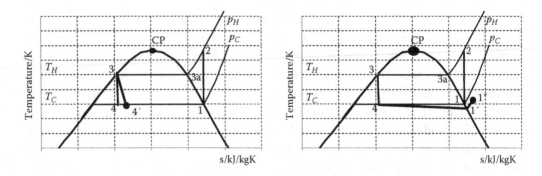

FIGURE 6.6 Left: Conditions around the expansion device. Right: Steps during evaporation process.

because in the line between throttling device and evaporator the temperature is lower than the environment, so that the refrigerant takes up some heat from the environment. In real systems, the distance between throttling device and evaporator is kept as short as possible so that this effect is negligible. The left-hand side of Figure 6.6 indicates the difference between the theoretical process from 3 to 4 and the real process from 3 to 4′. Note that the difference has been exaggerated to visually clarify the difference.

6.2.4.4 Evaporation

The evaporation process (see Figure 6.6, right-hand side) in the evaporator starts in the liquid-vapor region with the pressure given by the low-pressure isobar p_C. After the throttling process, an amount of vapor is available, the vapor that results from the flashing process. Due to the evaporation of liquid, the amount of vapor increases, and with it the velocity of the refrigerant also increases. With the velocity, the pressure drop per length unit also increases. This is assuming that the flow passage area for the refrigerant remains constant, which generally is the case but not always. From the pressure drop point of view, it would be better to enlarge the flow passage area as the flow progresses. This is possible for some heat exchanger types, such as the multipass shell-and-tube heat exchangers with evaporation in the tubes, for which the number of tubes per pass can be adjusted to allow for a more constant velocity through the whole evaporator, but for most other types is not feasible. This velocity increase also has a positive effect: As the velocity increases, the heat transfer between refrigerant and heat exchanger wall also increases, which has a positive effect for the performance of the heat exchanger. The pressure drop can be maintained at a low level by carefully designing the evaporator so that a constant flow passage area is not necessarily a problem. Due to the pressure drop, the process does not follow the low-pressure isobar p_C, line 4 to 1 but line 4 to 1″. Generally, the pressure drop is small so that the two lines coincide. The evaporation process shall always show some fluctuations. If the system is controlled so that the evaporator just delivers saturated vapor, then it is possible that liquid enters the compressor unless a liquid separator is installed at the evaporator outlet. This is the reason why mostly the system is controlled so that a small superheating process remains available at the outlet of the evaporator. The process moves along the isobar to state 1″. The refrigerant leaves the evaporator at state 1″. In liquid overfeed systems with a liquid vapor separator, the vapor leaves the separator practically at saturated conditions, so that the distance 1′ to 1″ can be considered to be zero.

6.2.5 Causes of Losses and Possible Improvements

The inlet condition of the compressor has been attained with state 1″ so that the whole cycle has been discussed. The discussion of the practical process in comparison with the theoretical process in the previous sections makes it possible to define actions that can lead to an improvement of cycle efficiency. Figure 6.7 compares the theoretical and practical cycles. The differences have been exaggerated to accentuate the differences. Three conditions are externally imposed on the system:

- The temperature of the heat source is imposed by the low-temperature heat source self. The temperature difference between the heat source flow and the evaporating refrigerant follows from an optimization loop in which the required surface area and the energy costs play a major role. This temperature difference fixes the evaporating temperature required for the cycle T_C.

- The goal of a heat pump is to deliver heat to a process flow that needs to be heated. The temperature level of this flow is imposed by the process that includes this heating step. So, the temperature of the heat sink is imposed by the application. Also, for the condenser, the temperature difference between the heat sink flow and the condensing refrigerant follows from an optimization loop in which the required surface area and the energy costs play a major role. This temperature difference fixes the condensing temperature required for the cycle T_H.

- The second goal of the heat pump is to deliver a certain heating power to the application so that $\dot{Q}_{out}$ is imposed. The T-s diagram indicates the specific value of the state variables (thus per kilogram refrigerant). The required mass flow of refrigerant through the cycle can be derived from the enthalpy difference between states 2″ and 3: $\dot{m}_{ref} = \dot{Q}_{out} / (h_{2″} - h_3)$.

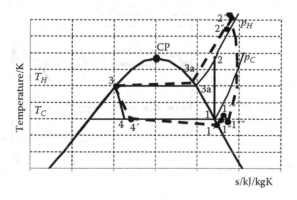

FIGURE 6.7 Comparison of theoretical and practical cycles in a T-s diagram.

Evaporating temperature, condensing temperature and heating capacity are imposed externally to the heat pump. Further calculations require the selection of the refrigerant to be used in the cycle.

6.2.5.1 Pressure-Enthalpy Diagram

The *T-s* diagram is useful to understand the processes of the cycle, to illustrate the effect of some parameters and to compare different processes with each other. To quantify heat and work amounts, in this diagram areas need to be integrated or enthalpies can, less accurately, be read from the diagram. The ln *p-h* diagram gives work and heat as straight lines, and for the design of heat pumps is much handier to use. Figure 6.8 gives the same theoretical and practical cycles in the ln *p-h* diagram as in Figure 6.7 drawn in the *T-s* diagram. A logarithmic scale has been selected for the pressure so that a handy temperature scale results in the liquid-vapor region where isothermal lines are horizontal just like the isobars. The first law of thermodynamics (Equation 6.1) and the COP (Equation 6.7) can be directly read from the diagram because the relevant variables are given by straight lines in this diagram. In the liquid-vapor region, isobars and isothermal lines coincide. In Figure 6.8, the real compression process line is also given in a comparable way as discussed for this process step in the *T-s* diagram.

In the *T-s* diagram and in the ln *p-h* diagram, the state variables are expressed per kilogram refrigerant. By dividing the required heat pump duty $\dot{Q}_{out}$ by $\Delta h_{cond} = h_2 - h_3$, the mass flow per time unit (kg/s) that circulates through the cycle can be determined.

The COP of a heat pump (Equation 6.7) can now be expressed as a function of enthalpy values only.

$$COP = \frac{\dot{Q}_{out}}{\dot{W}_{comp}} = \frac{\dot{m}_{ref} \times (h_2 - h_3)}{\dot{m}_{ref} \times (h_2 - h_1)} = \frac{h_2 - h_3}{h_2 - h_1} \qquad (6.15)$$

6.2.5.2 Impact of Compressor

In Figure 6.7, it is visualized that the work to be delivered to the compressor is significantly increased. The compressor work for the theoretical cycle is given by the area that results

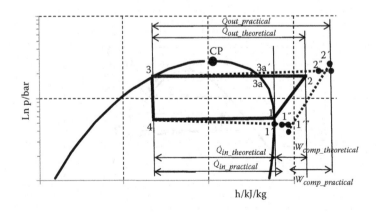

FIGURE 6.8 Comparison of theoretical and practical cycles in a ln *p-h* diagram.

from subtracting the area under line 2-3a-3 from the area under line 4-1. For the real cycle, the same applies, only now the area under line 2''-3a'-3 is subtracted from the area under line 4-1'-1'', and additional extra work is required. Now, on top of the enclosed area, the area under the compression line also needs to be taken into account. Figure 6.8 illustrates that the compressor work of the practical cycle ($w_{comp\_practical} = h_{2''} - h_{1''}$) is significantly larger than the compressor work for the theoretical cycle ($w_{comp\_theoretical} = h_2 - h_1$). To obtain a high COP, the compressor work should be as small as possible, so the indicated area should be reduced as much as possible:

- The pressure difference between $p_C = p_1$ and $p_{1''}$ should be as small as possible; thus, the pressure loss in the suction line, suction chamber and suction valve of the compressor should be as small as possible.

- The entropy increase during compression should be as small as possible. Valve losses, piston friction and pumping losses must be small, implying that the vapor flow needs to be optimized to prevent large losses. This is the task of the compressor designer. Changes of flow direction should be prevented as much as possible.

- The pressure difference between $p_{2'}$ and p_H must be as small as possible. Thus, the flow resistance in discharge valve, discharge chamber and discharge line to the condenser must be as small as possible.

From these points, it becomes clear that the compressor has a tremendous effect on the performance of the whole heat pump cycle.

The quality of a compressor is expressed by its isentropic efficiency:

$$\eta_{is\_comp} = \frac{w_{is}}{w_{practical}} = \frac{h_{2s} - h_1}{h_2 - h_1} \tag{6.16}$$

where h_{2s} is the enthalpy at compressor discharge if the process would take place isentropically. Using this definition, the COP of a heat pump can be expressed as

$$COP = \eta_{is\_comp} \frac{h_2 - h_3}{h_{2s} - h_1} \tag{6.17}$$

indicating that the COP of a heat pump is linearly dependent on the isentropic efficiency of the compressor used in the cycle.

The mass flow displaced by a compressor follows from its geometrical volume displacement and the volumetric efficiency of the compressor. The volumetric efficiency is defined as follows:

$$\eta_{vol\_comp} = \frac{\dot{V}_{comp\_in}}{V_{suction\_cavity} \times n} = \frac{\dot{m}_{ref} v_1}{\dot{V}_{displacement}} \tag{6.18}$$

with $\dot{V}$ the volume flow and n the rotational speed of the compressor. Here, v_1 is the specific volume of the refrigerant at the compressor inlet. When the volumetric efficiency is known, then the mass flow of refrigerant circulating through the cycle can be calculated.

$$\dot{m}_{ref} = \frac{\dot{V}_{displacement} \times \eta_{vol\_comp}}{v_1} \qquad (6.19)$$

where $\dot{V}_{displacement}$ is the geometrical volume displacement of the compressor associated with the size of the compressor inlet cavities and its rotational speed. The compressor power requirement follows from its isentropic efficiency and the technical compressor work for an isentropic compression process.

$$\dot{W}_{comp} = \dot{m}_{ref} \times \frac{1}{\eta_{is\_comp}} p_1 v_1 \frac{\kappa}{\kappa-1} \left[\left(\frac{p_2}{p_1} \right)^{\frac{\kappa-1}{\kappa}} - 1 \right] \qquad (6.20)$$

with κ the isentropic exponent of the refrigerant ($\kappa = c_p/c_v$).

6.2.5.3 Impact of the Throttling Process

When determining the work needed to drive the cycle, the throttling process must be considered in a correct way (see the left-hand side of Figure 6.9). Assume that, instead of following an isenthalpic line from state 3 to state 4, the process takes place in an adiabatic reversible (isentropic) way from state 3 to state 4'. The heat input to the cycle is then given by area 4'-1-8-6-4' and the required work by the area 4'-1-2-3a-3-4'. The work made available during the expansion process has totally been used as part of the energy required to drive the compressor.

Assume in another case that it is possible to cool the refrigerant liquid from the condenser down to state 5, still at the high pressure p_H. The isobar in this region coincides practically with the liquid saturation line of the refrigerant. Then, the subcooled liquid is expanded from high p_H to low pressure p_C. The state of the liquid remains practically at

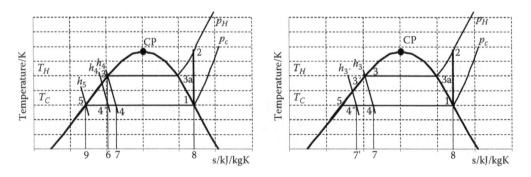

FIGURE 6.9 Left: Quantification of the impact of the throttling process in a T-s diagram. Right: Effect of subcooling after the condenser in a T-s diagram.

the same conditions (state 5) because the two pressure lines coincide in this region. The liquid is so subcooled that no vapor is formed during the flashing process. There is also no work delivered during this flashing process. After this flashing, the liquid is evaporated in the evaporator, making use of the heat delivered by the heat source. The cycle is now represented by 5-1-2-3a-3-5, the heat uptake in the evaporator by 5-1-8-9-5, while the heat delivered in the condenser remains unchanged. The required work (the difference between the two) is now smaller.

Both the adiabatic reversible expansion and the subcooling down to state 5 are hypothetical processes. The real process is isenthalpic from state 3 to 4. By moving the start point for the evaporation process from 4' to 4, the heat uptake from the heat source $\Delta \dot{Q}_{in}$ is reduced by the area 4'-4-7-6-4'. States 2-3a-3 remain in the same place, indicating that the heat delivered to the application ($\dot{Q}_{out}$) remains unchanged. The first law of thermodynamics, previously introduced in Equation 6.1, indicates that the reduction in heat uptake is compensated by an increase of the required compressor power input:

$$-\Delta \dot{Q}_{in} = \Delta \dot{W}_{comp} \tag{6.21}$$

Consider state 5 to be the reference state for the enthalpy so that $h_5 = 0$. Because there is no work exchange during the subcooling from state 3 to 5, the enthalpy change $h_3 - h_5$ is given by the surface under line 3-5: 3-5-9-6-3. Because $h_5 = 0$, h_3 is given by the area 3-5-9-6-3. The expansion from 3 to 4 is isenthalpic, that is, $h_3 = h_4$, so that also h_4 is given by the area 3-5-9-6-3.

Evaporation from 5 to 4' is given by the enthalpy change $h_{4'} - h_5$ given by the area 5-4'-6-9-5. Because $h_5 = 0$, h_4' is given by the area 5-4'-6-9-5. Previously, it was shown that $\Delta \dot{Q}_{in} = h_4 - h_{4'}$. This difference is obtained by subtracting h_4 (area 3-5-9-6-3) from h_4' (area 5-4'-6-9-5), which is area 3-5-4'-3. So, the extra work given in Equation 6.21 corresponds to the area enclosed by the saturation line, the isentropic line departing from the condenser outlet condition and the isothermal line for the evaporating condition in a T-s diagram. This work corresponds with the reduction of the heat uptake by the cycle $\Delta \dot{Q}_{in}$, as indicated by Equation 6.21. The expansion energy is totally dissipated in the refrigerant, reducing the capacity to absorb energy from the heat source. The work input required by the cycle is, for an isenthalpic expansion, given by the area 5-1-2-3a-3-5. The enclosed area of the cycle represents the required work only when the cycle takes place in a reversible way. The irreversibility associated with the expansion process illustrates the work required is enlarged and no longer represented only by the enclosed area. The COP of the cycle, Equation 6.7, depends on the slope of the saturated liquid line from state 3 to state 5 and so on the thermodynamic properties of the applied refrigerant.

Previously, it was assumed that the refrigerant leaves the condenser exactly in a saturated liquid state, state 3: All vapors are condensed, and the liquid is at saturation conditions. If the liquid is subcooled before the throttling process, as given in the right-hand side of Figure 6.9, to state 3', then the work input required by the cycle remains unchanged as follows from the previous discussion. The heat uptake by the evaporator, however, increases and is now given by the area 4"-1-8-7'-4". The COP (Equation 6.7) increases as follows from the first law of thermodynamics (Equation 6.1): The delivered heat is increased while $\dot{W}_{comp}$ is maintained. This requires that the subcooling in the condenser can be used for the heating of the process flow. In that case, it is evident that more heat is exchanged with the sink, and at the same time the irreversible throttling process is shorter.

The degree of subcooling that can be attained is limited. Nevertheless, it has previously been illustrated that when a pure refrigerant is used and the process fluid to be heated has a significant glide, some degree of subcooling is always possible and should be considered. The minimum approach temperature can then apply not only at the vapor saturation line but also at the heat sink stream inlet. Further subcooling is sometimes possible by making use of a heat exchanger between the liquid leaving the condenser and the vapor leaving the evaporator. In this case, the throttling losses are reduced, but the subcooling takes place outside the heat exchanger that produces the useful effect, the condenser. The superheating at the condenser inlet is then larger, so that in fact $\dot{Q}_{out}$ also increases.

6.2.5.4 Effect of the Evaporating Temperature

The compressor displaces the refrigerant through the cycle. The volume of its cavities (at the compressor inlet) determines how much refrigerant can be circulated through the cycle. The specific volume of the refrigerant at the inlet of the compressor times the mass flow determines the volume flow displaced by the compressor and therefore its size. If the evaporating temperature is lowered and with it the suction pressure p_C while all other conditions remain unchanged, then it is clear that the specific heat of evaporation, $\Delta h_{evap} = h_1 - h_4$, is reduced (see the left-hand side of Figure 6.10). The specific volume increases, and because of the larger pressure ratio p_H/p_C, the volumetric efficiency of the compressor is reduced. There are three negative effects being caused by the lower evaporating temperature:

- The specific heat uptake is decreased, $\dot{Q}_{in\_a} > \dot{Q}_{in\_b}$.

- Due to the increase in specific volume, the mass flow through the compressor is reduced.

- Due to a reduction of volumetric efficiency, the volume flow through the compressor is decreased.

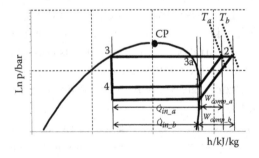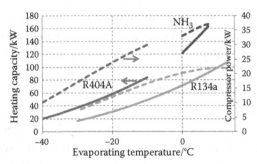

FIGURE 6.10 Left: Effect of lower evaporating temperature in a ln p-h diagram. Right: Effect of evaporating temperature on the heat pump capacity and power consumption for a specific compressor (heat delivery at 50°C).

These combined effects make the heat uptake by the cycle $\dot{Q}_{in}$ significantly decrease as the evaporating temperature decreases. The right-hand side of Figure 6.10 shows data for a specific compressor illustrating this effect. From the ln p-h diagram in the left-hand side of Figure 6.10, it appears that the work to be delivered to the cycle per circulating kilogram of refrigerant increases, $w_{comp\_b} > w_{comp\_a}$. At the same time, the mass flow through the compressor is reduced. Considering a reducing evaporating temperature first, the increase in required work has the largest impact, after which the reduction of the mass flow has the largest impact. The right-hand side of Figure 6.10 also shows the required compressor shaft power (dotted lines) and illustrates, for R134a, that for lower evaporating temperatures, first the required power slightly increases and then starts decreasing. Note that the right-hand side of Figure 6.10 applies for a specific compressor and so for a constant volume displacement. Other compressors may show a different behavior. From the ln p-h diagram in the left-hand side of Figure 6.10, it is clear that the discharge temperature of the compressor increases as the evaporating temperature decreases ($T_b > T_a$) due to the larger pressure ratio.

6.2.5.5 Effect of Condensation Temperature

If the condensation temperature is increased while all the other conditions are maintained, then from the ln p-h diagram in the left-hand side of Figure 6.11, both the work to be delivered to the cycle and the discharge temperature increase. Expansion takes place from a higher pressure level and the specific heat delivered in the condenser reduces. Due to the larger pressure ratio, the volumetric efficiency of the compressor reduces so that the mass flow through the cycle reduces. There are two effects:

- The heat uptake in the evaporator is reduced per circulating kilogram refrigerant, $\dot{Q}_{in\_a} > \dot{Q}_{in\_b}$, so that the required specific compressor power increases.

- The volumetric efficiency of the compressor reduces, so the mass flow of refrigerant also reduces.

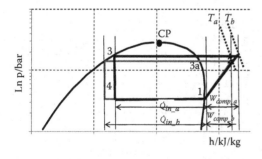
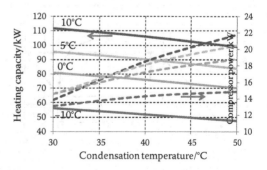

FIGURE 6.11 Left: Effect of higher condensing temperature in a ln p-h diagram. Right: Effect of condensing temperature on the heat pump capacity and power consumption for a specific compressor. Refrigerant is R134a.

The right-hand side of Figure 6.11 gives values of the heat pump capacity as a function of the condensation temperature. The increase in power requirement (dotted lines) as the condensation temperature increases can also be identified in the figure. The reduction in mass flow in the considered range does not have a large impact, so there is no power requirement decrease as the temperature rises as for the evaporator. The right-hand side of Figure 6.11 gives data for a specific compressor.

For example, for an evaporating temperature of 10°C, when the condensation temperature increases from 40°C to 50°C the heating capacity reduces by 7% while the specific work increases by 17%. The COP of the cycle where this compressor is used reduces from 5.7 to 4.5. At high condensation temperature, the discharge temperature can become so high that two-stage operation must be applied to keep the discharge temperature within acceptable limits.

6.2.5.6 Effect of Subcooling

When discussing the effect of throttling, it has already been remarked that subcooling of the liquid in the condenser leads to an increase of the heat uptake from the heat source. This is illustrated in the ln p-h diagram in the left-hand side of Figure 6.12. The heat removed from the condensed liquid $\Delta h_{subcooling}$ results in an increase of the heat delivered to the application of the heat pump. This subcooling is generally limited because the condensation temperature is selected as close as possible to the temperature of the process flow that needs to be heated. Depending on the temperature glide of this stream, the subcooling can attain a few degrees Kelvin. The right-hand side of Figure 6.12 illustrates the effect of subcooling for some refrigerants for a heat pump with an evaporating temperature of 55°C and condensation temperature of 120°C.

6.2.5.7 Effect of Superheating

Superheating can be divided into two parts: one part in the evaporator, the other in the suction line between evaporator and compressor. If the heat pump operates at high heat source temperatures, then it is possible that heat is lost in the suction line so that the superheating

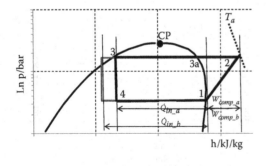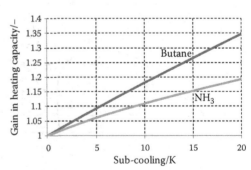

FIGURE 6.12 Left: Effect of subcooling at the outlet of the condenser on the heat pump capacity. Right: Heating capacity gain as a function of the subcooling at the condenser outlet for a heat pump with an evaporating temperature of 55°C and condensation temperature of 120°C.

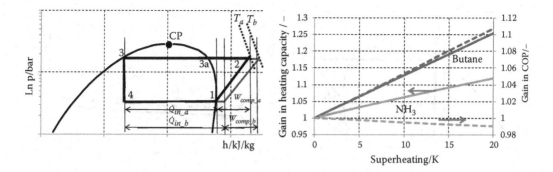

FIGURE 6.13 Left: Effect of superheating at the inlet of the compressor on the heat pump capacity. Right: Heating capacity and COP gain as a function of the superheating at the compressor inlet for a heat pump with an evaporating temperature of 55°C and condensation temperature of 120°C.

is reduced. The superheating that takes place in the evaporator is generally required for control purposes and increases the area requirement for the heat exchanger because the heat transfer coefficient of superheated vapor is significantly lower than the heat transfer coefficient of evaporating refrigerant. The increased temperature leads to an increase of the specific volume of the vapor. For the same volume displacement of the compressor, the mass flow reduces because less mass can be contained in the compressor cavity. Also, the discharge temperature becomes higher, as indicated in the left-hand side of Figure 6.13.

Because the isentropic lines become flatter as the temperature increases, the specific compressor work increases as the superheating increases. Generally, the increase in heat delivered to the application stream is larger than the increase in compressor work, so that superheating leads to higher heating capacities and COP. The gain in heating capacity and COP (dotted lines) is given in the right-hand side of Figure 6.13 for some refrigerants for a heat pump with an evaporating temperature of 55°C and condensation temperature of 120°C.

From this figure, it appears that the COP of NH_3 cycles slightly decreases as the superheating increases, while it increases for butane. So, depending on the slope of the isentropic lines in the vapor region for the considered refrigerant, superheating may lead to an improvement or deterioration of the heat pump COP. The impact on the heating capacity is positive and significant.

6.2.5.8 Combined Subcooling and Superheating

As discussed previously, subcooling of the condensate can be combined with superheating of the suction gas by exchanging heat between these two streams in a heat exchanger. This is illustrated in a ln p-h diagram in the left-hand side of Figure 6.14. If the heat exchanger is well insulated (externally adiabatic), then $h_4 - h_{4'} = h_3 - h_{3'} = h_{1'} - h_1 = \Delta h_{hex}$. The specific heat of liquid is larger than that of vapor, so that $T_3 - T_{3'} < T_{1'} - T_1$. The mass flows in both sides of the heat exchanger (at least for a single-stage cycle) are identical. Δh_{hex} comes as the enthalpy increase $h_{2'} - h_2$ as a gain in the heating capacity of the cycle. The subcooling Δh_{hex} takes place outside the condenser and delivers no extra heating capacity. Opposite

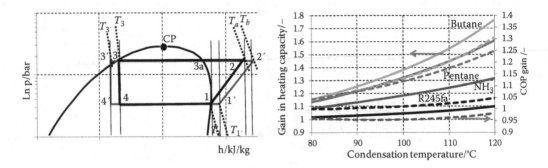

FIGURE 6.14 Left: A ln *p-h* diagram showing combined subcooling after condenser and superheating before a compressor making use of a heat exchanger. Right: Effect of internal refrigerant heat exchanger on heating capacity and COP (dotted lines) gain as a function of the condensation temperature. Evaporating temperature is kept at 55°C. Minimum-temperature approach is 5 K.

this gain, an increase in specific compressor work can take place. For some refrigerants, the isentropic lines in the vapor region have similar slope, so that the compressor work remains unchanged, but for other refrigerants, such as ammonia, the slope of the isentropic lines significantly decreases as the distance to the saturation line increases. In this case, the specific compressor power increases and reduces the advantages of the larger enthalpy change in the condenser.

The right-hand side of Figure 6.14 shows for some refrigerants how the application of a combined heat exchanger between the liquid line and the vapor line has an impact on heating capacity and COP. The application of such a heat exchanger allows for the realization of the superheating outside the evaporator so that the surface area available for evaporation can be optimally used.

The COP of ammonia cycles remains practically unchanged, while the heating capacity gain is significant. The other fluids all have both COP and heating capacity advantages. R245fa shows the smallest effect.

6.2.6 Sizing the System

Considering what has been previously discussed in relation to the imposed conditions of source and sink, it is possible to draw the optimum cycle in a ln *p-h* diagram. From this diagram, the heating capacity per circulating kilogram refrigerant can be determined. Because the heat pump duty is normally a requirement, the mass flow of refrigerant that needs to be circulated can then easily be calculated. Also, the specific volume of the refrigerant at compressor inlet, state 1, can be determined from the diagram so that the volume flow that the compressor needs to displace is known. With this known, a compressor can then be selected from the documentation of compressor manufacturers. Figure 6.15 shows the application range of semihermetic compressors presently available in the market. The figure shows that manufacturers offer mainly three compressor types for these applications – scroll, piston and screw – and that the application ranges partially overlap.

Important selection criteria for the compressor are its isentropic and volumetric efficiencies. Figure 6.16 shows that different compressors can have significantly different isentropic

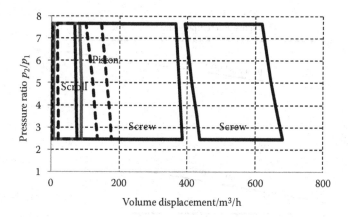

FIGURE 6.15 Application range of semihermetic compressors for vapor compression cycle applications. The application range of scroll compressors is indicated by the grey line; and of piston compressors with dotted lines.

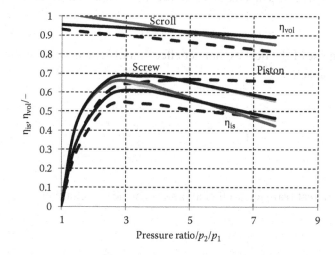

FIGURE 6.16 Examples of isentropic and volumetric efficiencies for some of the semihermetic compressors given in Figure 6.15. The isentropic efficiency includes the electric motor losses; the electric motor is included in the same casing as the compressor.

efficiencies. It also shows that efficiencies are generally lower than 70%. Note that the efficiency of some of the offered compressors is even significantly lower than shown in Figure 6.16. The volumetric efficiency shows an almost-linear dependence of the pressure ratio. Equations 6.19 and 6.20 in combination with the efficiencies shown in Figure 6.16 allow for the selection of a compressor for application in a specific cycle.

As previously discussed, a minimum-temperature driving force (approach) is required in the two heat exchangers so that the heat flow can take place. This is illustrated in Figure 6.17.

Condensation will only take place if a minimum temperature difference is available. The condensation temperature results from the surface area and overall heat transfer coefficient of the heat exchanger. A first approximation of the condensation temperature can be

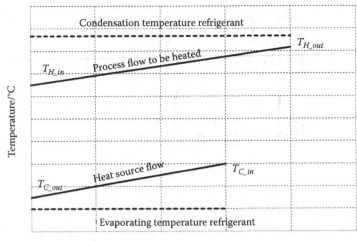

FIGURE 6.17 Effect of finite temperature differences between working fluid of the cycle and external fluid flows.

obtained making use of the heat exchanger effectiveness and assuming that the refrigerant has infinite heat capacity.

$$T_{cond} = \frac{T_{H\_out} - T_{H\_in}e^{-\frac{(UA)_{cond}}{\dot{m}_H c_{p\_H}}}}{1 - e^{-\frac{(UA)_{cond}}{\dot{m}_H c_{p\_H}}}} \tag{6.22}$$

with U the overall heat transfer coefficient and A the heat exchanger area. Similarly for the evaporator:

$$T_{evap} = \frac{T_{C\_out} - T_{C\_in}e^{-\frac{(UA)_{evap}}{\dot{m}_C c_{p\_C}}}}{1 - e^{-\frac{(UA)_{evap}}{\dot{m}_C c_{p\_C}}}} \tag{6.23}$$

The mass flow is known, and the specific enthalpy change in condenser and evaporator can be determined from the ln p-h diagram. With these data, the heat flow delivered by the condenser $\dot{Q}_{out}$ and the heat flow taken from the heat source in the evaporator $\dot{Q}_{in}$ can be calculated and be used to size condenser and evaporator. The expansion device is designed to allow for the required mass flow with the available pressure difference between high-pressure and low-pressure sides of the cycle. The refrigerant lines are sized taking into account the refrigerant mass flow and its local density. The most difficult aspect in the calculation of the heat pump is the determination of the applicable heat transfer coefficients. The discrepancy between the different correlations presented in the literature can be quite large, so that correlations applicable for the specific conditions encountered in the heat pump should be adopted.

Example 6.1

A butane (R600) vapor compression heat pump cycle consists of

- A reciprocating compressor with isentropic efficiency of 65% and volumetric efficiency of 80%;
- A process stream cooled condenser with UA = 100.0 kW/K and process stream flow rate of 15 kg/s (c_p = 4.16 kJ/kgK). The process stream flow enters the condenser with 80°C;
- An evaporator with UA = 91.5 kW/K and water flow rate of 33 kg/s (c_p = 4.18 kJ/kgK). Water enters the evaporator with 50°C.

The heating capacity delivered by the condenser is 1000 kW. Use the thermodynamic data of butane (R600) (use, for instance, the work of Lemmon et al., 2013) where needed. Assume saturation at compressor inlet and condenser outlet.

Calculate the following:

1. An approximation of the condensation temperature.
2. An approximation of the evaporating temperature.
3. The COP of the cycle.
4. The geometrical volume displacement of the compressor.

ANALYSIS

The temperature change of the process fluid follows from the heating capacity of the heat pump:

$$\dot{Q}_{out} = \dot{m}_H c_{p\_H}(T_{H\_out} - T_{H\_in})$$

so that

$$T_{H\_out} = T_{H\_in} + \frac{\dot{Q}_{out}}{\dot{m}_H c_{p\_H}} = 80 + \frac{1000}{15 \times 4.16} = 96.0°C$$

The condensation temperature follows from Equation 6.22.

$$T_{cond} = \frac{T_{H\_out} - T_{H\_in} e^{-\frac{(UA)_{cond}}{\dot{m}_H c_{p\_H}}}}{1 - e^{-\frac{(UA)_{cond}}{\dot{m}_H c_{p\_H}}}} = \frac{96 - 80 \times e^{-\frac{100}{15 \times 4.16}}}{1 - e^{-\frac{100}{15 \times 4.16}}} = 100.0°C$$

Assume $T_{evap} = 35°C$, then, with REFPROP, the enthalpy at the different states of the cycle can be derived:

State	T (°C)	h (kJ/kg)	s (kJ/kgK)
1	35	635.17	2.4274
2s	100	694.69	2.4274
2	102.1	726.74	2.5133
3	100	463.03	1.8066
4	35	463.03	1.8688

where h_2 has been obtained from

$$h_2 = h_1 + \frac{h_{2s} - h_1}{\eta_{is\_comp}} = 635.17 + \frac{694.69 - 635.17}{0.65} = 726.74 \text{ kJ/kg}$$

Then, the heat uptake from the source flow can be obtained from

$$\frac{\dot{Q}_{in}}{\dot{Q}_{out}} = \frac{\dot{m}_{ref} \times (h_1 - h_4)}{\dot{m}_{ref} \times (h_2 - h_3)}$$

or

$$\dot{Q}_{in} = \frac{(h_1 - h_4)}{(h_2 - h_3)} \times \dot{Q}_{out} = \frac{635.17 - 463.03}{726.74 - 463.03} \times 1000 = 574.4 \text{ kW}$$

The source outlet temperature follows from

$$T_{C\_out} = T_{C\_in} - \frac{\dot{Q}_{in}}{\dot{m}_C c_{p\_C}} = 50 - \frac{574.4}{33 \times 4.18} = 45.8°C$$

Now, the evaporation temperature can be calculated from Equation 6.23:

$$T_{evap} = \frac{T_{C\_out} - T_{C\_in} e^{-\frac{(UA)_{evap}}{\dot{m}_C c_{p\_C}}}}{1 - e^{-\frac{(UA)_{evap}}{\dot{m}_C c_{p\_C}}}} = \frac{45.8 - 50 \times e^{-\frac{91.5}{33 \times 4.18}}}{1 - e^{-\frac{91.5}{33 \times 4.18}}} = 41.4°C$$

This indicates that the assumed evaporating temperature was not correct, so that the calculation has to be repeated with the new value obtained for the evaporating temperature. With REFPROP, the enthalpy at the different states of the cycle can be derived:

State	T (°C)	h (kJ/kg)	s (kJ/kgK)
1	41.4	644.23	2.4332
2s	100	696.85	2.4332
2	101.5	725.18	2.5091
3	100	463.03	1.8066
4	41.4	463.03	1.8572

where h_2 has been obtained from

$$h_2 = h_1 + \frac{h_{2s} - h_1}{\eta_{is\_comp}} = 644.23 + \frac{696.85 - 644.23}{0.65} = 725.18 \text{ kJ/kg}$$

Then, the heat uptake from the source flow can be obtained from

$$\dot{Q}_{in} = \frac{(h_1 - h_4)}{(h_2 - h_3)} \times \dot{Q}_{out} = \frac{644.23 - 463.03}{725.18 - 463.03} \times 1000 = 691.2 \text{ kW}$$

The source outlet temperature now becomes

$$T_{C\_out} = T_{C\_in} - \frac{\dot{Q}_{in}}{\dot{m}_C c_{p\_C}} = 50 - \frac{691.2}{33 \times 4.18} = 45.0 \text{°C}$$

Now, the evaporation temperature can be calculated from Equation 6.23:

$$T_{evap} = \frac{T_{C\_out} - T_{C\_in} e^{-\frac{(UA)_{evap}}{\dot{m}_C c_{p\_C}}}}{1 - e^{-\frac{(UA)_{evap}}{\dot{m}_C c_{p\_C}}}} = \frac{45.0 - 50 \times e^{-\frac{91.5}{33 \times 4.18}}}{1 - e^{-\frac{91.5}{33 \times 4.18}}} = 39.7 \text{°C}$$

A next iteration gives a good approximation of the evaporating temperature. With REFPROP, the enthalpy at the different states of the cycle can be derived:

State	T (°C)	h (kJ/kg)	s (kJ/kgK)
1	39.7	641.83	2.4316
2s	100	696.25	2.4316
2	101.5	725.55	2.5101
3	100	463.03	1.8066
4	39.7	463.03	1.8601

where h_2 has been obtained from

$$h_2 = h_1 + \frac{h_{2s} - h_1}{\eta_{is\_comp}} = 641.83 + \frac{696.25 - 641.83}{0.65} = 725.55 \text{ kJ/kg}$$

Then, the heat uptake from the source flow can be obtained from

$$\dot{Q}_{in} = \frac{(h_1 - h_4)}{(h_2 - h_3)} \times \dot{Q}_{out} = \frac{641.83 - 463.03}{725.55 - 463.03} \times 1000 = 681.1 \text{ kW}$$

The source outlet temperature now becomes

$$T_{C\_out} = T_{C\_in} - \frac{\dot{Q}_{in}}{\dot{m}_C c_{p\_C}} = 50 - \frac{681.1}{33 \times 4.18} = 45.1°C$$

Now, the evaporation temperature can be calculated from Equation 6.23:

$$T_{evap} = \frac{T_{C\_out} - T_{C\_in} e^{-\frac{(UA)_{evap}}{\dot{m}_C c_{p\_C}}}}{1 - e^{-\frac{(UA)_{evap}}{\dot{m}_C c_{p\_C}}}} = \frac{45.1 - 50 \times e^{-\frac{91.5}{33 \times 4.18}}}{1 - e^{-\frac{91.5}{33 \times 4.18}}} = 39.8°C$$

Because the difference between the previous and the present value of the evaporating temperature is only 0.1 K, the evaporating temperature is taken as 39.8°C.

The COP of the cycle can now be obtained:

$$COP = \frac{\dot{Q}_{out}}{\dot{W}_{comp}} = \frac{\dot{Q}_{out}}{\dot{Q}_{out} - \dot{Q}_{in}} = \frac{1000}{1000 - 681.1} = 3.14$$

The mass flow through the cycle is obtained from

$$\dot{m}_{ref} = \frac{\dot{Q}_{out}}{h_2 - h_3} = \frac{1000}{725.55 - 463.03} = 3.81 \text{ kg/s}$$

The geometric volume displacement follows then from

$$\dot{V}_{displacement} = \frac{\dot{m}_{ref} v_1}{\eta_{vol\_comp}} = \frac{3.81 \times 0.07046}{0.80} = 0.51 \text{ m}^3/\text{s} = 1835 \text{ m}^3/\text{h}$$

This value is higher than the maximum volume flow shown in Figure 6.15 for semi-hermetic compressors, indicating that several compressors should be used or a larger open-type compressor should be considered.

6.2.7 Multistage Operation

The impact of the condensation temperature has previously been discussed in relation to Figure 6.11. If the condensation temperature is increased while the evaporating temperature is maintained, the following results:

- The heating capacity per circulated kilogram of refrigerant decreases.

- The specific compressor work increases.

- The displaced volume reduces due to the reduced volumetric efficiency, which is a function of the pressure ratio.

- The mass flow reduces because the displacement volume reduces while the specific volume at the compressor inlet remains unchanged.

- The discharge temperature increases.

Under extreme conditions, the heating capacity of the system equals the compressor energy input, so that the COP of the heat pump reduces to 1.0. Long before such conditions are attained, the technical and economic limits have already been attained. These limits are illustrated in Figure 6.18.

The first technical limit encountered is the compressor discharge temperature – conditions of line 1-2 in Figure 6.18. Most heat pump compressors are oil lubricated, and the refrigerant in the compressor contains oil. At temperatures between 160°C and 200°C,

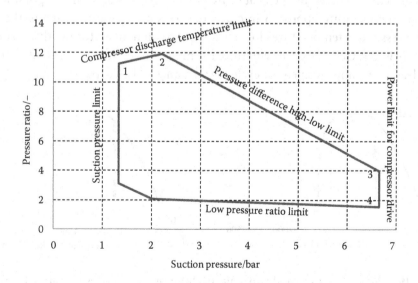

FIGURE 6.18 Limits of operation for a positive-displacement compressor.

the oil will chemically change and will contaminate the refrigerant. The oil will not burn because there is no oxygen present in the system. Some refrigerants decompose at high temperatures. Specifically, ammonia, in the presence of some metals, is sensitive to decomposed oil contamination. Also, the properties of the material of the valves of piston compressors may change, making them more susceptible to failure. Rotary compressors will not experience this problem. A practical limit for the discharge temperature is 130°C to 150°C.

The second technical limit is the pressure difference between above and below the piston in piston compressors or between inlet and outlet of the screw compressor in positive-displacement compressors (conditions in line 2-3 in Figure 6.18). The mechanism is designed for a certain maximum operating pressure difference. In centrifugal machines, 'surge' will occur for certain values of the pressure difference across the compressor.

The third technical limit is the maximum power that the crank mechanism can handle (conditions in line 3-4 in Figure 6.18). The right-hand side of Figure 6.11 shows that when the suction pressure is high (10°C evaporation) and the pressure ratio increases (condensation temperature increases), the required shaft power significantly increases so that, at some point, the power limit for the crank mechanism is attained.

The economic limit is normally attained previously to the technical limits. The cycle efficiency significantly reduces with an increase of the pressure ratio. When one of the previous limits is attained, then two-stage operation is required. Figure 6.19 illustrates this in a ln p-h diagram.

Comparison of the left (single-stage) and right (two-stage) sides of Figure 6.19 shows that the specific heating power is reduced, that the discharge temperature of the compressor is significantly lower and that the combined work of the two compressors w_{comp_L} + w_{comp_H} can be smaller than the work of the single compressor w_{comp}. The schematics of the process shown in Figure 6.19 are given in Figure 6.20. The discharge vapor of the first compressor stage can be cooled by the heat source that delivers heat to the evaporator. Taking a minimum-temperature approach into account, this can bring the inlet of the second-stage compressor to a temperature close to the saturation temperature so that its discharge temperature is low. Generally, this will solve the high-temperature problem of the cycle but will not lead to an improvement of cycle performance. Also, expansion takes place from

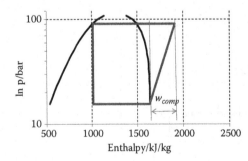
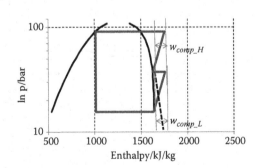

FIGURE 6.19 Single-stage (left) and two-stage (right) heat pump cycles operating between 40°C (evaporation) and 120°C (condensation).

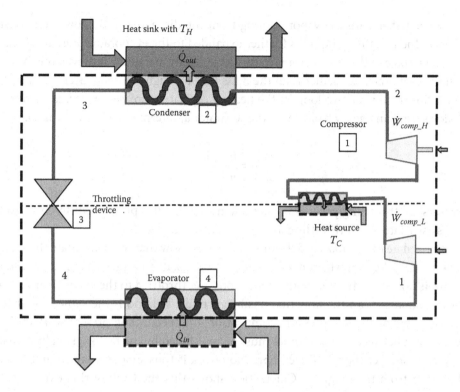

FIGURE 6.20 Schematics of a two-stage heat pump cycle with interstage cooling with heat source stream.

high to low pressure so that the expansion losses are large. The heat available to be delivered to the process flow is even decreased so that the COP of the cycle even reduces from 3.26 (single stage) to 3.00 (two stage) for the example operating conditions of Figure 6.19.

A two-stage cycle with a higher COP is shown in Figure 6.21. Expansion takes place in two steps. The high-pressure liquid from the condenser is expanded to the intermediate pressure. The flash gas formed during the expansion process is directly removed by the high-pressure

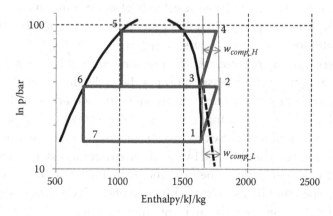

FIGURE 6.21 Two-stage heat pump cycle with interstage flash cooler and expansion from two pressure levels in a ln p-h diagram.

compressor together with the vapor coming from the discharge of the low-pressure stage compressor. The remaining liquid is further expanded to the evaporator pressure level. The specific heat uptake in the evaporator is in this cycle clearly increased; however, the mass flow through the evaporator is smaller than the mass flow through the condenser. It depends on the flow ratios within the two loops if the heating capacity increases or decreases. The flow ratio follows from an energy balance for the separator at intermediate pressure level.

$$\frac{\dot{m}_L}{\dot{m}_H} = \frac{h_3 - h_5}{h_2 - h_6} \tag{6.24}$$

The irreversible throttling process is here taken over smaller pressure differences so that the losses associated with throttling are significantly reduced.

The required heating capacity determines the mass flow that must circulate through the top cycle so that 1 kg of refrigerant circulates between states 4 and 5. In the separator, part of the flow is taken at saturated liquid conditions and throttled to the evaporator pressure. This flow is smaller than the flow that has been condensed and is the same flow that has to be compressed by the low-pressure compressor. The specific compressor work indicated by the diagram between states 1 and 2 must be multiplied by the factor given by Equation 6.24, which is smaller than 1. The compression work is thus smaller than in first instance could be read from the diagram. Calculation shows that the COP of this two-stage heat pump cycle is higher (3.78) than for the corresponding single-stage heat pump cycle (3.26) for the conditions of Figure 6.19 and Figure 6.21, a gain of 16% while the compressor discharge temperature is significantly reduced.

Figure 6.22 illustrates the schematics of the two-stage heat pump cycle shown in Figure 6.21. The hot discharge vapor from the low-pressure compressor is injected under the liquid level in the separator and so mixed with the liquid refrigerant that the vapor is brought to practically saturated conditions in the liquid-vapor separator. In this case, the suction vapor of the high-pressure compressor is practically at saturation conditions. In this cycle, the throttling of high-pressure liquid into the separator is liquid level controlled.

The interstage pressure level must be selected to attain minimum energy consumption by the cycle. This optimum pressure level is generally close to a pressure that creates the same pressure ratio for both compressors: $p_H/p_{int} = p_{int}/p_C = \sqrt{p_H/p_C}$. In the logarithmic scale of the ln p-h diagram, this intermediate pressure lies exactly halfway between p_H and p_C. When optimizing the energy consumption of the cycle, this is the starting value, and slightly higher and slightly lower intermediate pressures should be evaluated to identify the optimum value. The energy savings obtained by executing the heat pump cycle in two stages should be compared with the extra investments required for two-stage operation.

Two-stage operation can be obtained by installing two compressors. Piston compressors can include the two stages in the same machine. Also, for screw compressors a side port makes two-stage operation in a single machine possible. If two compressors are used for the two stages, then the compressors can be of different types.

Larger temperature differences between heat source and heat sink can also be obtained making use of two cascaded heat pump cycles. In this case, it is possible to select the most

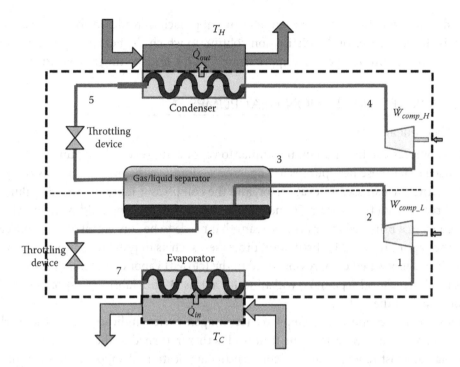

FIGURE 6.22 Schematics of two-stage heat pump cycle with interstage flash cooler and expansion from two pressure levels.

appropriate refrigerant for each of the temperature levels. The problem of the cascaded cycles is that an additional temperature driving force is required between the condenser of the low-temperature cascade cycle and the evaporator of the high-temperature cascade cycle.

When designing a heat pump, some parameters must be selected: For instance, what is the minimum-temperature approach in the evaporator and condenser, and what is the allowable pressure drop in these components? This means that the size of these components must be determined. Larger equipment requires larger investment but normally saves energy. Optimization of both effects is required by considering several alternative solutions. Table 6.1 gives an example of how the payback period (PBP) of a specific heat

TABLE 6.1   Example of Optimization of the Minimum-Temperature Approach of the Evaporator of a Vapor Compression Heat Pump (Condenser of the Distillation Column) Integrated with a Specific Distillation Heat Pump, Zhao (2011)

$\Delta T_{reboiler}$ (K)	$\Delta T_{condenser}$ (K)	p_{evap} (bar)	$\dot{W}_{comp}$ (kW)	Reboiler Cost (k€)	Condenser Cost (k€)	Compressor Cost (k€)	Installation Cost (k€)	Electricity Cost (k€)	PBP (years)
7	3	15.7	2.32	933.6	1647.7	2269.4	8037.4	1212.5	2.71
7	4	15.3	2.42	933.6	1306.0	2358.4	7618.9	1261.6	2.60
7	5	14.8	2.51	933.6	1090.5	2448.1	7410.3	1310.6	2.55
7	6	14.4	2.61	933.6	941.0	2536.2	7312.0	1359.7	2.55
7	7	14.0	2.70	933.6	830.6	2629.8	7280.8	1409.3	2.56
7	8	13.6	2.79	933.6	745.4	2721.2	7291.1	1458.3	2.59
7	9	13.2	2.89	933.6	677.5	2814.1	7332.4	1507.9	2.64
7	10	12.9	2.96	933.6	621.9	2907.6	7395.3	1557.5	2.69

pump depends on the minimum-temperature approach considered for the evaporator (which is the condenser of the distillation column to which the heat pump is applied). In this case, a minimum-temperature approach of 5 to 7 K leads to the shortest PBP.

6.3 VAPOR RECOMPRESSION HEAT PUMPS

6.3.1 Theoretical Cycle

Vapor recompression heat pumps are similar to vapor compression heat pumps. The major difference is that these heat pumps are open cycles and use the process fluid as the working fluid. The cycle misses an evaporator, and the compressor takes the working fluid as a vapor directly from the process. In most applications, the working fluid is water, which is removed from a brine, whey or other stream that needs to be concentrated. In some cases, this is the vapor produced in distillation processes, such as in propylene-propane splitters. Often, the oil-lubricated compressors used in the (closed) vapor compression systems cannot be used because oil separators will allow for at least 3-ppm lubricant vapor passing into the condenser of the system.

Vapor recompression heat pumps are (semi)open cycles, indicating that the working fluid only passes the heat pump one time and is then returned to the process from which it originated. Most reported studies concern the application of vapor recompression systems for waste-water treatment or seawater desalination (Pinder, 1968; Liang et al., 2013; Zhou et al., 2014; Li et al., 2015) and for heat pump–assisted distillation columns (Brousse et al., 1985; Modla and Lang, 2013; Jana, 2014). A particular case of vapor recompression applied to distillation is the internally heat-integrated distillation column. Instead of using a single-point heat source and sink, the whole rectifying section of a distillation column becomes the heat source, while the stripping part of the distillation column acts as a heat sink – thus providing a higher potential for energy savings (Kiss and Olujic, 2014).

The basic working principle scheme is given in Figure 6.23. In the distillation column in the left-hand side, the top product vapor (state 1) is compressed to a higher pressure (state 2) so that it condenses at the required heating temperature level (state 3). The condenser of the heat pump is simultaneously the reboiler of the column. Part of the condensate is expanded back to column pressure to create the required reflux; the rest is the column distillate product. Because generally there will be no exact match between the heat requirement for the reboiler and the heat released by condensation of the top product, a second (shown on top) water-cooled condenser is also required.

Similarly for the waste-water treatment plant on the right side, the vapor produced in the evaporator (state 1) is compressed to a sufficiently higher pressure level (state 2) so that it condenses in the evaporator at a higher temperature (state 3) and in this way evaporates the liquid in the flash chamber. The distillate can eventually be used to preheat the feed flow of the evaporator, as also illustrated in the same figure.

Because one of the temperature driving forces has been removed (there is no heat exchanger between top product vapor flow and an external medium), this heat pump shows theoretically the largest COP (ratio between heat delivery to the reboiler and compressor driving power requirements). The performance is nevertheless limited by the latent heat of

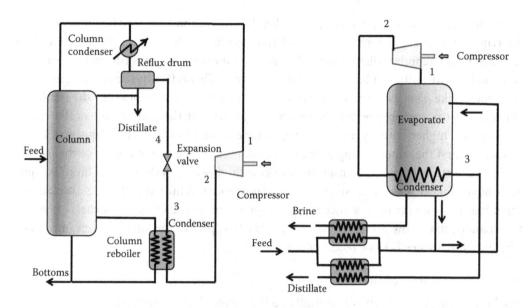

FIGURE 6.23 Vapor recompression heat pump–assisted distillation (left) and vapor recompression heat pump-based waste-water treatment plant (right).

the working fluids and by the compression and expansion losses. These last losses become dominant when a large pressure ratio has to be bridged between column and condensation pressure levels. For distillation columns, this is related to the temperature lift between the saturation temperatures of bottom and top products.

Figure 6.24 illustrates the processes in pressure-enthalpy and temperature-entropy diagrams. The mass flow through the compressor comprises the top product stream and the reflux stream. The reflux flow is obtained by flashing part of the condensate to the pressure level in the top of the column. Because part of this flow flashes directly into vapor, the condensate flow required for the reflux is larger than the reflux plus distillate flow. In the example represented in Figure 6.21, because the outlet vapor fraction is 0.197, the flow is 1.197 times the required total flow.

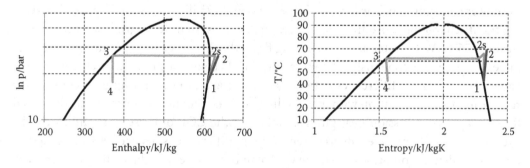

FIGURE 6.24 A ln p-enthalpy (left) and T-s diagram (right) of the vapor recompression open cycle. Line 1-2s applies for reversible compression while line 1-2 indicates a non-reversible compression process.

Vapor of the process fluid at state 1 is taken by the compressor, which compresses it to the required discharge pressure. The discharge pressure is determined by the conditions required in the reboiler (distillation column) or evaporator of the waste-water treatment plant added with sufficient temperature driving force. Theoretically, the compression process would take place isentropically from state 1 to state 2s, but in practice the isentropic efficiency of the compressor will be much smaller so that the compressor outlet conditions show a higher entropy and discharge temperature (state 2 in the diagrams). In the condenser, first the superheating is removed, followed by the condensation of the process fluid. At the outlet of the condenser, the working fluid (which is the process fluid) is liquefied but at a higher pressure at state 3. If the condensate is required at the same pressure as the initial pressure (as for instance to create the column reflux), the condensate needs to be expanded to the low pressure (state 4). In this case, part of the liquid flashes into vapor, which eventually needs to be recompressed.

6.3.2 System Components

The heat pump system itself is essentially built up from three components:

- A compressor that increases the pressure of the top product stream

- A condenser that operates as the reboiler in distillation columns or as the evaporator in waste stream purification flows

- A throttling valve that reduces the pressure back to the process level

The compressor has the same function in this cycle as in the vapor compression cycle, so that what has been discussed for this compressor also applies for vapor recompression systems. Specific for vapor recompression is that the working fluid is the process flow, which generally will have purity requirements. Often, oil-free compressors are required to prevent contamination with the lubricant of the compressor.

The effect of the condensing temperature on the performance of vapor compression heat pumps has been previously discussed. Similar conclusions can be drawn for vapor recompression cycles: for a specific compressor: the higher the condensation, the higher the power consumption. The left-hand side of Figure 6.25 illustrates the impact of increasing the discharge pressure in a vapor recompression system to upgrade a waste flow of steam at 75°C. With water as the operating fluid, the discharge temperature significantly increases with the condensing temperature and becomes a problem even for oil-free compressors. Higher differences between condensing and evaporating temperature will require multistage operation with intercooling between the stages, as previously discussed for the vapor compression heat pumps. Note that the COP of these heat pumps can be extremely high. If only a few degrees temperature difference is needed between evaporator and condenser conditions, then the COP can become very high, leading to extremely short payback times. Shen et al. (2014) have recently proposed the use of water-injected twin-screw compressors for these applications. Although their study has been only theoretical and water-injected screw compressors in practice are only used for compressed air

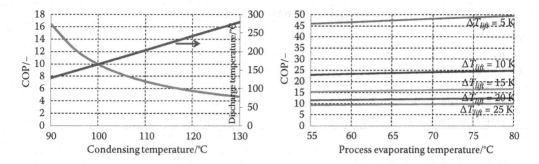

FIGURE 6.25 Left: Effect of condensing temperature on COP and compressor discharge temperature. Suction at 75°C; process fluid is water, and the isentropic efficiency of the compressor is assumed to be 70%; Right: COP of vapor recompression heat pumps with water as working fluid as a function of the suction temperature (process evaporating temperature) and ΔT_{lift} (difference between condensation temperature and suction temperature). The compressor efficiency is assumed to be 70%.

applications at much lower operating temperatures (Prins, 2001), this is an interesting route to exploit.

Because the vapor recompression cycle is open, the throttling losses are normally not a problem and do normally not affect the COP of the cycle. However, if (part of) the product flow needs to be returned to the process at a lower pressure, such as when a reflux is required in distillation columns, the vapor produced during the expansion process needs to be removed again by the compressor so that it then has an impact in the mass flow through the compressor. Subcooling of the liquid condensate before expansion will reduce the amount of vapor produced and so the impact of throttling. When the discharge temperature of the compressor is low, application of combined subcooling and superheating as previously discussed for vapor compression heat pumps will make subcooling of the condensate feasible. If the differential pressure and mass flows are large enough, then most probably hydraulic power recovery turbines could be an interesting option for converting the available energy into electric power.

6.3.3 Operating Conditions and Temperature Lift

The working medium that is used is the process fluid. For distillation columns, the suction pressure of the compressor is determined by the pressure in the top of the column. The discharge pressure is related to the saturation pressure of the top product at a temperature slightly higher than the reboiler temperature. In the example illustrated in the left-hand side of Figure 6.24, the pressure ratio created by the compressor is 1.50. When larger pressure ratios have to be bridged, two or multistages are possible with intercooling between the stages. As discussed in the previous section, the rotating equipment will suffer from high temperatures. Large temperature lifts lead to large pressure ratios, and these will lead to large entropy production during compression and during the throttling process. The system efficiency reduces with the pressure ratio, as illustrated in the left-hand side of Figure 6.25. This figure shows that, for the illustrated conditions, the COP is still 4.7 for a temperature lift (difference between condensing and evaporating temperatures) of 55 K.

TABLE 6.2　Examples of Vapor Recompression Plants (IEA-IEST, 2014)

Application	$T_{evaporation}$ (°C)	$T_{condensation}$ (°C)	Fluid (-)	p_H/p_C (-)
Metallurgical process: copper wire production from liquid copper	60	62	Water	1.03
Alcohol distillation (two stage)	75	88/110	Water	1.7/2.1
Propylene-propane splitter	≈15	≈27	Propylene	1.4
Production of dimethyl-terephthalate	103	118/134	Water	1.6/1.6

6.3.4 Stage of Development

The vapor recompression heat pump has been extensively demonstrated in large-scale plants. This technology has been commercially available for many years, and a number of companies offer this type of heat pump. Both small-scale (food-processing industry) and large-scale (process industry) systems have been demonstrated. Table 6.2 shows some of the example plants reported by the International Energy Agency–Industrial Energy Technology and Systems (IEA-IEST, 2014). Boot et al. (1998) gave a more extensive overview of processes in which mechanical vapor recompression has been successfully applied.

6.3.5 Performance

The right-hand side of Figure 6.25 shows how vapor recompression heat pumps perform depending on suction temperature level and discharge pressure level, which is indicated by the temperature lift. The figure makes clear that these heat pumps can attain extremely high COP values if small temperature differences can be attained over the condenser. The results apply for water as the working fluid, and it should be remarked that large temperature lifts lead to very high discharge temperatures, which might limit the application for such cases. The COP values will be different for different fluids and different operating conditions but will remain high.

6.3.6 Part-Load Operation

Part-load operation can easily be attained by controlling the compressor. Varying the rotational speed is an efficient option, but depending on the compressor type, various alternative control strategies will be available. Changes in operating temperatures and in operating pressures are for some of the compressor types no problem (except for modified power requirements). If centrifugal compressors are applied, a surge control will be required to prevent damage of the system. For instance, Brousse et al. (1985) showed the working principle of this type of control.

6.4 COMPRESSION RESORPTION HEAT PUMPS

6.4.1 Technology Description

Compression resorption heat pumps (CRHPs) are vapor compression heat pumps that work with mixtures such as ammonia-water. Because the cycle combines parts of an absorption heat pump with components from a vapor compression cycle, the cycle is also called a hybrid heat pump. The condenser is now an absorber in which mass transfer plays a major role; the evaporator is now a desorber. Because in absorption heat pump cycles absorption normally takes place at the low-pressure side of the cycle and here the absorption takes place at the high-pressure side of the cycle, the absorber is generally referred to as a resorber. The evaporation in the desorber is incomplete, so that solution recirculation between the desorber and the resorber or wet compression is possible. This is illustrated in Figure 6.26 (left), in which the wet vapor leaving the desorber directly enters the compressor, and in the right-hand side of Figure 6.26 liquid and vapor are separated. The vapor is compressed by the compressor, while the liquid is brought to the same pressure level, making use of a pump. When all the solution is recirculated, the cycle is called the Osenbrück cycle, after the name of its inventor. The name *hybrid wet compression cycle* is used for those cycles for which all the solution is sent to the compressor, avoiding the use of a solution pump.

The use of a mixture allows for lower pressure levels and for condensation and evaporation at gliding temperatures, which can result in higher efficiency. Wet compression has the effect of suppressing vapor superheating. If the technical problems surrounding wet compression are solved, it can also improve the heat pump efficiency. Comparison between Figures 6.1 and 6.26 shows that the process sequence is comparable with the processes encountered in vapor compression heat pumps. The operation of vapor compression cycles that make use of nonazeotropic mixtures as working fluids is similar, but generally these fluids do not show large heat-of-mixing effects such as absorption processes show, and the evaporation process continues until saturated vapor conditions. Figure 6.27 illustrates the difference between a mixture used in compression resorption heat pump cycles and a mixture used in vapor compression cycles with nonazeotropic mixtures. The figure shows, for a pressure of 5 bar, how the liquid enthalpy varies with the concentration: While for ammonia-water the enthalpy of mixing plays a large role for R32-R134a, the enthalpy of mixing is negligible. The application of nonazeotropic mixtures in vapor compression heat pump cycles has extensively been discussed, for instance, by Vorster and Meyer (2000) and Radermacher and Hwang (2005).

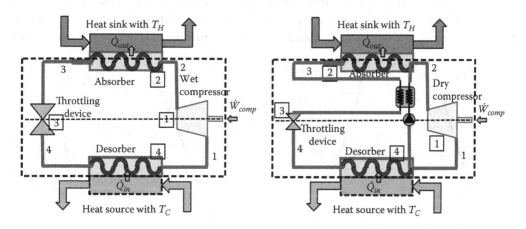

FIGURE 6.26 Compression resorption heat pump: left: wet compression; right: with solution recirculation (Osenbrück cycle).

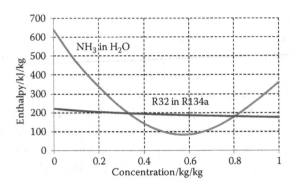

FIGURE 6.27 Saturated liquid enthalpy lines for a compression resorption working fluid (ammonia-water) and for a nonazeotropic working fluid for vapor compression cycles (R32-R134a) at a pressure of 5 bar.

Itard (1995, 1998) and Itard and Machielsen (1994) have demonstrated that next to the temperature-entropy diagram, also the temperature-enthalpy diagram, as the pressure-enthalpy diagram for vapor compression heat pumps, facilitates the quantification of the energy flows within the cycle. Both diagrams are shown in Figure 6.28 for an ammonia-water mixture with 35% ammonia. Because all processes take place with the same average concentration, the different processes can be drawn in these diagrams. In the diagrams, three isobaric lines are also drawn in the two-phase region. Because both absorption and desorption take place at practically constant pressure, these lines indicate the temperature profile of the mixture as the processes progress in these heat exchangers.

The compressor rises the pressure of a nonazeotropic gas-liquid mixture (for instance, ammonia-water) so that its temperature also rises, process 1-2 in Figure 6.26. The resorber, process 2-3 in Figure 6.26, delivers heat to the process stream that needs to be heated. As the gas is absorbed, absorption heat is released. During the resorption process (high-pressure side of the heat pump), the temperature changes are as indicated in Figure 6.29. The concentration of the ammonia-water mixture is selected so that this temperature glide corresponds to the temperature glide of the process stream that needs to be heated. In this way, the entropy losses during the heat release process are minimized. Figure 6.29 illustrates a

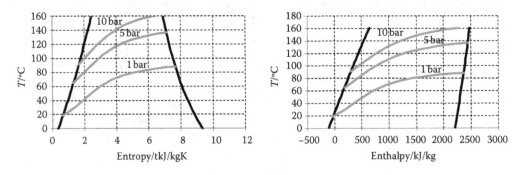

FIGURE 6.28 Temperature-entropy (left) and temperature-enthalpy (right) diagram for ammonia-water solution with 35% ammonia showing isobaric lines in the liquid vapor region.

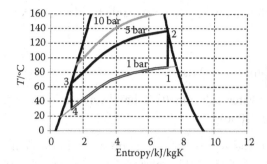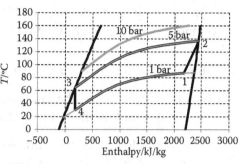

FIGURE 6.29 Wet compression heat pump cycle in temperature-entropy and pressure-enthalpy diagram. Compression process is assumed isentropic; expansion process is isenthalpic.

wet compression process in the temperature-entropy (left) and in the temperature-enthalpy diagram (right) between 1 and 5 bar. The outlet conditions of the compressor have been selected to be exactly at saturated vapor conditions. Van de Bor et al. (2014) have demonstrated that for such operating conditions the maximum COP is attained.

In this case, the COP is 8.3. This can easily be confirmed from Figure 6.29. The power required for the compression process is given by $\dot{W}_{comp} = \dot{m}_{working\_fluid} \times (h_2 - h_1)$ and will not significantly change if the desorption process stops at a lower vapor quality. The heat delivered to the process flow that needs to be heated is given by $\dot{Q}_{out} = \dot{m}_{working\_fluid} \times (h_2 - h_3)$, so that the larger this enthalpy difference is, the larger the COP. At the outlet of the resorber, state 3 in Figure 6.29, all vapor has been absorbed and saturated mixture leaves the resorber. The flow is then throttled to suction pressure level by means of a throttling device, process 3-4 in Figure 6.29. The mixture is then directed into the desorber. In the desorber, process 4-1 in Figure 6.29, the source flow cools as it delivers heat to the heat pump working fluid.

The conditions after the separator in the compression resorption cycle with liquid recirculation (Osenbrück cycle) cannot be shown in a temperature-enthalpy diagram for a constant mixture concentration. After separation, the liquid phase has a much lower ammonia concentration, while the vapor entering the compressor has a much higher ammonia concentration. Representation of these steps will require the use of an enthalpy-concentration diagram for the working fluid as normally used for the processes in absorption heat pumps. Also, in this case some of the advantages of the cycle will be lost because the vapor will be superheated: The compressor discharge temperature significantly increases, and with the superheating, the entropy losses associated with it will reduce the efficiency of the cycle.

6.4.2 System Components

The system is essentially built up from four components: a (wet) compressor, a resorber, a throttling device and a desorber. The compressor has the same function in this cycle as in the vapor compression cycle, so that what has been discussed for the vapor compression cycle also applies for compression resorption heat pumps. Specific for these heat pumps is that the compressor operates in the two-phase regime, which puts specific requirements on the compressor. Hulten and Berntsson (2002) listed the options available for the compressor of compression resorption heat pumps:

- Oil-free (centrifugal or oil-free screw compressors) use;

- Cooling by injection of the ammonia-water solution (this has been exploited by Zaytsev, 2003, and Infante Ferreira et al., 2006);

- Use of an insoluble lubricant (this will lead to the accumulation of an oil layer in the surfaces of the desorber and resorber, reducing their performance, and has been exploited by Brunin, 1995, and Brunin et al., 1997);

- Use of a soluble oil (there is no oil soluble in ammonia-water that can withstand the high operating temperatures encountered in the compressors of these cycles).

Several attempts have been undertaken to come to specific compressor designs. Itard (1998) investigated the application of a liquid ring compressor. Zaytsev (2003) modified a screw compressor to make it suitable for operation in an ammonia-water compression resorption heat pump.

The effect of the condensing temperature on the performance of vapor compression heat pumps has been previously discussed. As can be derived from Figure 6.29, similar conclusions can be drawn for compression resorption cycles: For a specific compressor, the higher the resorption temperature is, the lower the heating capacity and the higher the power consumption will be.

6.4.3 Operating Conditions and Temperature Lift

The working fluid of the heat pump will be one of the mixtures generally applied in liquid sorption heat pump/refrigeration systems. Due to its favorable thermodynamic performance, ammonia-water is often preferred, but other mixtures may perform better for some specific operating conditions. If ammonia-water is considered, different overall ammonia-water concentrations will lead to different pressure levels and to different performance.

The operating conditions are imposed by the process stream that needs to be heated and the temperature driving forces across the heat exchangers. The temperature at the resorber outlet should be slightly above the inlet temperature of the process stream that needs to be heated. From this restriction and the selected overall concentration, the high-pressure level follows. State 4 (Figure 6.29) has the same enthalpy as state 3 and should have a temperature slightly lower than the temperature of the source outlet temperature. The operating low-pressure level is then fixed. State 1, the compressor inlet condition, results from the inlet temperature of the source medium and the temperature difference across the heat exchanger wall. State 2 finally results from the isentropic efficiency of the compressor. The conditions of state 2 impose the maximum temperature that can be reached at the process stream outlet.

Van de Bor et al. (2014) have proposed an equilibrium model for the compression-resorption heat pump to obtain an estimate of the performance of these heat pumps under the conditions imposed by the sink and source streams. Starting with the resorber outlet (assumed to be a saturated liquid) temperature, the resorber pressure and resorber inlet conditions are determined as discussed next.

The resorber outlet temperature is given by a minimum approach temperature added to the inlet temperature of the process stream that needs to be heated:

$$T_3 = T_{process\_stream\_in} + \Delta T_{approach} \tag{6.25}$$

The operating pressure then follows from the saturated liquid condition ($x = 0$):

$$p_{resorber} = f(T_3, x = 0, w_{ave}) \tag{6.26}$$

with w_{ave} the average ammonia-water concentration and x the vapor fraction. Considering that the highest COP is attained when resorption starts at saturated vapor conditions, the temperature and enthalpy at the inlet of the resorber follow from

$$T_2 = f(p_{resorber}, x = 1, w_{ave}) \tag{6.27}$$

$$h_2 = f(p_{resorber}, x = 1, w_{ave}) \tag{6.28}$$

T_2 should be larger than $T_{process\_stream\_out} + \Delta T_{approach}$. If this is not the case, then the minimum-temperature approach should be increased, and the steps from Equation 6.25 should be repeated. The outlet enthalpy of the resorber is obtained from

$$h_3 = f(p_{resorber}, x = 0, w_{ave}) \tag{6.29}$$

As previously discussed, the process in the expansion valve is isenthalpic. Therefore, the inlet enthalpy of the desorber is given by

$$h_4 = h_3 \tag{6.30}$$

If the available temperature glide at the process side of the resorber is larger than the available glide in the process side of the desorber, then the pressure at the desorber inlet is obtained as follows:

$$T_4 = T_{source\_stream\_out} - \Delta T_{approach} \tag{6.31}$$

$$p_{desorber} = f(T_4, h_4, w_{ave}) \tag{6.32}$$

The conditions at the compressor outlet for the theoretical case of 100% isentropic efficiency should be on the same entropy line as the inlet condition.

$$s_2 = f(p_{resorber}, x_2 = 1, w_{ave}) \tag{6.33}$$

This is used as a start condition to determine the conditions at the desorber outlet.

$$h_{1\_init} = f(p_{desorber}, s_2, w_{ave})\qquad(6.34)$$

Iterations are carried out for the entropy at the compressor inlet s_1 until the values for the isentropic enthalpy obtained from the two following calculations converge:

$$h_1 = f(p_{desorber}, s_1, w_{ave})\qquad(6.35)$$

$$h_1 = \frac{h_{2s} - \eta_{is}h_2}{1 - \eta_{is}}\qquad(6.36)$$

with

$$h_{2s} = f(p_{resorber}, s_1, w_{ave})\qquad(6.37)$$

and η_{is} is the compressor isentropic efficiency, which in the first instance could be taken as 70%. The corresponding desorber outlet temperature is

$$T_1 = f(p_{desorber}, h_1, w_{ave})\qquad(6.38)$$

Alternatively, when the available glide for the desorber is larger than for the resorber, the desorber pressure is determined as a function of the desorber outlet temperature and (assumed) entropy.

$$T_1 = T_{source\_stream\_in} - \Delta T_{approach}\qquad(6.39)$$

$$s_{1\_init} = f(p_{resorber}, s_2, w_{ave})\qquad(6.40)$$

so that

$$h_{1\_init} = f(T_1, s_{1\_init}, w_{ave})\qquad(6.41)$$

The entropy is iterated until the same enthalpy at the desorber outlet is obtained from Equations 6.36 and 6.41. The desorber pressure follows then from

$$p_{desorber} = f(T_1, h_1, w_{ave})\qquad(6.42)$$

If this temperature is higher than the process temperature minus the required temperature driving force, the temperature driving force at the inlet of the desorber should be increased.

6.4.4 Stage of Development

According to Nordtvedt (2005), only a few experimental or demonstration compression resorption heat pumps have been reported that operated with wet compressors. These studies are summarized in Table 6.3.

Alternatively, dry compression combined with liquid recirculation can be considered. This variant of the cycle has been successfully applied in a few projects by Hybrid Energy, a start-up company from the Institute for Energy Technology (Norway) that has been active since 2004. Table 6.4 gives a summary of projects realized during the period 2011–2015.

Figure 6.30 illustrates some results reported by van de Bor et al. (2015), and it shows the expected compressor discharge temperatures and makes visible the disadvantage of the

TABLE 6.3   Reported Experimental Studies in Which Wet Compression Has Been Applied with Ammonia-Water as the Working Fluid

Reference	Compressor Type Applied	$\dot{Q}_{out}$ (kW)	$T_{application}$ (°C)	T_{source} (°C)
Malewski (1987–1988)	Screw	500	60–80	35
Bergmann and Hivessy (1990)	Screw	1000	15–85	25–5
Torstensson and Nowacki (1990)	Scroll	1.4	35–60	16–3
Itard (1998)	Liquid-ring	13	40–53	44–38
Zaytsev (2003)	Screw	19	76–92	70–65

TABLE 6.4   Reported Hybrid Energy Compression Resorption Heat Pump Systems in Which Dry Compression with Liquid Recirculation Has Been Applied with Ammonia-Water as the Working Fluid

Application	Compressor Type Applied	$\dot{Q}_{out}$ (kW)	$T_{application}$ (°C)	T_{source} (°C)
Dairy	Piston	350	55–85	50–15
Dairy	Piston	1200	55–85	45–22
Slaughterhouse	Piston	650	50–87	49–40
Slaughterhouse	Piston	500	15–86	25–17
Sludge heating biogas	Piston	1100	60–72	40–20

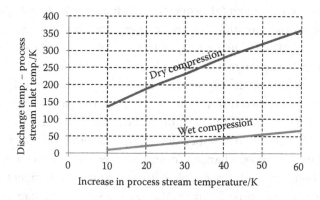

FIGURE 6.30   Compressor outlet temperatures for dry and wet compression resorption cycles.

dry compression solution: State 2 can reach very high temperatures, up to 400°C, depending on the operating conditions. Such high temperatures deliver lubrication problems as the refrigerant will start decomposing. It is evident that, under dry operating conditions, the heat pumps cannot operate under the imposed operating conditions. As already mentioned, an additional problem of CRHP systems that compress dry vapor and make use of oil-lubricated compressors is that some oil vapor will be transported to the heat exchangers, and it will accumulate at the heat exchangers' surface, reducing the performance of the heat exchangers with time. It should be noted that these problems do not exist under wet compression conditions. Van de Bor et al. (2015) showed that the COP of the wet CRHP is generally higher in comparison to that of the dry compression, but that the level of this advantage strongly depends on the operating conditions.

6.5 TRANSCRITICAL VAPOR COMPRESSION HEAT PUMPS

6.5.1 Technology Description

Transcritical heat pumps (TCHPs) are practically identical to vapor compression heat pumps. The only difference is that the high-pressure side of these heat pumps operates in the supercritical region of the working fluid, so that the cycle does not have a condenser but instead uses a dense gas heat exchanger to reject heat to the application. The consequence of operating in the supercritical region is that the process does not take place at constant temperature and that a large temperature glide applies. In recent years, transcritical CO_2 heat pumps have become popular because CO_2 – as a refrigerant – has low environmental impact, and heat rejection in this cycle takes place with a significant temperature glide. A few review articles have been published in which the last developments have been reported (Austin and Sumathy, 2011; Sarkar, 2012; Ma et al., 2013). The critical temperature (31°C) of CO_2 makes the heat pump suitable for space heating and hot water production. Process stream heating will generally make use of sources with temperatures that are higher than the critical temperature of CO_2, so that a transcritical CO_2 cycle cannot be applied. Sarkar et al. (2007) investigated alternatives to CO_2 as working fluid for transcritical cycles, but up to now CO_2 is the only fluid used. CO_2 has a number of properties that make it suitable as a working fluid for heat pumps, such as the high latent heat of vaporization, good transport properties and good environmental properties, such as a low global warming potential and zero ozone-depletion potential. The low temperature associated with the critical pressure of CO_2 implies that when using CO_2 as refrigerant, a TCHP cycle is needed for the condenser/gas cooler to exchange heat with a heat sink. A transcritical vapor compression heat pump consists at least of an evaporator, a compressor, a dense gas heat exchanger and an expansion valve. A temperature profile exists along the gas heat exchanger; however, due to the supercritical nature of this part of the heat pump, no phase change takes place. The evaporator, where the fluid is subcritical, has zero temperature glide, and the fluid undergoes a liquid/vapor phase change. Transcritical vapor compression heat pumps can achieve relatively high lifts with reasonable efficiency as long as temperature glides match the glides of the source and sink. Figure 6.31a shows a schematic diagram of a transcritical CO_2 cycle. To operate a transcritical CO_2 cycle, a large pressure difference between

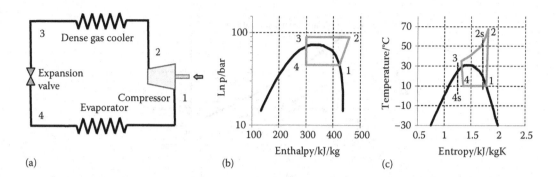

FIGURE 6.31 Schematic diagram of transcritical CO_2 (a) and corresponding ln p-h diagram (b) and T-s diagram (c).

evaporating pressure and heat rejection pressure has to be overcome, so that the compression work of such a cycle is typically large.

Furthermore, expansion losses that occur during the isenthalpic expansion process in a transcritical CO_2 vapor compression heat pump cycle (TCHP) can make up to 40% of the total losses in the system (Yang et al., 2005). In the T-s diagram shown in Figure 6.31c, the throttling losses can be readily identified. An ideal expansion process would be an isentropic one, from state 3 to state 4s. The isenthalpic throttling process occurs from state 3 to state 4, increasing the entropy and thus creating irreversibilities. Due to the large compression work and throttling losses, the COP of transcritical CO_2 systems is generally low. In recent years, efforts have increased to find technological innovations to improve the energetic efficiency of CO_2 cycles.

6.5.2 Optimal Heat Rejection Pressure in Transcritical Cycles

As mentioned by Ma et al. (2013), in traditional subcritical cycles, the enthalpy value at the condenser outlet is only a function of temperature (or pressure). If the evaporation and condensation temperatures are given, then the performance of the cycle is basically fixed. In the supercritical region, temperature and pressure are independent variables. When the outlet temperature of the dense gas cooler is constant, the enthalpy of the working fluid still changes with the discharge pressure of the system. The higher the pressure, the steeper are the isotherms. Therefore, the heating capacity increases with an increase of the pressure. At the same time, the compressor work increases as the discharge pressure increases. So, when other parameters are constant, as the discharge pressure is increased there is a maximum value for the heat pump COP. The corresponding discharge pressure is called the optimal high pressure.

Liao et al. (2000) derived a correlation for optimal discharge pressure based on evaporator temperature and gas cooler outlet temperature. Later, Sarkar et al. (2004) also developed correlations for optimal pressure, system COP and optimal gas cooler inlet temperature based on the gas cooler outlet temperature and evaporator temperature. Cecchinato et al. (2010), however, have shown that when the outlet temperature of the process stream that needs heating is the controlled variable, Liao and Sarkar's prediction methods can lead to significant deviations from the optimum (up to 30%). Figure 6.32 has been adapted from

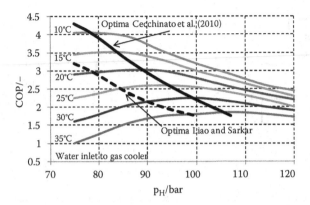

FIGURE 6.32 Transcritical heat pump efficiency as a function of the cycle high pressure and water inlet temperature to gas cooler.

Cecchinato et al. (2010) and illustrates, for a specific dense gas cooler, that fixing the cycle outlet conditions instead of the outlet conditions of the external fluid leads to different optima.

6.5.3 Ways to Improve the Performance of Transcritical Cycles

One of the technologies with potential for CO_2 heat pump system improvement is expansion work recovery. Two lines of research have been developed in recent years: the use of expanders and the use of ejectors. To improve the COP of a TCHP, the expansion valve can be replaced by an ejector or expander to recover expansion work, which reduces both compressor work and throttling losses. In addition, an ejector has the advantages of simplicity and availability compared to an expander.

An ejector typically consists of four main components: a nozzle for the primary or motive fluid, a suction chamber for the secondary fluid, a mixing chamber and a diffuser. In Figure 6.33, the refrigerant flow, pressure and velocity profile inside an ejector are shown. Its operation is based on the Venturi effect. A low-pressure zone is created at the exit of the primary nozzle by accelerating and expanding a high-pressure primary fluid to supersonic conditions from point 3 to point 4, thereby drawing in and entraining another (secondary) fluid from point 10 to point 5. The two fluid flows become mixed in a mixing section, making use of the momentum of the primary fluid. Finally, the mixed fluid flow is discharged at an elevated pressure using a diffuser from point 6 to point 7. The sudden increase in pressure in the constant area mixing section indicates the presence of a shock wave where the refrigerant flow decreases from supersonic velocity to subsonic velocity.

Figure 6.34 shows a schematic diagram of a transcritical ejector expansion cycle and its corresponding *p-h* diagram. The ejector expansion heat pump cycle consists of two loops: a primary and a secondary loop. The primary loop is circulated by a compressor through a gas cooler (or condenser if subcritical), an ejector and a separator (states 1, 2, 3, 4, 6 and 7), whereas the secondary loop circulates through the expansion valve, evaporator, ejector and separator (states 8, 9, 10, 5, 6 and 7). The primary and secondary loops come together at the mixing section and diffuser of the ejector (states 6 and 7).

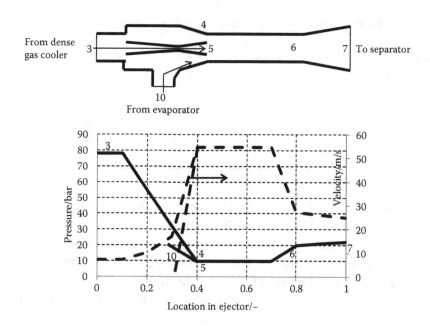

FIGURE 6.33 Refrigerant flow, pressure and velocity profile inside an ejector.

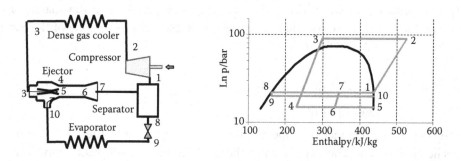

FIGURE 6.34 Schematic diagram of transcritical ejector–expansion heat pump cycle and its corresponding *p-h* diagram.

The use of an ejector provides two benefits: work recovery, which is related to COP improvement, and flash gas bypass, which is related to evaporator size reduction (Sarkar, 2012). The COP improvement stems from the fact that the compression stage in the ejector expansion heat pump cycle is smaller than in the TCHP cycle, from state 1 to point 2 instead of from point 10 to 2a (not indicated in the figure but slightly to the right of state 2). The flash gas bypass is the result of the separation of the two-phase mixture into a vapor flow and a liquid flow in the separator (from state 7 to states 1 and 8). Therefore, the specific enthalpy that can be utilized for evaporation in the ejector expansion heat pump cycle is larger than that of a TCHP cycle, from state 9 to state 10 instead of from a state with the enthalpy of state 3 but suction pressure to state 10.

Weerstra (2014) compared experimentally the two cycles and reported that the ejector expansion heat pump cycle generally showed a COP improvement over the conventional

TABLE 6.5   Examples of CO_2 Transcritical Heat Pump Plants (IEA-IEST, 2014)

Application	Compressor Type Applied	$\dot{Q}_{out}$ (kW)	$T_{application}$ (°C)	T_{source} (°C)
Slaughterhouse (add-on to refrigeration plant)	Piston	800	30–90	20–14
Cafeteria	Piston	53	20–80	12–6
Noodle production	Piston	72	20–90	10–5
Air drying process	Piston	110	80–120	35–9
Chicken product manufacturing	Piston	80	20–65	20–14

single-stage transcritical vapor compression cycle operating between the same external conditions. This improvement could be as large as 30% but generally lower. Compared to a two-stage compression system, the ejector expansion heat pump cycle shows comparable performance if it is optimally configured. Weerstra (2014) further reported that the performance of the ejector, and thus the performance of the complete system, is highly dependent on the operating conditions at the outlet of the gas cooler (i.e., the inlet conditions of the motive nozzle of the ejector).

6.5.4 Stage of Development

Carbon dioxide TCHPs have been extensively demonstrated in recent years in small scale, mainly food-processing plants. This technology has been commercially available for only a few years. Most plants are less than 10 years old. Table 6.5 shows some of the example plants reported by IEA-IEST (2014).

6.6 STIRLING HEAT PUMPS

6.6.1 Technology Description

A Stirling machine is a device that operates on a closed regenerative thermodynamic cycle, with cyclic compression and expansion of the working fluid at different temperature levels. The flow is controlled by volume changes, so that there is a net conversion of heat to work or vice versa. The first Stirling machine was patented in 1816 by Robert Stirling and was used throughout the nineteenth century. After this period, the internal combustion engines became a better and cheaper alternative for the Stirling machine, reducing Stirling machine use. It was not until the late 1930s, that Philips Electric Company in Eindhoven, the Netherlands, began new research that became the basis of all present interest in Stirling engines. This interest is mainly triggered by the favorable characteristics of the Stirling cycle:

- Working fluids that have no environmental impact

- Silent and vibration-less operation (Organ, 2014, reported that in fact this is only partially true)

- High efficiencies compared to other heat engines (in practice, the efficiency of the realized machines is not significantly higher than from alternative cycles)

- Mechanical simplicity

A Stirling heat pump is a Stirling engine that works in reverse mode. The working principle of a Stirling engine has been discussed by, among others, Walker et al. (1982). The working principle as discussed by these authors is here converted to the working principle of a heat pump. Figure 6.35 is a cross section of an idealized Stirling heat pump. It consists of a cylinder containing a displacer and a piston, which are independently driven with the same crank shaft.

The space above the displacer, called the expansion space, is maintained at a temperature close to the heat source temperature. The space between the displacer and the piston is called the compression space and is maintained at a temperature close to the temperature of the process stream that needs to be heated. The volume of the two spaces varies cyclically with the motion of the piston and displacer, as shown in Figure 6.36. The

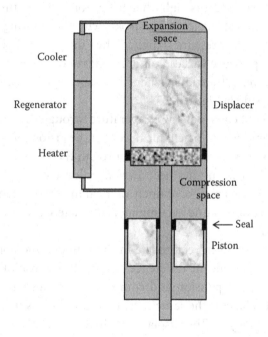

FIGURE 6.35 Schematic of a Stirling heat pump.

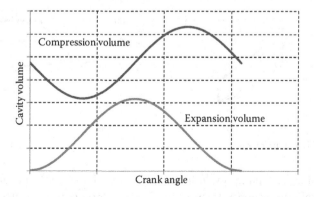

FIGURE 6.36 Volume variation with crank shaft rotation in a Stirling heat pump.

volume variations of the expansion space lead those of the compression space by about a quarter of a cycle (90° of crank rotation). The two spaces are coupled by a duct, which includes the heat exchangers of the cycle so that the pressure is approximately the same in the two spaces.

As the volumes of the spaces vary, some of the working fluid passes from one space to the other. The connecting duct contains three heat exchangers: the heat delivery heat exchanger (heater), the regenerator and the heat uptake heat exchanger (cooler). The heater is a heat exchanger with small flow passage areas for the working fluid, while the process stream that needs to be heated flows through fine tubes. The cooler is similarly equipped with small flow passage areas for the working fluid, while the heat source medium flows through small-diameter tubes. In operation, the working fluid is concentrated in the cold expansion space, and the pressure is high when the piston is rising (the compression stroke). Conversely, the displacer is at the top of its stroke, with the working fluid concentrated in the compression space so the pressure is higher when the piston is descending (the expansion stroke). The cyclic pressure variation is approximately sinusoidal and so phased with the piston motion to develop the kidney-shaped pressure-volume or indicated diagram, as shown in the left-hand side of Figure 6.37.

There is a net transfer of energy at low temperature through the cooler to the heat pump. Part of the energy is converted to mechanical work at the piston for useful application, and the remainder is rejected from the heat pump through the heater at application temperature level. The regenerator is another heat exchanger located between the heater and the cooler. It consists of a porous matrix of finely divided material, often wire mesh, and may be thought of as a thermodynamic accumulator alternately accepting and rejecting heat from the working fluid.

The displacer and piston both reciprocate in the cylinder and appear to have much in common, but there are significant differences to justify the use of separate names. The open duct connecting the expansion and compression spaces ensures that the pressure above and below the displacer is the same except for minor pressure losses in the regenerator and other heat exchangers. The displacer is therefore essentially a no-work element,

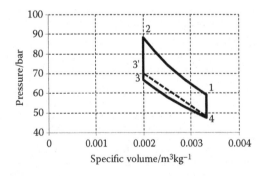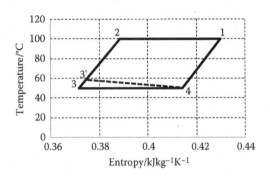

FIGURE 6.37 Stirling heat pump cycle in the pressure-volume (left) and in the temperature-entropy diagram (right) for xenon as the working fluid. The heat pump operates between 50°C and 100°C.

simply displacing fluid from the compression to expansion spaces or vice versa. The displacer seal is subject only to a small pressure difference, and some leakage across the seal is tolerable, although for proper operation of the heat pump, the main flow of working fluid must always be directed through the heat exchangers.

Although the displacer has no significant pressure difference, it experiences the full temperature difference between T_{source} at the top side and T_{sink} at the bottom side. To minimize thermal conduction, the displacer is therefore a long, thin-wall element of relatively low structural strength and light in weight. The piston sustains the full pressure difference between the working space pressure on the top face and the crankcase pressure on the bottom side. This pressure difference across the piston is substantial and varies cyclically; therefore, the piston must be a strong structural element. At the same time, there is no great temperature difference across the piston, so that it can be designed without regard to thermal conduction between the upper and lower faces. In simple terms, a displacer has 'high ΔT, zero Δp', whereas a piston has 'high Δp, zero ΔT'.

6.6.2 Basic Thermodynamics

As example, an ideal Stirling heat pump cycle using xenon as working fluid is shown in Figure 6.37 assuming an ideal cycle behavior on the pressure-volume (left) and on the temperature-entropy (right) diagrams: cycle 1-2-3-4. With reference to Figure 6.37, starting from state 1 (all fluid in the hot compression space at its maximum volume and temperature), the compression piston moves upward to state 2, reducing the volume to a fraction of its starting value, while the temperature is kept constant by transferring heat to the process stream that needs to be heated. Displacer and compression pistons then move simultaneously downward: The working fluid emerges into the expansion space after having been cooled to state 3 (minimum temperature) by the regenerator porous matrix. The displacer moves further downward to state 4, while temperature is kept constant by introducing low-grade heat from the heat source. Now, both the piston and displacer move simultaneously upwards, causing the fluid to absorb heat from the regenerator at constant volume and thus emerging in the compression space at the maximum temperature and closing the cycle.

The compression ratio v_1/v_2 is 1.67, and the temperature ratio T_1/T_4 is 1.15. Useful heat Q_{out} is equal to compression work, while the ratio of compression W_{comp} to expansion work W_{exp} coincides with the temperature ratio T_1/T_4.

$$Q_{out} = mRT_H \ln\frac{v_1}{v_2} \tag{6.43}$$

with Q_{out} the useful heat, m the mass of working fluid in the system and T_H the heat delivery temperature ($= T_1 = T_2$). The compression work is obtained from

$$W_{comp} = mRT_H \ln\frac{v_1}{v_2} \tag{6.44}$$

While the expansion work can be obtained from

$$W_{exp} = mRT_C \ln \frac{v_3}{v_4} \tag{6.45}$$

With T_C the heat delivery temperature (= $T_3 = T_4$). Because $v_1 = v_4$ and $v_2 = v_3$,

$$\frac{W_{comp}}{W_{exp}} = \frac{T_H}{T_C} \tag{6.46}$$

The COP of the ideal Stirling heat pump follows from

$$COP = \frac{mRT_H \ln \dfrac{v_1}{v_2}}{mRT_H \ln \dfrac{v_1}{v_2} + mRT_C \ln \dfrac{v_3}{v_4}} = \frac{T_H}{T_H - T_C} \tag{6.47}$$

which is exactly the COP of a Carnot cycle: Assuming reversible work exchange, the Stirling cycle is strictly ideal (equivalent to a Carnot heat pump). According to this equation, the heat pump COP is independent of the working fluid employed to operate the system. The power is fixed solely by the temperature of the sink and source between which the system is placed. What happens between the hot and cold reservoirs does not influence the heat capacity of the cycle.

This simple thermodynamic model used to determine the dynamic stability of the Stirling cycle is not able to simulate the characteristics of the heat exchangers. To generate a more comprehensive simulation model, an extended thermodynamic dynamic model needs to be used. Martini (1983) made a categorization with five distinct methods for Stirling cycle analysis: a zeroth-order (rough estimation), first-order (approximation), decoupled (second-order), nodal (third-order) and multi–first-order analysis.

Around 1960, Schmidt developed an analysis of the actual Stirling machine (Walker et al., 1994). It included swept volumes of expansion and compression spaces, variable phase angles and clearance volume in the working space. The results from this analysis follow the trends indicated by experimental data; however, the thermal efficiency is close to the Carnot value, which is usually a factor 2 to 3 higher than the real efficiency. This efficiency trend is mainly due to the fact that the Schmidt theory assumes isothermal processes (Figure 6.38). The regeneration process is assumed to be perfect and friction losses are neglected. Despite the optimistic prediction of efficiency the Schmidt cycle is a useful analytical tool providing a fair degree of realism about the operation of Stirling machines. A correction factor is usually applied to account for the losses.

The isothermal assumption is that the gas in the expansion space is at the source temperature (cooler temperature) and the gas in the compression space is at the heater temperature (heat sink temperature). This assumption is illustrated in Figure 6.38. It can clearly be seen that in the compression space/heater and the cooler/expansion space, the

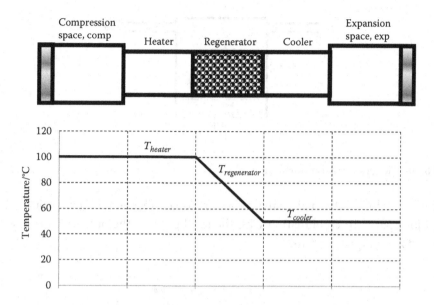

FIGURE 6.38 Ideal isothermal model. The heat pump operates between 50°C and 100°C.

temperature is kept constant. The temperature varies linearly over the regenerator (Urieli and Berchowitz, 1984). These assumptions imply that there are no temperature driving forces in the heat exchangers. In this model, the Stirling heat pump consists of only five components. This means that in this model the remaining spaces (ducts, etc.) are included in the spaces of the five components. The isothermal assumption is used to obtain a simple relation for the working gas pressure. The pressure is a function of volume variations. By assuming that the volumes of the working spaces vary sinusoidally, a solution is easy to obtain. According to Urieli and Berchowitz (1984), the sinusoidal behavior of the volumes is a good approximation of the real volume variation in Stirling machines.

In the isothermal model, some specific assumptions are made:

- Each component is homogeneous, with the working gas in it represented by its instantaneous mass m, temperature T, pressure p and volume V.

- Pressure is constant in the whole system.

- No leakages of mass are taken into account: Mass of the working gas is constant.

- Ideal gas law applies.

- The heat pump frequency is constant.

- Cyclic steady state is obtained.

- Kinetic and potential energies of the working gas are neglected.

To find the set of equations that describe the system, a generalized component model can be used. This model is shown in Figure 6.39. If this model is used for all the five

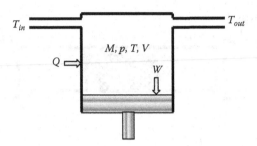

FIGURE 6.39 Generalized component model.

components in this isothermal analysis (heater, cooler, regenerator, expansion space and compression space), in combination with the ideal gas law, the pressure variation can be determined from Equation 6.48.

$$p = m \times R \left(\frac{V_{comp}}{T_{comp}} + \frac{V_{heater}}{T_{heater}} + \frac{V_{regenerator} \times \ln\left(\frac{T_{heater}}{T_{cooler}}\right)}{T_{heater} - T_{cooler}} + \frac{V_{cooler}}{T_{cooler}} + \frac{V_{exp}}{T_{exp}} \right)^{-1} \quad (6.48)$$

The set of equations of the isothermal analysis that results is as follows:

Compression work:

$$Q_{comp} = W_{comp} = \oint p \frac{dV_{comp}}{d\theta} d\theta \quad (6.49)$$

with θ the crank angle.

Expansion work:

$$Q_{exp} = W_{exp} = \oint p \frac{dV_{exp}}{d\theta} d\theta \quad (6.50)$$

The net work input is:

$$W_{Stirling\_cycle} = W_{comp} + W_{exp} \quad (6.51)$$

and the COP:

$$COP = \frac{Q_{comp}}{W_{Stirling\_cycle}} \quad (6.52)$$

Schmidt found closed-form solutions to this set of equations for the special case of sinusoidal volume variations.

6.6.2.1 Second-Order Analysis

The second-order analysis is derived from the Schmidt analysis and includes several losses (Figure 6.40). It assumes adiabatic expansion and compression spaces.

The losses that can be implemented are as follows:

1. Heat transfer losses

   - Shuttle heat transfer, which is caused by the displacer. The displacer has an oscillatory movement from the hot expansion space to the cold compression space; it thereby transfers heat by conduction.

   - Pumping losses. A gap is present between the displacer and the cylinder wall, which is sealed at the cold part of the displacer. As the pressure changes in the heat pump, gas flows in and out of the gap. As the gas is cooled at the cold part, extra heat has to be added to compensate this.

   - Hysteresis heat transfer. This is heat transfer from the working gas to the cylinder due to temperature gradients. The temperature gradients are caused by the pressure variations in the compression space.

2. Fluid friction losses

   - The flow friction losses are mainly due to a pressure drop in the regenerator. The heater and cooler also cause a small pressure drop.

   - Viscosity of the working gas.

   - Gas spring hysteresis losses.

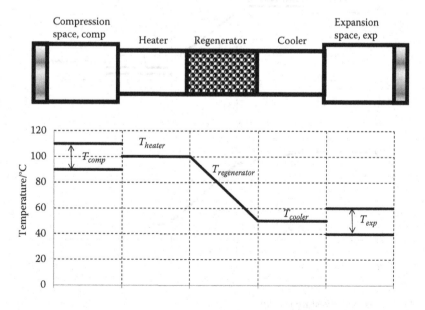

FIGURE 6.40   Ideal adiabatic model.

3. Mechanical friction losses

- The mechanical friction losses are due to seal friction. This is hard to compute in a reliable way and must be measured.

6.6.2.2 Third-Order Analysis

The third-order analysis has the highest analytical level and is an attempt to create a realistic and complete understanding of the complex processes occurring in a Stirling machine (Figure 6.41). The model uses control volumes or nodes. Three conservation laws and the equation of state are solved for each element. The three conservation laws to be solved are energy, mass and momentum. These equations are in a complex differential form and are therefore solved numerically. Empirical equations for fluid-friction and heat transfer effects must be added. A mathematical stable method must be found for numerical solution of the differential equations to calculate the pressure, temperature and mass distribution in the heat pump.

Several third-order analyses have been developed. Urieli and Berchowitz (1984) developed a third-order model that is still used in many research projects.

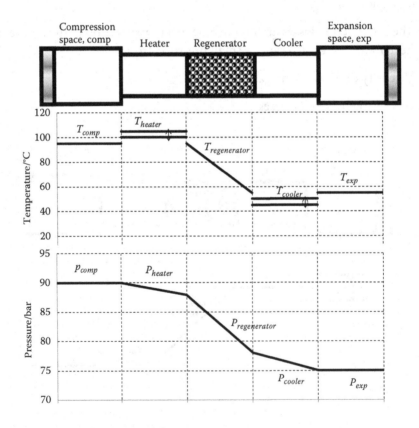

FIGURE 6.41 Quasi-steady-state model.

6.6.3 Thermodynamic Losses

The higher isochoric heat capacity of the compressed fluid (0.120 kJ/kgK vs. 0.111 kJ/kgK) prevents a reversible regeneration of the heat available between $T_H = T_1$ and $T_C = T_4$; see Figure 6.37 for the state numbers. Even for an infinite heat transfer surface $(u_2 - u_3) > (u_1 - u_4)$, so that the regeneration heat $Q_{regeneration}$ is determined by $(u_1 - u_4)$:

$$Q_{regeneration} = |m \times (u_4 - u_1)| \tag{6.53}$$

For this reason, the working fluid cools only down to 3′ instead of 3 as indicated in Figure 6.37. The performance of idealized cycles is not a meaningful index of the actual merits of real cycle configurations. A thorough analysis of the losses of a Stirling heat pump, as in the third order analysis, is beyond the scope of this book. A simplified method is proposed which takes into account the main causes of losses which are typical of Stirling heat pumps. With reference to Figure 6.37, the following nonidealities can be further introduced.

Additionally to the effect of the higher heat capacity of the compressed fluid already discussed above, a minimum-temperature approach will be needed between the gas flow and the temperature attained by the regenerator bed. As a consequence, the gas flow is not at T_H as the compression process starts and also not at T_C as the expansion starts. In the diagrams of Figure 6.37 state 1 moves to lower temperature (along the same isochoric line) and state 3′ moves to higher temperatures. By defining a regenerator effectiveness which indicates the departure of the actual regenerator from an ideal regenerator, it is possible to identify the operating conditions at the outlet of the regenerator.

$$\eta_{regenerator} = \frac{u_1^1 - u_4}{u_1 - u_4} \tag{6.54}$$

and

$$u_1^1 = u_4 + \eta_{regenerator} \times (u_1 - u_4) \tag{6.55}$$

where the regenerator effectiveness can be in the range 80% to 90%. Note that again the internal energy change on the low-pressure side of the cycle determines the internal energy change in the high-pressure side.

If heat should be delivered at temperature T_H, then the compression needs to be performed at a slightly increased temperature, taking into account the temperature driving force required to transfer the heat from working fluid to the process stream in the heater. Also, at the heat source side, expansion needs to take place at a slightly decreased temperature to allow for the finite size of the cooler. This effect is similar to the effect of temperature driving forces in the heat exchangers of vapor compression cycles.

Pressure losses, which are known to be important in Stirling heat pumps, reduce the expansion work and increase the compression work. Angelino and Invernizzi (1996) have proposed the use of isothermal compression/expansion efficiencies to take these losses into account.

$$\eta_{comp\_isothermal} = \frac{W_{comp\_isothermal}}{W_{comp\_real}} \tag{6.56}$$

and

$$\eta_{exp\_isothermal} = \frac{W_{exp\_real}}{W_{exp\_isothermal}} \tag{6.57}$$

Mechanical losses can eventually be taken into account by including these losses in the efficiencies.

6.6.4 Heat Pump Cycle Performance

The cycle discussed previously that includes the identified thermodynamic losses is referred to as a 'real' cycle. Angelino and Invernizzi (1996) have compared the COP of the real Stirling cycle with the reversible cycle COP, Equation 6.47, by calculating the ratio between the two COPs (second law efficiency as previously introduced). They assumed isothermal compression and expansion efficiencies of 85% and a minimum-temperature approach at the outlet of the regenerator of 10 K. They obtained, for the conditions discussed, a second law efficiency of 42%. Higher compression ratios and higher temperature ratios have a positive effect on the second law efficiency because the losses then become relatively smaller. It can be expected that the efficiencies are comparable to the efficiencies previously reported for vapor compression cycles so that second law efficiencies will be in the same range as for vapor compression heat pumps.

Haywood (2004) has experimentally studied a Stirling heat pump with air as working fluid and has identified a significant deviation between the expected indicated diagram for the real cycle as proposed by Angelino and Invernizzi (1996) and the measured diagram. He also has confirmed that the quality of the seals plays a major role in the performance of the cycle. One of Haywood's (2004) experiments is reproduced in Figure 6.42 to illustrate an experimental cycle and the effect of leakages.

6.6.5 Working Fluid Selection

A Stirling heat pump is able to use a large variety of working fluids. In the nineteenth century, most Stirling engines used air at ambient pressures. Air is a cheap and readily available fluid. The disadvantage of using air is the small power density. As the technology evolved, the demand for a fluid that could be used in a small engine with a high power

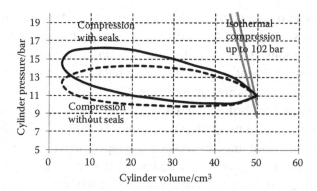

FIGURE 6.42 Stirling heat pump cycle in an experimental pressure-volume diagram with and without seals.

output increased. The fluids should have the following thermodynamic, heat transfer and gas dynamic properties (Thombare and Verma, 2008):

- High thermal conductivity (heat has to be transferred to all the cross sections of the flow)

- High specific heat capacity (high heat transfer capability)

- Low viscosity (viscosity is related to friction)

- Low density (related to the working fluid mass in the system)

The selection of the working fluid is also related to the operating pressures and fluid safety. Table 6.6 illustrates the differences for the proposed working fluids. Helium and hydrogen are the predominant working fluids for Stirling cycles. Hydrogen has the highest efficiency over the whole range of rotational speeds and reaches the highest power density. A disadvantage of hydrogen is the high flammability when mixed with air. The mixture can spontaneously explode when it is exposed to a spark, heat or even sunlight. Because of safety reasons, hydrogen is not used for stationary application. Some leakage of working fluid cannot be prevented, enabling hydrogen gas to build up around the Stirling heat

TABLE 6.6   Properties of Different Working Fluids for Operating Conditions between 50°C and 100°C

Fluid	Operating Pressures (bar)	c_{v_high}/c_{v_low} (-)	c_v (kJ/kgK)	Density (kg/m³)
Air	65–157	1.02	0.74	70–140
Carbon dioxide	68–149	1.12	0.89	165–330
Helium	69–164	1.00	3.2	10–20
Hydrogen	69–167	1.01	10.4	5–10
Nitrogen	67–165	1.02	0.75	70–140
Xenon	67–147	1.12	0.12	500–1000

pump, which poses a potential hazard. For stationary applications, helium is mostly used since helium has the second-highest power density, and space limitations are less abundant for stationary applications. Also, the criteria for safe operation and storage are promoters for the use of helium. Berchowitz et al. (2008) proposed CO_2 as a working fluid. Fluid properties of various working fluids are listed in Table 6.6.

Because the medium that is used is a gas (generally a noble gas), the working fluid has no temperature limitations in its applications. However, the moving parts will need lubrication and need to be maintained at acceptable temperatures. To prevent fouling of the regenerator, sliding surfaces are covered with low-friction materials such as PTFE (polytetrafluoroethylene) or graphite so that the use of oil as a lubricant can be prevented in the working fluid side of the machine. Because the compression space is at heater temperature, most probably the piston rings will limit the structural integrity of the moving parts. The impact of the quality of the seals is demonstrated in Figure 6.42.

The density has been selected so that the lowest pressure is 65 to 70 bar. A compression ratio of 2 has been assumed. The table shows that the mismatch in specific heat capacity between the low-pressure and high-pressure regeneration process is really only a problem for carbon dioxide and xenon, being close to 1 for the other fluids. For the imposed compression ratio, the table shows that the pressure ratio is similar for the considered fluids.

6.6.6 Stage of Development

The Stirling principle has been known now for a long time, since 1816. Although large research and development efforts have been devoted to this system, no mass-scale commercial introduction has been observed. Organ (2014) indicated that although some demonstration projects have been executed, the expectations have always been too high, so that the practical performance was disappointing and projects did not receive a follow-up. Ultra-low temperature applications (−80°C) as refrigeration Stirling cycles are being commercialized by Stirling Ultracold (USA), which makes use of the free piston Stirling technology from Global Cooling. These machines are hermetically sealed, are driven by an integrated linear motor and use helium (10 g) as the working fluid. An ethane thermosiphon is used to distribute the cooling effect. All moving parts are supported by non-contacting gas bearings, and no oil is used for the lubrication. Stirling Cryogenics BV (the Netherlands) manufactures Stirling refrigeration cycles for liquefaction purposes: air, methane, nitrogen and oxygen liquefying systems. These machines make use of the crank shaft–driven concept illustrated in Figure 6.35 with piston and displacer being independently driven. The working fluid is again helium. The cooling capacity of the cycle at −196°C is 1 to 4 kW. These are tailor-made machines that require maintenance every 6,000 hours of operation.

Based on the work of Angelino and Invernizzi (1996), the efficiency of a Stirling heat pump increases with the compression ratio for fluids operating close to the critical point. Because a higher compression ratio leads to a higher pressure ratio, leakage flows will increase, while the model by these authors neglects the leakage flows. Figure 6.43 has been derived from the original data of Angelino and Invernizzi (1996) and makes clear that xenon performs better than carbon dioxide. If real heat pump cycles could reach COPs as

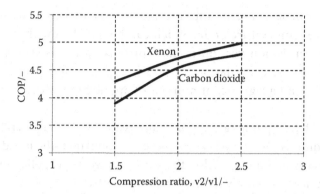

FIGURE 6.43 The COP for a Stirling cycle heat pump that operates between 50°C and 104°C (based on Angelino and Invernizzi, 1996). Compression and expansion efficiencies are 85%, and the regenerator temperature approach is 10 K.

given in Figure 6.43, then the second law efficiency of the heat pump would be 47% to 58% of Carnot (assuming 5 K between reservoirs and cycle), which would be slightly better than the average vapor compression cycle. Including the impact of leakage losses would give a better prediction of the performance of the practical cycles.

6.7 CONCLUDING REMARKS

This chapter provided an overview of the alternative mechanically driven technologies available for the implementation of heat pumps: vapor compression, vapor recompression, compression resorption, transcritical and Stirling heat pumps. The chapter discussed the methods for the prediction of the performance of these cycles when the externally imposed operating conditions are prescribed: heat sink in- and outlet temperatures and corresponding heating capacity and heat source inlet temperature.

Vapor compression cycles have been the most extensively applied heat pumps, and most components are affordable because they are produced in large amounts. When the application temperatures become high, then there is only limited experience, and mainly the operation of compressors becomes critical. The efficiency of these heat pumps can be improved by applying a heat exchanger between the liquid line (outlet condenser) and the suction line (inlet compressor), by applying two-stage operation with an interstage flash separator or by recovering part of the expansion work with, for instance, an ejector. This last aspect has only been discussed in the frame of the TCHPs because in these cycles the throttling losses are the largest losses but can also be applied for common vapor compression cycles.

Vapor recompression cycles have the smallest number of components, and because they do not require an evaporator, their performance is the highest of the heat pumps discussed. These cycles should be preferred whenever they can be applied. There is significant industrial experience with these heat pumps. Unfortunately, their application is limited because the working fluid is the process fluid and by the fact that generally a lubricant-free compressor will be required, which are less efficient or extremely expensive.

Compression resorption heat pumps are specifically attractive when the application flow shows a large temperature glide, which can be matched with the resorber of this cycle. The wet compression version is most promising but is still not commercially available. The solution recirculation version has been offered by a single company for about 10 years, and a significant number of applications exist, mainly in the food sector.

Transcritical CO_2 heat pumps are vapor compression cycles that operate partially under supercritical conditions. These cycles have become popular in the last decade because of their small environmental impact. Also, these cycles have been applied in the last 10 years in a significant number of mainly food-processing plants.

Stirling heat pumps have always been considered promising. While all the previous cycles include isobaric heat transfer processes, in Stirling cycles the pressure varies as heat transfer takes place, and ideally the processes in the heat exchangers should take place at constant temperature. The performance prediction methods differ significantly from the methods used for the other cycles. Although promising, these cycles have only commercial applications related to very low-temperature refrigeration applications.

List of Symbols

A	Heat exchanger area	m^2
c_p	Specific heat at constant pressure	$kJkg^{-1}K^{-1}$
c_v	Specific heat at constant volume	$kJkg^{-1}K^{-1}$
h	Enthalpy	$kJkg^{-1}$
m	Mass	kg
$\dot{m}$	Mass flow	kgs^{-1}
n	Rotational speed	s^{-1}
p	Pressure	kPa
Q	Heat	kJ
$\dot{Q}$	Heat flow	kW
R	Gas constant	$kJkg^{-1}K^{-1}$
s	Entropy	$kJkg^{-1}K^{-1}$
T	Temperature	K
u	Internal energy	$kJkg^{-1}$
U	Overall heat transfer coefficient	$Wm^{-2}K^{-1}$
v	Specific volume	m^3kg^{-1}
V	Volume	m^3
$\dot{V}$	Volume flow	m^3s^{-1}
w	Technical work	$kJkg^{-1}$
w_{ave}	Average solution concentration	$kgkg^{-1}$
W	Work	kJ
$\dot{W}$	Power	kW
x	Quality (vapor fraction)	–

Greek Symbols

Δh	Enthalpy change	kJkg^{-1}
$\Delta T_{approach}$	Smallest temperature difference	K
ΔT_{lift}	Temperature difference between heat sink and heat source	K
κ	Isentropic exponent	–
η	Efficiency	–
θ	Crank angle	–

Subscripts

C	Source (cold)
comp	Compressor
cond	Condenser
evap	Evaporator
exp	Expansion
H	Sink (hot), high (pressure side)
hex	Heat exchanger
in	At inlet
init	Initial value
int	Intermediate
is	Isentropic
L	Low (pressure side)
ln	Logarithmic
out	At outlet
ref	Refrigerant
s	Isentropic
vol	Volumetric

Abbreviations

COP	Coefficient of performance
CP	Critical point
CRHP	Compression resorption heat pump
PBP	Payback period
PTFE	Polytetrafluoroethylene
TCHP	Transcritical heat pump
VC	Vapor compression
VRC	Vapor recompression

REFERENCES

Angelino G., Invernizzi C., Potential performance of real gas Stirling cycle heat pumps, *International Journal of Refrigeration*, 19 (1996), 390–399.

Austin B. T., Sumathy K., Transcritical carbon dioxide heat pump systems: A review, *Renewable and Sustainable Energy Reviews*, 15 (2011), 4013–4029.

Berchowitz D. M., Janssen M., Pellizzari R. O., CO_2 Stirling heat pump for residential use, International Refrigeration and Air Conditioning Conference at Purdue, Purdue University, 2008, Paper 2352, 1–8.

Bergmann G., Hivessy G., Mean features and operational experience of the hybrid heat pump pilot plant. In *Proceedings of the 3rd Workshop Research Activities on Advanced Heat Pumps*, 24–26, Technical University of Graz, Austria, 1990.

Boot H., Nies J., Verschoor M.J.E., de Wit J.B., *Handbook industriële warmtepompen* (in Dutch), Kluwer bedrijfsinformatie, Blaricum, the Netherlands, 1998.

Brousse E., Claudel B., Jallut C., Modelling and optimization of the steady state operation of a vapour recompression distillation column, *Chemical Engineering Science*, 40 (1985), 2073–2078.

Brunin O., Feidt M., Hivet B., Comparison of the working domains of some compression heat pumps and a compression-absorption heat pump, *International Journal of Refrigeration*, 20 (1997), 308–318.

Brunin O., Pompe a chaleur a compression-absorption; etude et realisation experimentale (in French), PhD thesis, Universite Henri Poincaré–Nancy, France, Mechanique Energetique, 1995.

Cecchinato L., Corradi M., Minetto S., A critical approach to the determination of optimal heat rejection pressure in transcritical systems, *Applied Thermal Engineering*, 30 (2010), 1812–1823.

Haywood D., Investigation of Stirling-type heat pump and refrigerator systems using air as the refrigerant, PhD thesis, University of Canterbury, New Zeeland, 2004.

Hulten M., Berntsson T., The compression/absorption heat pump cycle – Conceptual design improvements and comparisons with the compression cycle, *International Journal of Refrigeration*, 25 (2002), 487–497.

Infante Ferreira C. A., Zamfirescu C., Zaytsev D., Twin screw oil-free wet compressor for compression-absorption cycle, *International Journal of Refrigeration*, 29 (2006), 556–565.

International Energy Agency, Application of industrial heat pumps, *IEA industrial energy related systems and technologies, Annex 13 and IEA heat pump programme annex 35*, Report No. HPP-AN35-1, ISBN 978-91-88001-92-4, IEA Heat Pump Centre, Borås, Sweden, 2014.

Itard L. C. M., Wet compression versus dry compression in heat pumps working with pure refrigerants or non-azeotropic mixtures, *International Journal of Refrigeration*, 18 (1995), 495–504.

Itard L. C. M., Wet compression-resorption heat pump cycles: Thermodynamic analysis and design, PhD thesis, Delft University of Technology, the Netherlands, 1998.

Itard L. C. M., Machielsen C. H. M., Considerations when modelling compression/resorption heat pumps, *International Journal of Refrigeration*, 17 (1994), 453–460.

Jana A. K., Advances in heat pump assisted distillation column: A review, *Energy Conversion and Management*, 77 (2014), 287–297.

Kiss A. A., Olujic Z., A review on process intensification in internally heat-integrated distillation columns, *Chemical Engineering and Processing: Process Intensification*, 86 (2014), 125–144.

Lemmon E., Huber M., McLinden M., *NIST Standard reference database 23: Reference fluid thermodynamic and transport properties REFPROP*, Version 9.1, National Institute of Standards and Technology, Gaithersburg, MD, 2013.

Li Y., Wub H., Liang XG., Rong C., Chen H., Experimental study of waste concentration by mechanical vapor compression technology, *Desalination*, 361 (2015), 45–52.

Liang L., Han D., Ma R., Peng T., Treatment of high-concentration wastewater using double-effect mechanical vapor recompression, *Desalination*, 314 (2013), 139–146.

Liao S.M., Zhao T.S., Jakobsen A., A correlation of optimal heat rejection pressures in transcritical carbon dioxide cycles, *Applied Thermal Engineering*, 20 (2000), 831–841.

Ma Y., Liu Z., Tian H., A review of transcritical carbon dioxide heat pump and refrigeration cycles, *Energy*, 55 (2013), 156–172.

Malewski W. F., Integrated absorption and compression heat pump cycle using mixed working fluid ammonia and water. In *Proceedings of the Institute of Refrigeration*, 1987–1988, 4–1, London, 1988.

Martini W. R., *Stirling engine design manual*, CR-182290, NASA, Houston, TX, 1983.

Modla G., Lang P., Heat pump systems with mechanical compression for batch distillation, *Energy*, 62 (2013), 403–417.

Nordtvedt S. R., Experimental and theoretical study of a compression/absorption heat pump with ammonia/water as working fluid, PhD thesis, Norwegian University of Science and Technology, Trondheim, 2005.

Organ A. J., *Stirling cycle engines: Workings and design*, Wiley, New York, 2014.

Pinder K. L., Direct contact vapor recompression evaporation desalination process economic assessment, *Desalination*, 4 (1968), 45–54.

Prins J., Experimental compressed air unit. In *Twin screw compressor leakage flows*, Delft University of Technology, the Netherlands, 2001.

Radermacher R., Hwang Y., *Vapor compression heat pumps with refrigerant mixtures*, CRC Press, Taylor & Francis, Boca Raton, FL, 2005.

Sarkar J., Ejector enhanced vapor compression refrigeration and heat pump systems – A review. *Renewable and Sustainable Energy Reviews*, 16 (2012), 6647–6659.

Sarkar J., Bhattacharyya S., Gopal M. R., Optimization of a transcritical CO_2 heat pump cycle for simultaneous cooling and heating applications, *International Journal of Refrigeration*, 27 (2004), 830–838.

Sarkar J., Bhattacharyya S., Gopal M.R., Natural refrigerant-based subcritical and transcritical cycles for high temperature heating, *International Journal of Refrigeration*, 30 (2007), 3–10.

Shen J., Xing Z., Wang X., He Z., Analysis of a single-effect mechanical vapor compression desalination system using water injected twin screw compressors, *Desalination*, 333 (2014), 146–153.

Stolk A., Koudetechniek A1 – Koudeopwekking, Lecture notes i77A (in Dutch), Werktuigbouwkunde en Maritieme Techniek, TU Delft, the Netherlands, 1990.

Thombare D. G., Verma S. K., Technological development in the Stirling cycle engines, *Renewable and Sustainable Energy Reviews*, 12 (2008), 1–38.

Torstensson H., Nowacki J. E., *Sorptions/kompressionvärmepump*, *Slutrapport*, Studsvik Energy, Nyköping, Sweden, 1990.

Urieli I., Berchowitz D. M., *Stirling cycle engine analysis*, Hilger, Bristol, UK, 1984.

van de Bor D. M., Infante Ferreira C. A., Kiss A. A., Optimal performance of compression resorption heat pump systems, *Applied Thermal Engineering*, 65 (2014), 219–225.

van de Bor D. M., Infante Ferreira C. A., Kiss A. A., Low grade waste heat recovery using heat pumps and power cycles, *Energy*, 89 (2015), 864–873.

Vorster P. P. J., Meyer J. P., Wet compression versus dry compression in heat pumps working with pure refrigerants or non-azeotropic binary mixtures for different heating applications, *International Journal of Refrigeration*, 23 (2000), 292–311.

Walker G., Fauvel R., Gustafson R., van Bentham J., Stirling engine heat pumps, *International Journal of Refrigeration*, 2 (1982), 91–97.

Walker G., Reader G., Fauvel O. R., Bingham E. R., *The Stirling alternative. Power systems, refrigerants and heat pumps*, Gordon and Breach, Philadelphia, 1994.

Weerstra J., Dynamic behavior of CO_2 refrigeration system with integrated ejector, MSc thesis, Delft University of Technology, the Netherlands, 2014.

Yang J., Ma Y., Li M., Guan Q. Exergy analysis of transcritical carbon dioxide refrigeration cycle with an expander, *Energy*, 30 (2005), 1162–1175.

Zaytsev D., Development of wet compressor for application in compression-resorption heat pumps, PhD thesis, Delft University of Technology, the Netherlands, 2003.

Zhao X., Heat pumps in separation processes, MSc thesis, Delft University of Technology, the Netherlands, 2011.

Zhou Y., Shi C., Dong G., Analysis of a mechanical vapor recompression wastewater distillation system, *Desalination*, 353 (2014), 91–97.

CHAPTER 7

Thermally Driven Heat Pumps

7.1 INTRODUCTION

This chapter discusses the characteristics of thermally driven heat pumps. These cycles are often preferred by users because of the absence of moving parts (except for circulation pumps, which require limited maintenance). Their efficiency is lower than the efficiency of mechanically driven heat pumps, but because the systems are driven by thermal energy, fewer energy conversion steps are required. Considering a coefficient of performance (COP) for a heat pump when driven by primary fuel making use of a burner with efficiency of 0.85 to be 2.0, then the total efficiency is 1.70. Assuming the energy conversion efficiency of an average power plant to be 0.42, the COP of the corresponding mechanically driven heat pump is 4.05. In the discussions in this chapter, this relation to primary energy use should be taken into account: A COP of 2.0 for a thermally driven heat pump is comparable to the COP of mechanically driven heat pumps.

The chapter starts with a discussion of liquid-vapor absorption heat pumps, the most frequently applied thermally driven systems in practice. The working fluids in these cycles are ammonia-water mixtures, with ammonia as the refrigerant, and water-lithium bromide mixtures, with water as the refrigerant. Other mixtures have been proposed and investigated but are not yet applied on a commercial scale. Absorption has been a topic discussed in the refrigeration courses at the Delft University of Technology in the Netherlands for many years (Stolk, 1990). The topics discussed in these courses form the basis of the section about liquid-vapor absorption: determination of cycles based on equilibrium conditions and effect of irreversibilities and deviation from equilibrium in the different components of the cycle. Sorption cycles were extensively discussed by Niebergall (1981) in the German language, which in fact was a reprint of the original 1959 version. Herold et al. (1996) published a dedicated book that also focused on absorption heat pumps. Bogart (1981) dedicated a whole book specifically to ammonia-water absorption systems.

Recently, Li et al. (2014) reviewed the progress in the development of solid-vapor sorption cycles. A few years earlier, a review was published by Demir et al. (2008) so that recent progress can be identified. Adsorption cycles are the topic of the next section. Commercial applications of these cycles are more limited but are clearly expanding. Wang et al. (2014)

published a book specifically dedicated to adsorption cycles where the topic is more extensively discussed.

Finally, thermally driven ejector systems are discussed. These cycles are often used as open cycles as 'thermal vapor recompression' heat pumps, making use of steam as the driving force. In this case, water is the working fluid. Power (1993) published a dedicated book for steam jet ejectors. In closed systems, different refrigerants can be applied.

7.2 LIQUID-VAPOR ABSORPTION HEAT PUMPS

7.2.1 Reversible Cycle and Efficiency

This section introduces and discusses liquid absorption heat pump cycles, which are cycles with a large number of implemented heat pumps. Absorption heat pumps are based on the same principle as vapor compression heat pumps: evaporation and condensation of the working fluid within a closed cycle. These cycles are nevertheless driven in a different way: The mechanical compressor is substituted by a thermal compressor. This thermal compressor consists of an absorber, which has a function comparable to the suction phase of a mechanical compressor, and a desorber (often called generator) with a function comparable to the discharge phase of a mechanical compressor. In intermittent cycles, as the solid sorption systems, the frequency is much smaller than for piston compressors, a few cycles per day instead of 20 per second. A continuously operating machine is as such comparable to a turbocompressor.

The working principle schematic is very much comparable with the schematic for the vapor compression cycle (see Figure 7.1). The same four main components can again be identified: thermal compressor (1), condenser (2), throttling device (3) and evaporator (4). Only the driving part (the compressor) deviates from the vapor compression cycle. The separation between high- and low-pressure parts goes from throttling device to driving part.

The following concerns the different components of vapor compression and liquid absorption:

- Mechanical work is added to the vapor compression heat pump to increase the pressure of the vapor of the working fluid. Energy is added as heat $\dot{Q}_{des}$ to the liquid solution in the desorber of the absorption heat pump. Only a relatively small amount of mechanical energy is needed to drive the solution pump $\dot{W}_{pump}$.

- Heat $\dot{Q}_{des}$ is supplied to the desorber of the thermal compressor at relatively high temperature. At the absorber, heat is delivered at the application temperature level $\dot{Q}_{abs}$. The energy balance becomes

$$\dot{Q}_{in} + \dot{Q}_{des} + \dot{W}_{pump} - \dot{Q}_{abs} - \dot{Q}_{cond} = 0 \tag{7.1}$$

The energy consumption by the pump is normally small and can be neglected. Note that the total heat delivered by the heat pump is the sum of the heat delivered by the condenser and the heat delivered by the absorber:

$$\dot{Q}_{out} = \dot{Q}_{abs} + \dot{Q}_{cond} \tag{7.2}$$

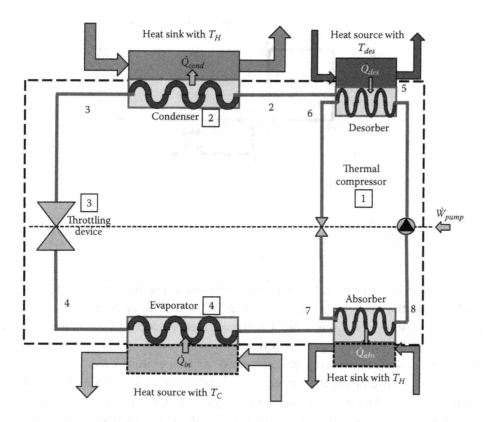

FIGURE 7.1 Liquid-vapor absorption heat pump with its main components: thermal compressor, condenser, throttling valve and evaporator. The thermal compressor includes an absorber, a pump, a desorber and a throttling device.

The COP of these heat pumps is given by

$$COP = \frac{\dot{Q}_{out}}{\dot{Q}_{des} + \dot{W}_{pump}} = \frac{\dot{Q}_{abs} + \dot{Q}_{cond}}{\dot{Q}_{des} + \dot{W}_{pump}} \qquad (7.3)$$

This COP is not directly comparable with the COP of a vapor compression heat pump. The efficiency of the conversion of primary energy into electricity should be accounted for. For the average Dutch power station, the conversion rate is 0.42, so that the COP of vapor compression cycles should first be multiplied with this factor before comparing it to the result of Equation 7.3.

- The separation between high and low pressure in the thermal compressor moves along the solution pump and the solution throttling device. High and low pressure are separated at three locations (see Figure 7.1).

Comparison with other processes is the easiest when the process is drawn in a state diagram, for instance, a T-s diagram. This is not so simple for an absorption heat pump because there are two loops that partially overlap: the solution loop and the refrigerant

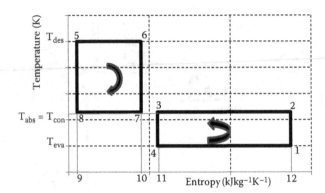

FIGURE 7.2 Ideal liquid-vapor absorption heat pump in T-s diagram, composed of a refrigerant loop 1-2-3-4 and a solution loop 5-6-7-8.

loop. It is not possible to include both in the same diagram. Niebergall (1981) proposed a simplification that makes visualization possible. This is illustrated in Figure 7.2, which shows two Carnot cycles. Cycle 1-2-3-4 is the refrigerant loop and is comparable to the refrigerant loop for vapor compression cycles.

Process 4-1 represents the evaporation process: At temperature $T_C = T_{evap}$, the evaporation heat $\dot{Q}_{in}$ is removed from the source. Process 2-3 is the condensation process: At temperature $T_H = T_{cond}$, the condensation heat $\dot{Q}_{cond}$ is added to the process flow that needs to be heated. Process 3-4 indicates an isentropic (adiabatic reversible) expansion. Process 1-2 represents an isentropic pressure increase. This process is coupled with the solution loop. The solution loop is given by 5-6-7-8. Process 5-6 represents the desorption process: At high temperature T_{des}, the desorption heat $\dot{Q}_{des}$ is added. Process 7-8 is the absorption process. This process takes place at temperature $T_H = T_{abs}$, while the absorption heat $\dot{Q}_{abs}$ is delivered to the process stream that needs to be heated. Process 6-7 is an isentropic expansion process from desorber pressure level to absorber pressure level. Process 8-5 is an isentropic pressure increase of the solution. Processes 8-5 and 1-2 are in reality connected, but this cannot be represented in the T-s diagram. The two loops concern different fluids, so that different entropy scales apply. Only the isothermal lines are in the correct position. Isentropic and isobaric lines are only schematically represented. Desorption and condensation take place at the same pressure, but this cannot be correctly indicated in this (schematic) diagram. The same applies for absorption and evaporation, which also take place at the same pressure. The advantage of this diagram is that the Carnot COP can be calculated in a similar way as for the vapor compression heat pump. The solution cycle can be considered as a power cycle that drives the refrigerant cycle. The solution cycle operates clockwise as all power cycles, while the refrigerant cycle operates anticlockwise as all heat pumps. The desorption heat $\dot{Q}_{des}$ is represented by the area of the rectangle 5-6-10-9-5, while the absorption heat $\dot{Q}_{abs}$ is indicated by the area 8-7-10-9-8. From this cycle results the heat represented by the enclosed area 5-6-7-8-5. This heat is delivered to the refrigerant cycle. The evaporation heat $\dot{Q}_{in}$ is given by the area 4-1-12-11-4; the condensation heat $\dot{Q}_{cond}$ is given by the area 2-3-11-12-2. This cycle requires the heat represented by the enclosed area 1-2-3-4-1. In an ideal Carnot process, the heat represented by the two enclosed areas is identical. The areas are not identical because the masses that flow through the two loops are not identical. Also, this is not visible from the *T-s* representation. The Carnot COP is given by

$$COP_{Carnot} = \frac{(8-7-10-9-8)\times f+(2-3-11-12-2)}{(5-6-10-9-5)\times f} \tag{7.4}$$

where f is the ratio of the solution and refrigerant mass flows. The heat delivery to the heat pump application is partially contributed by the absorber and partially by the condenser. The heat represented by the enclosed areas is comparable to the required compressor work in vapor compression heat pumps.

The cycle representation in Figure 7.2 also makes it possible to quantify the Carnot efficiency on the basis of the temperatures imposed to the cycle:

$$COP_{Carnot} = \eta_{Carnot\_power\_cycle} \times COP_{Carnot\_hp} = \left(1-\frac{T_H}{T_{des}}\right)\times\left(\frac{T_H}{T_H-T_C}\right) \tag{7.5}$$

Because T_{des} is normally relatively low, the efficiency of the power cycle is low, so that the COP of absorption heat pumps will always be quite low.

From this discussion, it is clear that a *T-s* diagram can only give a schematic representation of the processes occurring in an absorption heat pump cycle. For binary systems as used in these cycles, two diagrams are commonly applied: the $\log p - 1/T$ diagram and the enthalpy-concentration diagram. The first one simplifies the identification of the process limits. The second one allows determination of the enthalpy at different positions in the cycle so that energy flows can be quantified. Figure 7.3 shows a $\log p - 1/T$ diagram for a

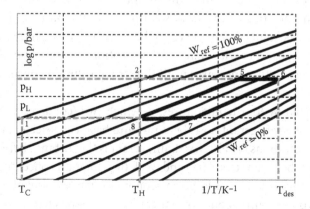

FIGURE 7.3 Liquid-vapor absorption heat pump in $\log p - 1/T$ diagram. Evaporation is indicated by state 1; condensation by state 2; desorption by line 5-6; absorption by line 7-8. The solution loop is shown with bold lines.

generic refrigerant-absorbent pair for an absorption heat pump cycle. The refrigerant concentration in the solution w_{ref} is given as a parameter. The top line is the saturation vapor pressure line for pure refrigerant (w_{ref} = 100%). The bottom line (w_{ref} = 0%) is the saturation line of the pure absorbent. The lines in-between give the vapor pressure above the solution with the given concentration. The y axis gives the total pressure, thus the sum of the refrigerant and absorbent vapor pressure. The vapor pressure of the absorbent must be small in comparison to the vapor pressure of the refrigerant; otherwise, the system cannot operate. The ideal absorbent has no vapor pressure. In practice, the vapor pressures indicated by the diagram are only slightly higher than for the refrigerant only.

The diagram gives the relation between pressure p, temperature T and concentration w_{ref}. The enthalpy h at the relevant process conditions cannot be read from this diagram. The absorption heat pump process can be drawn in the $\log p - 1/T$ diagram. An ideal process is first discussed. Evaporation takes place at source temperature T_C, and pure refrigerant (w_{ref} = 100%) evaporates, state 1 in Figure 7.3. This condition fixes the low pressure in the heat pump p_L. Condensation takes place at heat rejection temperature (application temperature for the heat pump T_H), while the refrigerant condenses as pure refrigerant (w_{ref} = 100%), state 2 in Figure 7.3. This state determines the high pressure of the cycle p_H. The lowest temperature in the absorber is generally identical to the condensation temperature because both heat flows reject heat to the process stream that needs to be heated. The pressure in the absorber is the same as in the evaporator p_L. This pressure applies for the whole low-pressure side of the heat pump. The conditions at the absorber outlet, state 8, are determined by pressure and temperature, so that from the diagram the concentration of the strong solution at the outlet of the absorber is known (w_s). The circulation pump brings this strong solution to the desorber. The pressure in the desorber is the same the pressure in the condenser p_H, which was determined by condensation temperature T_H and the pure refrigerant (w_{ref} = 100%) line. The lines for constant pressure p_H and constant concentration w_s meet at state 5, where the desorption process starts. The desorption end temperature T_{des} depends on the medium used for driving the heat pump, and together with the high-pressure level of the cycle p_H determines the condition at the desorber outlet, state 6. At the same time, the concentration of the weak solution w_w at the outlet of the desorber is also determined. This weak solution flows back to the absorber through the solution throttling device and enters the absorber with state 7. By absorption of vapor from the evaporator, the concentration of the solution increases until the strong solution concentration is attained again, state 8.

The process is fixed by four temperatures, of which two (condensation and end absorption) can coincide. If a refrigerant-absorbent mixture has been selected, then the operating pressures and concentrations are also fixed. Important is the distance between states 7 and 8, that is, the difference between strong (w_s) and weak (w_w) solution concentration, the so-called degassing width. If this difference is large, then only a limited amount of solution needs to be circulated to desorb a certain amount of refrigerant. The opposite is also true: If the difference is small, then a large amount of solution must be circulated to produce a certain amount of refrigerant. The smaller the amount of solution that needs to be heated, the higher the COP of the heat pump. A large degassing width leads in principle to higher efficiencies.

The amounts of heat involved with the different processes are not given by the log $p - 1/T$ diagram. However, the slope of the lines of constant concentration depends on the sum of the evaporation and absorption heat. This can be derived from Clapeyron's relation for ideal gases:

$$\Delta h_{LG} = T \times (v_G - v_L) \frac{dp}{dT} \tag{7.6}$$

Neglecting the specific volume of the liquid phase because it is very small in comparison to the vapor volume:

$$\Delta h_{LG} = T \times v_G \frac{dp}{dT} \tag{7.7}$$

so that

$$\frac{dp}{dT} = \frac{\Delta h_{LG}}{T \times v_G} \tag{7.8}$$

For an ideal gas, $v_G = \dfrac{RT}{p}$, and the equation becomes

$$\frac{dp}{dT} = \frac{\Delta h_{LG} \times p}{R \times T^2} \tag{7.9}$$

This can also be written as

$$\frac{d \ln p}{d \frac{1}{T}} = -\frac{\Delta h_{LG}}{R} \tag{7.10}$$

Integration then gives

$$\ln p = a_1 - a_2 \Delta h_{LG} \frac{1}{T} \tag{7.11}$$

with a_1 and a_2 integration constants. Equation 7.11 applies also for binary mixtures if the latent heat of evaporation Δh_{LG} is substituted by the latent heat of evaporation plus absorption, $\Delta h_{LG} + \Delta h_{abs}$.

$$\ln p = b_1 - b_2 (\Delta h_{LG} + \Delta h_{abs}) \frac{1}{T} \tag{7.12}$$

Mixtures with a large heat of absorption show a larger slope of the lines of constant concentration than mixtures with a small heat of absorption. If the latent heats of evaporation and absorption are temperature independent and if the absorbent has negligible vapor pressure, then the diagram shows straight lines. In practice, this is not true, so that the lines are slightly curved. Nevertheless, this curvature is generally small and mostly is neglected. Note that the scale of $1/T$ increases from right to left, so that in fact a rising temperature scale results.

In an enthalpy-concentration diagram for binary mixtures, it is possible to indicate energy flows (see Figure 7.4). For ammonia-water as a working mixture, this diagram has been published by Niebergall (1981) and more recently by the International Institute of Refrigeration (IIR, 1994). In this diagram, the solution enthalpy is given as a function of the refrigerant concentration in the absorbent solution. The following parameters play a role: concentration, pressure, temperature, specific heat of the refrigerant, specific heat of the absorbent, evaporation heat of the refrigerant and evaporation heat of the absorbent.

The calculation of the diagram is complex and generally validated with experimental data. The diagram in Figure 7.4 applies for a specific pressure of a specific refrigerant-absorbent mixture. The line at the bottom of the figure is the saturated liquid line (or boiling line). This line gives the enthalpy of a boiling solution as a function of the refrigerant concentration in the solution for the specific pressure for which the diagram applies. Below the saturated liquid line, the solution is in the liquid phase. Above this line, the solution is in the vapor-liquid region. The isothermal lines in the liquid region show a trend as indicated by the dotted lines. The line on top of the diagram is the saturated vapor line

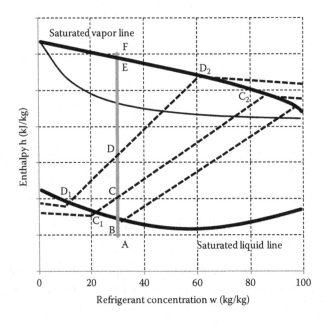

FIGURE 7.4 Enthalpy concentration diagram for a refrigerant-absorbent mixture at a given pressure. The dotted lines are isotherms. The vertical line shows the heating of a solution with given concentration from A to F.

(or condensation line). This line forms the top limit of the two-phase region in which both liquid and vapor coexist. Along this line, the vapor is just dry; above the line, the vapor is in a superheated state. The isothermal lines are straight lines in the two-phase region. In the superheated region, the isothermal lines show a negative slope because the heat capacity of the vapor at constant temperature is practically linearly dependent on the vapor composition.

Assume there is a solution in the liquid region with refrigerant concentration w_A and temperature T_A, given by state A in Figure 7.4. By heating the solution, its enthalpy rises. The concentration of the solution remains constant. The process follows a line of constant concentration, a vertical line in the diagram. In state B, the saturation line is attained. Further enthalpy increase leads to boiling of the solution. In state C, the system consists of liquid at state C_1 and vapor at state C_2. The line C_1-C_2 is the isothermal line that goes through state C. The mixture is composed of $(h_{C_2} - h_C)/(h_{C_2} - h_{C_1}) \times 100\%$ liquid and $(h_C - h_{C_1})/(h_{C_2} - h_{C_1}) \times 100\%$ vapor according to the mixing rule. The concentration of the liquid phase is decreased, while the concentration of the vapor phase is increased. Further enthalpy increase leads to state D. The vapor fraction is again increased. The vapor now has state D_2 while the liquid is at state D_1. The concentration of both liquid and vapor phases is further reduced. The concentration of the liquid phase reduces because refrigerant vapor is boiled off. The concentration of the vapor phase reduces because also vapor of the absorbent is released. Evaporation takes place at rising temperature and reducing concentration of the lighter component of the liquid phase. The average concentration of the total mixture remains constant w_A because the enthalpy increase takes place in a closed vessel maintained at constant pressure. This is a theoretical case in which volume increase takes place. In state E, all liquid has been vaporized. The vapor has then the concentration w_A. In state F, the vapor is superheated. The whole process takes place at constant pressure, the pressure for which the diagram applies. Assume that this pressure corresponds to the pressure in the evaporator and absorber of an absorption heat pump, pressure p_L. A similar diagram can be drawn for pressure p_H, the pressure in the desorber and condenser. Combining both diagrams in a single diagram results in Figure 7.5. At higher pressure, both saturated liquid and saturated vapor lines have higher enthalpies. In the liquid and vapor phases, the isothermal lines for the different pressures coincide. The intersection of the isotherms with the boiling and dew lines takes place at different positions, so that the isotherms for different pressures in the two-phase region do not coincide. See the dotted lines in Figure 7.5.

The isotherm is given by line D_1-D_2 for p_L and by C_1-C_2 for p_H. The position of the isotherms in the two-phase region can be obtained, for what concerns the ammonia-water mixture, making use of the auxiliary lines introduced by Bosnjakovic (see Niebergall, 1981). The auxiliary lines for p_L and p_H are also shown in Figure 7.5, which indicates how these lines are used to construct the isotherms.

In this combined diagram, it is possible to draw the processes taking place in an absorption heat pump (see Figure 7.6). The states 5 to 8 correspond with the states indicated in Figure 7.3, the log p – $1/T$ diagram. In the ideal case, the solution in state 5 is in equilibrium with vapor of 100% refrigerant in state 2. The ideal case does not take place for ammonia-water for which this diagram applies because this is only possible when the vapor pressure

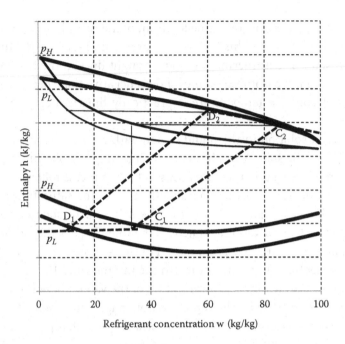

FIGURE 7.5 Enthalpy concentration diagram for a refrigerant-absorbent mixture showing an isothermal line at two different pressure levels.

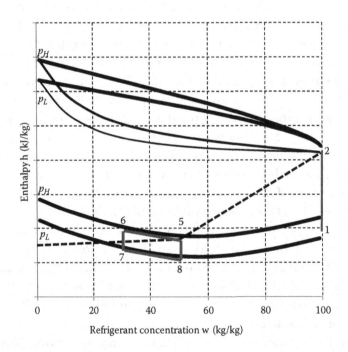

FIGURE 7.6 Enthalpy concentration diagram for a refrigerant-absorbent mixture showing an absorption heat pump cycle. During evaporation (line 1-2) and condensation (line 2-1), the refrigerant concentration is 100%. The desorption process is given by line 5-6, while the absorption process is given by line 7-8.

of the absorbent is negligible at the desorber temperature level. This is the case for salt absorbents such as LiBr and LiNO$_3$. The enthalpy-concentration diagram has a different shape for these absorbents, as discussed further in the chapter.

From state 8 at low pressure p_L, the pressure of the solution is increased by the solution pump to the high-pressure level p_H. There is no heat added to the flow, and the pumping work is relatively small and often negligible, so that the enthalpy of the solution practically remains unchanged. The operating condition changes from saturated liquid at low pressure to subcooled liquid at high pressure. By adding heat in the desorber, the saturated liquid condition at high pressure, state 5, is attained where desorption starts. Refrigerant vapor escapes from the solution, reducing its concentration and becoming weaker in refrigerant. The solution condition moves from state 5 to state 6. As indicated in the figure, the end of the desorption process is attained at the highest desorber temperature, state 6. This weak solution is expanded from p_H to p_L and sent to the absorber. The expansion process takes place along a line of constant enthalpy. State 6 leaves its saturated high-pressure condition and becomes a two-phase mixture at low pressure and with the same enthalpy. By heat removal in the absorber, the solution is cooled until saturated conditions are attained, state 7. The absorption process can then start from state 7 to state 8. In the enthalpy concentration diagram, the high concentrations of refrigerant are given on the right-hand side, while in the log p – $1/T$ diagram the high refrigerant concentrations are given in the left-hand side. In both diagrams, the enclosed area is not related to the energy input of the cycle. The enthalpy concentration diagram is most useful to develop energy and mass balances, which are needed for the design of heat pumps. This is illustrated further in the chapter.

7.2.2 Theoretical Cycle

The real process will deviate from the ideal cycle considered previously. The impact of these deviations can be both qualitatively and quantitatively determined making use of the log p – $1/T$ diagram. A qualitative evaluation is first presented to increase the understanding of these effects. The following nonidealities play a role in absorption heat pumps:

- Nonideal internal heat transfer

- Nonideal absorption

- Nonideal desorption

- Nonideal evaporation

- Nonideal external heat transfer

- Pressure drop

- Noncondensable gases

These nonidealities are discussed one by one considering that all except a specific nonideality remains unchanged.

7.2.2.1 Nonideal Internal Heat Exchange

Internal heat exchange increases the COP of the heat pump. In absorption heat pumps, there are two locations where internal heat exchange can be applied:

- Heat transfer between cold vapor from the evaporator and hot liquid from the condenser, heat exchanger hex1 in Figure 7.7. Such heat exchangers are also applied in vapor compression heat pumps.

- Heat transfer between cold solution from the absorber and hot solution from the desorber, heat exchanger hex2 in Figure 7.7.

The $\log p - 1/T$ diagram represents only equilibrium conditions. Subcooling and superheating cannot be indicated in the diagram. In hex1, the pressure remains unchanged and the concentration of the refrigerant flow also does not change. Nevertheless, because the temperatures change, the equilibrium vapor pressure also changes. States 1 and 2 move to each other. The corresponding saturation pressures are p_{Hs} and p_{Ls} (see the left-hand side of Figure 7.8).

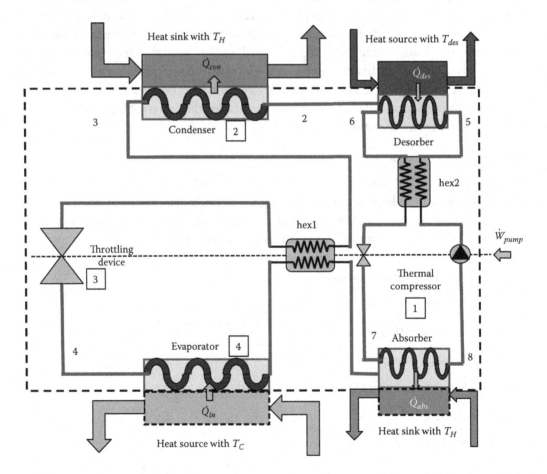

FIGURE 7.7 Schematic of liquid-vapor absorption heat pump system showing the location of the two commonly applied internal heat exchangers, hex1 and hex2.

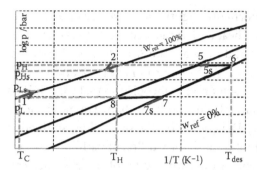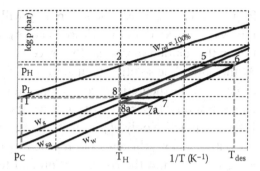

FIGURE 7.8   Effect of internal heat exchangers on absorption heat pump cycle (left). Effect of non-ideal absorption on absorption heat pump cycle (right).

The limits of the process do not change. The effect of heat exchanger hex1 is difficult to indicate in a log p – $1/T$ diagram because the flow becomes, respectively, subcooled and superheated. This heat exchanger always has a positive but limited effect in absorption heat pumps because it has no negative effects, such as for the vapor compression heat pump.

The same problem applies also for heat exchanger hex2. The temperature of the solution as it enters the desorber is higher. A much smaller subcooling than discussed previously will be attained before entering the desorber. It is even possible that desorption already started in the heat exchanger, so that state 5 moves to state 5s. The desorption trajectory from 5 to 5s takes place within the heat exchanger. The process remains the same.

The weak solution entering the absorber is now colder because the solution is cooled before being expanded. State 7 moves to 7s. The amount of vapor formed after expansion is now smaller; it can even occur that after expansion the solution enters the absorber under subcooled conditions. In the heat exchanger, there is no free refrigerant vapor that could be absorbed, so the subcooling can only be removed at the absorber. State 7s indicates the conditions at the absorber inlet. The process limits remain unchanged, only the solution enters the absorber under subcooled conditions. Heat flows and COP change, but this is not visible in log p – $1/T$ diagrams.

7.2.2.2 Incomplete Absorption

In the theoretical process, the pressure in the absorber is everywhere p_L, and the saturation pressure is also p_L. The situation in the real process is different, as indicated in the right-hand side of Figure 7.8.

The ideal absorption process is given by line 7-8. The solution concentration is then increased from w_w (weak) to w_s (strong), and the solution is cooled from T_7 to T_8. This is only possible with an infinite heat and mass exchanging surface of the absorber. In a real absorption process. a driving force is required so that the vapor from the evaporator can be absorbed by the weak solution. The liquid solution needs to be subcooled in relation to the operating pressure p_L. State 7 needs to move to state 7a, which is only possible by cooling the solution to T_{7a}. Absorption can only be initiated when the subcooling attains the value T_7-T_{7a}. The concentration of the solution starts increasing, and the heat of absorption is released so that heat removal must be guaranteed because otherwise the driving force for the process reduces. The subcooling will normally slightly decrease because the heat removal is slower than the release of absorption heat and at the same time the concentration of the solution increases. The process slows and stops at state 8a at the absorber outlet. At the outlet, the solution temperature is imposed by the environment (T_H), and it is the same as for the theoretical process, only now the solution is subcooled so that the attained concentration is lower: w_{sa} instead of w_s. The attained concentration change in the absorber is thus in practice only $w_{sa}-w_w$ instead of w_s-w_w. Depending on the subcooling attained at the outlet of hex2, the absorption process will follow a slightly different path. If the operating pressure and the concentration and temperature at the outlet of the absorber can be determined, then the subcooling at the absorber outlet can be calculated. In practical systems, this subcooling is in the range 3 to 20 K. It must be clear that a lower subcooling will provide the best performance of the cycle, but the largest surface area is required for the absorber.

7.2.2.3 Incomplete Rectification

A continuously operating absorption heat pump requires the use of a liquid solution of refrigerant and absorbent. Some absorbents have a nonnegligible vapor pressure at the operating temperatures of the desorber, so that an amount of absorbent vapor also is released during desorption. The presence of absorbent in the refrigerant loop has a negative effect on performance, and it should be prevented from entering the condenser. To prevent this as much as possible, the vapor leaving the desorber is purified in a rectifier. This purification is realized by partially condensing the vapor flow that leaves the desorber, creating a reflux. This partial condenser is called the rectifier; often, the solution flow coming from the absorber is used to remove the heat. The reflux is absorbent rich due to the difference in vapor pressure between refrigerant and absorbent. A few distillation trays or a packed bed with Raschig rings guarantee that the vapor that finally flows to the condenser is richer in refrigerant. Ideally, the vapor leaves the rectifier as pure refrigerant (i.e. w_{ref} = 100%). In practice, the rectification is not complete, so that the refrigerant concentration in the refrigerant loop w_b is less than 100%. The impact of incomplete rectification is illustrated in the right-hand side of Figure 7.9 in a $\log p - 1/T$ diagram. The left-hand side shows possible construction of desorber and rectifier.

The vapor that reaches the condenser has a concentration w_b that is lower than 100%. The temperature $T_H = T_8$ is imposed by the application of the heat pump. The condensation pressure is now decreased in comparison to the ideal case, and p_H moves to a slightly lower pressure level, indicated by the line through the marker in Figure 7.9. Because the evaporating temperature T_C is also fixed by the heat source, the evaporating pressure reduces from p_L to a slightly lower pressure. In this way, the advantage of a lower condensation pressure cancels out.

Because the end desorption temperature is fixed by the temperature driving heating temperature T_{des}, not only the weak solution concentration decreases but also the strong solution concentration decreases. The final result is that the concentration change in the absorber and desorber remains practically the same. The solution loop is now 5b-6b-7b-8b.

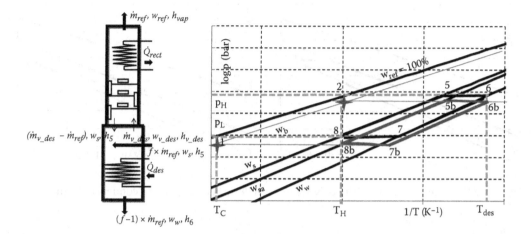

FIGURE 7.9 Effect of incomplete rectification on absorption heat pump cycle (right). Example of construction of rectification section on top of desorber (left).

7.2.2.4 Incomplete Evaporation

Due to the incomplete rectification, the refrigerant that flows through the condenser and evaporator is not pure and contains an amount of absorbent. Evaporation takes place with a gliding temperature instead of at constant temperature. The highest temperature is the temperature imposed by the external source T_C (see Figure 7.9). Evaporation starts at a lower temperature, the saturation temperature of pure refrigerant at the resulting lower evaporating temperature. Unless some action is undertaken, the amount of absorbent that remains in the evaporator will increase with time. The absorbent in the evaporator must be continuously or periodically drained to the absorber. The effect on the total cycle is as previously discussed for the incomplete rectification.

7.2.2.5 Incomplete External Heat Exchange

Because the external heat exchangers have a finite heat-exchanging area and heat transfer coefficients are limited, large temperature driving forces can occur at different parts of the cycle, resulting in the following:

- The driving heating medium is at a temperature that is higher than the end desorption temperature. This causes the weak solution concentration at the outlet of the desorber to increase from w_w to w_{we} (Figure 7.10, left-hand side).

- The process medium that needs to be heated has a temperature that is lower than the condensation temperature. This causes the condensation temperature to increase from T_H to T_{He} (Figure 7.10, right-hand side). The condensation pressure increases from p_H to p_{He}, as does the desorber pressure. Because the temperature at the desorber outlet is fixed, the weak solution concentration moves from w_w to w_{we}.

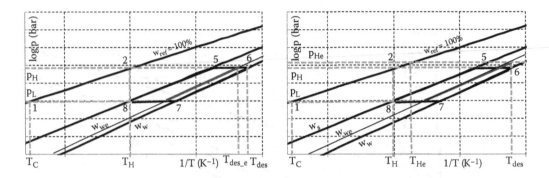

FIGURE 7.10   Effect of temperature driving force in desorber (left) and in condenser (right).

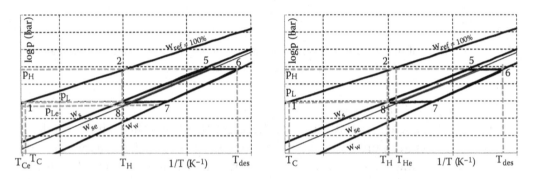

FIGURE 7.11   Effect of temperature driving force in evaporator (left) and in absorber (right).

- The evaporating temperature is lower than the temperature of the heat source. This causes the evaporating pressure to decrease from p_L to p_{Le} and so does the absorber pressure (Figure 7.11, left). Because the absorber outlet temperature is fixed T_H, the strong solution concentration moves from w_s to a lower concentration w_{se}.

- The process medium that needs to be heated has a temperature that is lower than the absorption end temperature. The absorption end temperature moves from T_H to T_{He} so that the strong solution concentration moves from w_s to a lower concentration w_{se} (Figure 7.11, right).

- The hot parts of the system will exchange heat with the environment, which is at a lower temperature. Heat exchange to the environment is not visible in the $\log p - 1/T$ diagram. The heat flow delivered by the cycle needs to be larger to compensate for the external heat transfer losses. The COP of the cycle is reduced.

- The parts that are colder than the environment will receive heat from the environment. Again, this cannot be visualized in the $\log p - 1/T$ diagram. The heat input from the source can decrease, which is advantageous for the size of the evaporator.

7.2.2.6 Pressure Drop

The pressure difference between absorber and desorber is covered by the circulation pump. This pump requires some technical work, but this is generally very small in comparison to

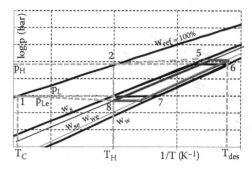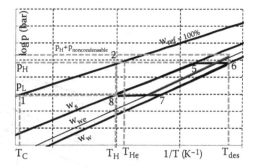

FIGURE 7.12   Effect of pressure drop between desorber and condenser and between evaporator and absorber (left) and effect of noncondensables (right).

the total energy used to drive the cycle. The pressure drop between desorber and absorber can be considered part of the throttling process. The same applies for the pressure drop between condenser and evaporator.

The pressure drop between the desorber and condenser, as well as between the evaporator and absorber, has a direct impact on the cycle performance (see Figure 7.12, left). The desorber pressure must be higher than the condensation pressure to compensate for the pressure drop. The end desorption temperature T_{des} is fixed, so that the concentration at the outlet of the desorber becomes higher, w_{we} instead of w_w. Also, the absorber pressure must be lower than the evaporation pressure to compensate for the pressure drop. The end absorption temperature T_H is fixed such that the concentration at the outlet of the absorber becomes lower, w_{se} instead of w_s. The actual concentration change reduces significantly due to contributions from both sides.

7.2.2.7 Presence of Noncondensable Gases

Air is the most commonly encountered noncondensable gas in absorption cycles. This air can be the remaining amount of gas in the system if the evacuation of the system has not taken place correctly. It can also leak into the parts of the system that operate below atmospheric pressure or during maintenance. When very high desorption temperatures are applied, it is also possible that the refrigerant dissociates, delivering noncondensable gases.

Air and other noncondensable gases will generally accumulate in both the condenser and the absorber. The condensation pressure increases and with it the desorber pressure (see Figure 7.12, right). The weak solution concentration at the outlet of the desorber becomes higher: w_{we}.

Everywhere in the heat pump where noncondensables are collected, the heat and mass transfer resistances will increase. These gases only play a role in the vapor phase. Diffusion of these gases in the liquid phase is negligible. The effect of noncondensables in the absorber is quite large because they negatively influence the mass transfer process.

The effects discussed are cumulative and lead to a significant reduction of the actual concentration change attained in the absorber. In reality, these effects have been exaggerated

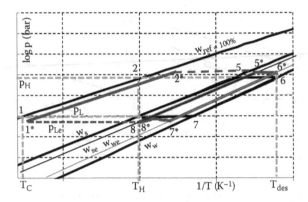

FIGURE 7.13   Combined effect of irreversibilities in log p – $1/T$ diagram.

to make the different effects clear. Figure 7.13 illustrates the combined effects in a log p – $1/T$ diagram.

The theoretical cycle is shown by 5-6-7-8 and 1-2, while the actual cycle is represented by 5*-6*-7*-8* and 1*-2*. The absorption process efficiency is defined as the ratio between the actual concentration change attained in the absorber and the concentration change theoretically expected from the imposed temperatures T_H and T_{des}:

$$\eta_{abs} = \frac{w_{se} - w_{we}}{w_s - w_w} \tag{7.13}$$

The indexes s and w indicate strong and weak solution, respectively (outlet and inlet of the absorber, respectively). The actual concentration change across the absorber has a large impact in the COP of the heat pump cycle. If it is small, then a large amount of solution must be circulated through the absorbent loop to produce the required amount of refrigerant needed for the refrigerant loop. The pumping power is then significantly increased but still quite small. The larger mass flow of absorbent implies a larger heat input in the desorber, which makes the COP of the cycle drop. The non-reversible part of the process increases. This effect can be partially compensated by applying a larger internal heat exchanger (hex2 in Figure 7.7) but cannot be cancelled. Energy and mass balances can be determined making use of enthalpy-concentration diagrams. Figure 7.14 illustrates a theoretical and an actual cycle in the enthalpy-concentration diagram.

7.2.3 Sizing the System

Considering what has been previously discussed in relation to the imposed conditions of source, sink and driving temperature level, it is possible to draw the theoretical cycle in a log p – $1/T$ and in an enthalpy concentration diagram. From this diagram, the heating capacity per circulating kilogram of refrigerant can be determined. Because the heat pump duty is normally a requirement, the mass flow of refrigerant that needs to be circulated can then easily be calculated. This is first illustrated with an example.

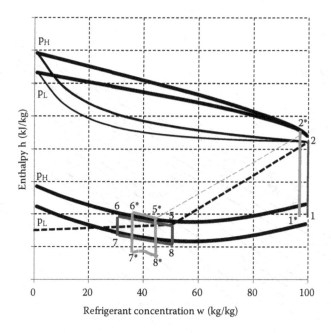

FIGURE 7.14 Combined effect of irreversibilities in enthalpy-concentration diagram.

7.2.3.1 Example with Ammonia-Water

An ammonia-water absorption heat pump heats up a process flow stream from 90°C to 115°C, while the condensation takes place at 100°C, and the strong solution leaves the absorber at 100°C. The heat pump makes use of a waste heat stream as the heat source and cools this stream from 50°C to 40°C, while the refrigerant evaporates at 35°C. Condensing steam at 250°C is used to drive the cycle so that the end desorption temperature reaches 245°C.

The heating capacity delivered by the condenser and absorber is 1000 kW. Use the thermodynamic data of ammonia-water (use, for instance, data from Lemmon et al., 2013) where needed. Assume the produced refrigerant vapor is pure ammonia, and there are saturation conditions at the outlet of condenser, evaporator, absorber and desorber.

Calculate the following:

1. The weak and strong solution concentration circulating through the solution loop.

2. The circulation ratio that follows from these concentrations.

3. The enthalpy in relevant parts of the system.

4. The heat and pumping power requirement of the heat pump.

5. The COP of the cycle.

7.2.3.1.1 Analysis   If equilibrium conditions can be assumed, then the operating conditions can be indicated in a $\log p - 1/T$ diagram. Figure 7.15 illustrates the operating conditions in a $\log p - 1/T$ of ammonia-water. The evaporating temperature determines

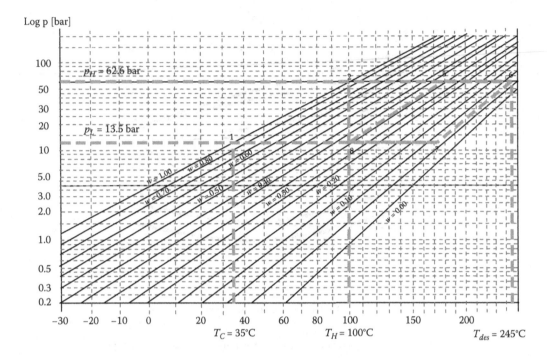

FIGURE 7.15   Heat pump process in log p – $1/T$ diagram for the ammonia-water pair.

the low-pressure level of the cycle (p_L = 13.5 bar), while the condensing pressure determines the high-pressure level of the cycle (p_H = 62.6 bar). From Figure 7.15, it is clear that
the concentration of the solution increases from 7.5% to 37% as the solution passes the
absorber.

A species balance around the desorber allows determination of the circulation ratio
of the solution in the absorbent loop. Figure 7.16 illustrates the conditions around the
desorber.

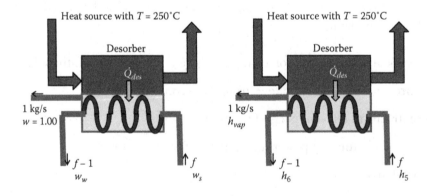

FIGURE 7.16   Species balance around the desorber (left) and energy balance around the desorber
(right).

Per kilogram refrigerant that is desorbed and delivered to the condenser, an amount f of solution must be fed to the desorber:

$$f \times w_s = (f - 1) \times w_w + 1 \times 1.00$$

It follows that f is given by

$$f = \frac{1 - w_w}{w_s - w_w} \tag{7.14}$$

Using the values previously obtained for the concentrations, this gives a circulation ratio of 3.1, which is favorable. This ratio is commonly significantly larger, indicating that the concentration change in the absorber is normally smaller than indicated in this example.

The right-hand side of Figure 7.16 allows for the calculation of the heat input required by the cycle. The energy balance gives

$$\dot{Q}_{des} + f \times \dot{m}_{ref} \times h_5 = (f - 1) \times \dot{m}_{ref} \times h_6 + \dot{m}_{ref} \times h_{vap}$$

so that

$$\dot{Q}_{des} = \dot{m}_{ref} \times \left[f \times (h_6 - h_5) + (h_{vap} - h_6) \right] \tag{7.15}$$

The pumping power can be calculated from

$$\dot{W}_{pump} = \frac{f \times \dot{m}_{ref}}{\rho_8} (p_H - p_L) \tag{7.16}$$

where ρ_8 is the density of the ammonia-water solution at the pump inlet (temperature = 35°C; concentration is 37%), according to REFPROP (Lemmon et al., 2013) equal to 860.54 kg/m³. The heat delivered by the condenser is obtained from

$$\dot{Q}_{cond} = \dot{m}_{ref} \times [h_{vap} - h_{2L}] \tag{7.17}$$

and the heat delivered by the absorber follows from the energy balance around the absorber:

$$\dot{Q}_{abs} = \dot{m}_{ref} \times [f \times (h_7 - h_8) + (h_{1V} - h_7)] \tag{7.18}$$

Because the heat pump delivers 1000 kW to the process flow, and this is the sum of the heat flow in the condenser and in the absorber, the required mass flow of refrigerant is

$$\dot{m}_{ref} = \frac{\dot{Q}_{abs} + \dot{Q}_{cond}}{[f \times (h_7 - h_8) + (h_{1V} - h_7)] + [h_{vap} - h_{2L}]} \tag{7.19}$$

The enthalpies can be read from the concentration-enthalpy diagram for the two relevant pressures, which are given in Figure 7.17. Alternatively, REFPROP can be used for this purpose. Table 7.1 gives a summary of the relevant data.

This allows for the calculation of the relevant parameters: $\dot{m}_{ref} = 0.312$ kg/s; $\dot{Q}_{des} = 619$ kW; $\dot{W}_{pump} = 5.5$ kW. The COP of the cycle can now be calculated:

$$COP = \frac{\dot{Q}_{abs} + \dot{Q}_{cond}}{\dot{Q}_{des} + \dot{W}_{pump}} \qquad (7.20)$$

$$COP = \frac{1000}{619 + 5.5} = 1.60$$

Note that if the pumping power is neglected, the COP becomes only 1% higher (1.615). This is the reason why the pumping power often is neglected in the calculation of the COP.

The condenser and absorber power can also be calculated:

$$\dot{Q}_{cond} = \dot{m}_{ref} \times [h_{vap} - h_{2L}] = 0.312 \times (1576 - 521) = 330 \text{ kW}$$

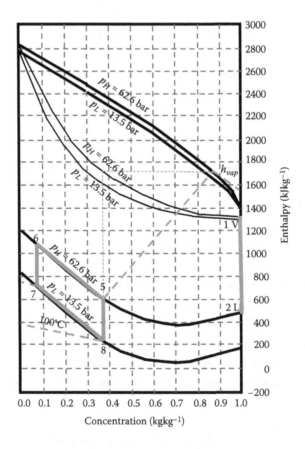

FIGURE 7.17   Heat pump process in ammonia-water enthalpy-concentration diagram.

TABLE 7.1   Conditions at Several Locations of the Equilibrium Heat Pump Cycle

State	T (°C)	Concentration (%)	Enthalpy (kJ/kg)
Vapor at absorber inlet, 1 V	35	100	1288
Liquid at condenser outlet, 2 L	100	100	521
Solution at desorber inlet, 5	180	37	600
Solution at desorber outlet, 6	245	7.5	1080
Solution at absorber inlet, 7	172	7.5	720
Solution at absorber outlet, 8	100	37	210
Vapor at outlet desorber, h_{vap}	245	100	1576

and

$$\dot{Q}_{abs} = \dot{m}_{ref} \times \left[f \times (h_7 - h_8) + (h_{1V} - h_7) \right] = 0.312$$
$$\times [3.1 \times (720 - 210) + (1288 - 720)] = 670 \text{ kW}$$

In this example, the flow that needs to be heated is water, which is heated from 90°C to 115°C. The flow can be heated in parallel or in series. In series, the flow should first pass the condenser because it has a constant temperature and then the absorber, which shows a glide. As previously discussed, a minimum temperature driving force (approach) is required in the two heat exchangers so that the heat flow can take place. This is illustrated in the left-hand side of Figure 7.18. The temperature difference is 5 K at the outlet of the condenser and at the inlet of the absorber. The mass flow that can be heated is 15.9 kg/s, and the flow is heated from 90°C to 115°C. Alternatively, the flow can circulate in parallel through the two heat exchangers: For this, 15.8 kg/s are heated in the condenser from 90°C to 95°C, and 2.1 kg/s are heated in the absorber from 90°C to 167°C. Bringing the two flows together gives a lower temperature of 103.4°C, and the required discharge temperature cannot be attained.

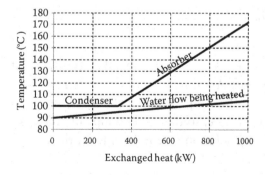
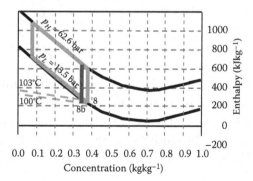

FIGURE 7.18   Left: Heating of the process flow by the heat pump with condenser and absorber connected in series. Right: Effect of subcooling at outlet of the absorber; the outlet condition moves from 8 to 8b.

It should be noted that if the vapor that leaves the desorber is not pure refrigerant, then the refrigerant will show a glide in the condenser so that a higher temperature can be attained at the outlet of the condenser for the same minimum temperature driving force.

7.2.3.1.2 Effect of Nonequilibrium   Figure 7.17 shows that if the vapor would just leave the desorber, the vapor concentration would be around 85%. Rectification can significantly increase the purity of the vapor.

The liquid at the outlet of the absorber needs to be subcooled; otherwise, the mass transfer process cannot take place. Assume the subcooling is 3 K so that the saturation conditions at the absorber outlet are at 103°C. The strong solution concentration is then in reality not 37% but only 33%.

The strong solution is pumped from absorber to desorber. The pumping power has been calculated previously. The enthalpy after the pump can be obtained from the specific pumping power:

$$h_{after\_pump} = h_{8b} + \frac{\dot{W}_{pump}}{f \times \dot{m}_{ref}} \tag{7.21}$$

This gives an enthalpy of $215 + 5.5/(3.1*0.312) = 220.7$ kJ/kg. The solution concentration remains unchanged after the pump. After the pump, the solution enters the solution heat exchanger (hex2 in Figure 7.7). The liquid flow through both sides of the heat exchanger is different: $f \times \dot{m}_{ref}$ through the strong solution side and $(f-1) \times \dot{m}_{ref}$ through the weak solution side. The specific heat of both flows is quite similar, so that the heat capacity at the strong solution side is the largest. The temperature rise in this side is smaller than the temperature rise in the weak solution side.

$$\dot{Q}_{hex2} = f \times \dot{m}_{ref} \times c_{p\_strong}(T_5 - T_{after\_pump}) = (f-1)$$
$$\times \dot{m}_{ref} \times c_{p\_weak}(T_6 - T_7)$$

Because $c_{p\_strong} \approx c_{p\_weak}$, $(T_5 - T_{after\_pump}) < (T_6 - T_7)$, indicating that the smallest temperature glide takes place at the pump side. Assuming a minimum temperature approach of 5 K the at cold side outlet, $T_7 - T_8$, the temperature in state 7 is $100 + 5 = 105$°C, while the weak solution concentration is 7.5%, so that an enthalpy of 400 kJ/kg applies. The enthalpy at the outlet of the strong solution side follows from the energy balance:

$$h_5 = h_{after\_pump} + \frac{(f-1) \times \dot{m}_{ref} \times (h_6 - h_7)}{f \times \dot{m}_{ref}} \tag{7.22}$$

Then, h_5 becomes

$$h_5 = 220.7 + \frac{(3.1-1) \times 0.312 \times (1080 - 400)}{3.1 \times 0.312} = 681.3 \text{ kJ/kg}$$

Figure 7.19 makes clear that the outlet of the heat exchanger at the strong solution side is above the saturation temperature, so the solution starts boiling within the heat exchanger. Operation under such conditions requires that the design of the heat exchanger takes into account the volume flow increase associated with the production of vapor. The velocity of the solution in the rest of the heat exchanger is then lower, leading to lower heat transfer coefficients. This is the reason why generally the heat exchanger is designed so that the strong solution leaves the heat exchanger just at saturated conditions. This can easily be attained by increasing the approach temperature. The advantage of partially boiling in the heat exchanger is that the required desorption heat is reduced: The enthalpy at the desorber inlet is now 681.3 kJ/kg instead of 600 kJ/kg.

Figure 7.19 clearly illustrates that the weak solution leaving the heat exchanger with state 7 is significantly subcooled, so that after the throttling device, at the inlet of the absorber, the solution enthalpy and temperature do not change. Table 7.2 shows the conditions in the real cycle.

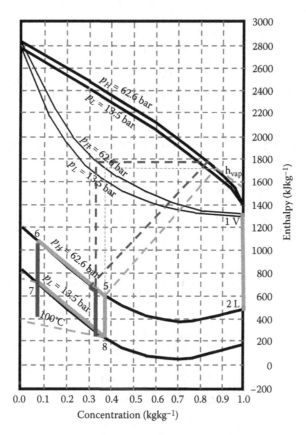

FIGURE 7.19 Enthalpy-concentration diagram showing the temperature change in the internal heat exchanger hex2 assuming a 5 K temperature approach at a cold outlet.

TABLE 7.2 Conditions at Several Locations of the Real Heat Pump Cycle

State	T (°C)	Concentration (%)	Enthalpy (kJ/kg)
Vapor at absorber inlet, 1 V	35	100	1288
Liquid at condenser outlet, 2 L	100	100	521
Solution at desorber inlet, 5	180	33	681
Solution at desorber outlet, 6	245	7.5	1080
Solution at absorber inlet, 7	172	7.5	400
Solution at absorber outlet, 8	100	33	215
Vapor at outlet desorber, h_{vap}	245	100	1576

Repeating the previous calculations for the real cycle conditions, the circulation ratio becomes 3.6:

$$f = \frac{1-w_w}{w_s - w_w} = \frac{1-0.075}{0.33-0.075} = 3.6$$

while the mass flow of refrigerant becomes 0.383 kg/s.

$$\dot{m}_{ref} = \frac{\dot{Q}_{abs} + \dot{Q}_{cond}}{[f \times (h_7 - h_8) + (h_{1V} - h_7)] + [h_{vap} - h_{2L}]}$$

$$= \frac{1000}{[3.6 \times (400-215) + (1288-400)] + (1576-521)} = 0.383 \text{ kg/s}$$

With these values, the power requirement by the desorber and pump can be calculated:

$$\dot{Q}_{des} = \dot{m}_{ref} \times [f \times (h_6 - h_5) + (h_{vap} - h_6)] = 0.383$$
$$\times [3.6 \times (1080-681) + (1576-1080)] = 740.1 \text{ kW}$$

$$\dot{W}_{pump} = \frac{f \times \dot{m}_{ref}}{\rho_8} (p_H - p_L) = \frac{3.6 \times 0.383}{860.54} (6260-1350) = 7.9 \text{ kW}$$

With these values, the COP can be obtained.

$$COP = \frac{\dot{Q}_{abs} + \dot{Q}_{con}}{\dot{Q}_{des} + P_{pump}} = \frac{1000}{740+7.9} = 1.34$$

The COP is significantly lower for the real cycle than for the cycle that is in equilibrium in all states.

7.2.3.1.3 Purification of the Refrigerant Vapor   The desorber is constructed is such a way that the vapor leaving it comes into direct contact with the strong solution entering the desorber. For this reason, the produced vapor is in equilibrium with the boiling strong solution, which corresponds to a vapor concentration as high as possible. Nevertheless, this vapor contains a large amount of absorbent vapor, as illustrated in Figure 7.19. To increase the purity of the vapor, a distillation section and a partial condenser (called the rectifier), which produces a reflux, are installed on top of the desorber. The vapor leaving the desorber and entering the distillation column has a concentration w_V and is in equilibrium with the strong solution concentration w_s. In each distillation column tray, equilibrium is attained between liquid and vapor. Vapor enters from the bottom, and reflux liquid enters from the top. Because each tray is adiabatic, the equilibrium conditions and temperature follow. A comprehensive discussion about the design of the distillation column and amount of heat that needs to be removed in the rectifier is included in the work of Niebergall (1981). Arora (2006) also discussed the topic but with less detail. From a global design point of view, it is important to quantify the required reflux (the circulating refrigerant flow is reduced) and the amount of heat that needs to be removed in the rectifier (reflux condenser). The mass and energy flows that play a role are illustrated in Figure 7.20, which shows the purification section above the desorber.

The ratio between the desorbed flow and the refrigerant flow follows from the species balance:

$$\dot{m}_{v\_des} w_{v\_des} = \dot{m}_{ref} w_{ref} + (\dot{m}_{v\_des} - \dot{m}_{ref}) w_s$$

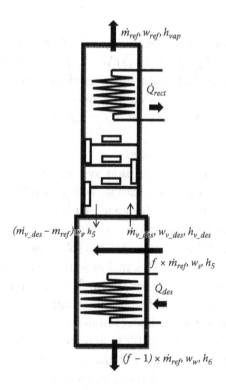

FIGURE 7.20   Desorber with purification section above it (distillation column and rectifier).

so that

$$\frac{\dot{m}_{v\_des}}{\dot{m}_{ref}} = \frac{w_{ref} - w_s}{w_{v\_des} - w_s} \tag{7.23}$$

For the conditions illustrated in Figure 7.19, $w_{v\_des} = 0.82$. If pure refrigerant should leave the purification section, then $w_{ref} = 1.0$, and the ratio becomes 1.37, so that a large amount of vapor needs to be condensed in the rectifier. The amount of heat that needs to be removed follows from an energy balance for the purification section.

$$\dot{m}_{v\_des}h_{v\_des} = \dot{m}_{ref}h_{vap} + (\dot{m}_{v\_des} - \dot{m}_{ref})h_5 + \dot{Q}_{rect}$$

so that

$$\dot{Q}_{rect} = \dot{m}_{v\_des}(h_{v\_des} - h_5)$$

$$-\dot{m}_{ref}(h_{vap} - h_5) = \dot{m}_{ref}\left[\frac{w_{ref} - w_s}{w_{v\_des} - w_s}(h_{v\_des} - h_5) - (h_{vap} - h_5) \right] \tag{7.24}$$

Taking the conditions of Table 7.2 and Figure 7.19 into account, for which $h_{v\_des} = 1770$ kJ/kg and $h_{vap} = 1310$ kJ/kg, $\dot{Q}_{rect} = 330.5$ kW. Because the produced vapor has a relatively low concentration, purification to 100% requires a large reflux condenser. Note that this heat can be used to heat the solution coming from the absorber. In that case, less subcooling will be attained at the inlet of the absorber.

7.2.3.2 Example with Water-Lithium Bromide

Water-lithium bromide is used in absorption systems with evaporating temperatures above 0°C. In general, higher COPs are feasible with this refrigerant-absorbent mixture than with ammonia-water. It has the advantage that the refrigerant is not contaminated with absorbent (the vapor pressure of lithium bromide is negligible), so no purification step is required. A disadvantage of this mixture is that it solidifies at a concentration of lithium bromide of around 70% (30% water). This makes some operating conditions impossible. For example, the conditions for the previous example are not feasible. For this reason, the evaporating conditions have been modified in this example.

A water-lithium bromide absorption heat pump heats up a process flow from 90°C to 115°C, while condensation takes place at 100°C, and the strong solution leaves the absorber at 100°C. The heat pump makes use of a waste heat stream as the heat source and cools this stream from 60°C to 55°C, while the refrigerant evaporates at 50°C. Condensing steam at 180°C is used to drive the cycle, so that the end desorption temperature reaches 175°C.

The heating capacity delivered by the condenser and absorber is 1000 kW. Use the thermodynamic data of water-lithium bromide (use, for instance, data from Kim and Infante Ferreira, 2006) where needed. Because, as a salt, lithium bromide has negligible vapor pressure, the produced refrigerant vapor is pure water. Assume that saturation conditions are attained at the outlet of the condenser, evaporator, absorber and desorber.

Calculate the following:

1. The weak and strong solution concentration circulating through the solution loop.

2. The circulation ratio that follows from these concentrations.

3. The enthalpy at relevant parts of the system.

4. The heat and pumping power requirement of the heat pump.

5. The COP of the cycle.

7.2.3.2.1 Analysis   If equilibrium conditions can be assumed, then the operating conditions can be indicated in a $\log p - 1/T$ diagram. Figure 7.21 illustrates the operating conditions in a $\log p - 1/T$ diagram of water-lithium bromide.

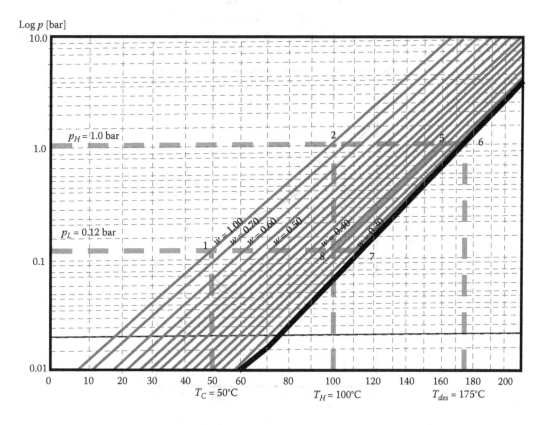

FIGURE 7.21   Heat pump process in water-lithium bromide $\log p - 1/T$ diagram. The bold line indicates the crystallization limit of LiBr.

The evaporating temperature determines the low-pressure level of the cycle (p_L = 0.12 bar), while the condensing pressure determines the high-pressure level of the cycle (p_H = 1.0 bar). From Figure 7.21, it is clear that the concentration of the solution increases from 32% to 38% as the solution passes the absorber. The bold line indicates the crystallization limit and shows that, for the present conditions, 175°C is the highest allowable temperature. Note that in Figure 7.21 the concentration is expressed in kilograms refrigerant per kilogram solution, and that most frequently the concentration is expressed in kilograms lithium bromide per kilogram solution.

Equations 7.14 to 7.22 can be used to quantify the relevant data, but these equations require that the enthalpy is known at several locations in the cycle. Equation 7.14 gives a circulation ratio of 11.3, which is significantly larger than for the ammonia-water heat pump. The relevant enthalpy values can be read from Figure 7.22. The enthalpy values for water vapor are not given in the figure and should be derived from, for instance, the work of Lemmon et al. (2013). Table 7.3 lists the relevant enthalpy values.

The mass flow of refrigerant results from Equation 7.19: $\dot{m}_{ref}$ = 0.194 kg/s. The power requirement by the desorber and pump can be calculated using Equations 7.15 and 7.16: $\dot{Q}_{des}$ = 526.8 kW and $\dot{W}_{pump}$ = 0.14 kW, with the density of the solution taken as 1400 kg/m³. With these values, the COP can be obtained making use of Equation 7.20:

$$COP = \frac{\dot{Q}_{abs} + \dot{Q}_{cond}}{\dot{Q}_{des} + \dot{W}_{pump}} = \frac{1000}{526.8 + 0.14} = 1.90$$

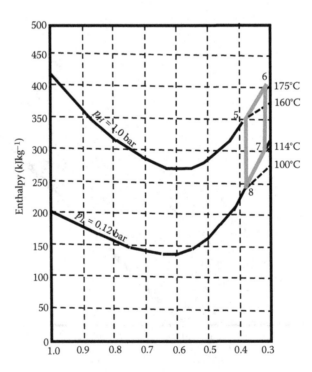

FIGURE 7.22 Heat pump process in water-lithium bromide enthalpy-concentration diagram. The two lines indicate the saturated liquid lines for p_L = 0.12 bar and p_H = 1.0 bar.

TABLE 7.3 Conditions at Several Locations of the Water–Lithium Bromide Heat Pump Cycle

State	T (°C)	Concentration (%)	Enthalpy (kJ/kg)
Vapor at absorber inlet, 1 V	50	100	2373
Liquid at condenser outlet, 2 L	100	100	200
Solution at desorber inlet, 5	160	38	355
Solution at desorber outlet, 6	175	32	400
Solution at absorber inlet, 7	114	32	300
Solution at absorber outlet, 8	100	38	240
Vapor at outlet desorber, h_{vap}	175	100	2607

Note that for the water-lithium bromide heat pump, the pumping power is only 0.01% of the delivered capacity, and the efficiency of this cycle is significantly larger than for the ammonia-water cycle.

7.2.4 Multieffect Operation

Making use of internal heat exchangers, it is possible to improve the COP of absorption heat pumps. Figure 7.23 has been derived from the work of Kim (2006) and Grossman and Zaltash (2001) and illustrates the impact of multieffects on performance. *Effect* stands here for refrigerant production locations; in a double-effect cycle, the refrigerant vapor coming out of the desorber is condensed in a second desorber at an intermediate-pressure level so that a second site for the production of refrigerant results.

Alternatively, the refrigerant produced in a single desorber can be partially consumed internally to make absorption possible at temperatures lower than the application level. In this case, the efficiency is lower than for the single-effect cycle, as illustrated in the figure. A single-effect heat pump will have a COP around 1.8, while a double-effect heat pump will have a COP of around 2.2. The triple-effect cycle shows even higher COPs but is not

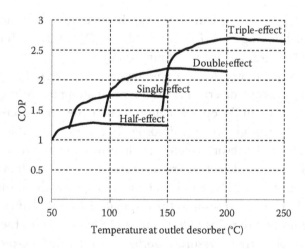

FIGURE 7.23 The COP of absorption heat pumps as a function of the number of effects with the end desorption temperature as a parameter.

available commercially, while single- and double-effect pumps are. In the next example, a double-effect cycle is discussed.

7.2.4.1 Double-Effect Example with Water–Lithium Bromide

The literature reports two configurations for double-effect absorption cycles: the series and the parallel configuration. The parallel configuration has a slight higher COP compared with the series configuration (Herold et al., 1996; Arun et al., 2001; Garousi Farshi et al., 2011); therefore, it is considered for this example. A double-effect water–lithium bromide absorption heat pump heats up a process flow from 90°C to 115°C, while condensation takes place at 100°C and the strong solution leaves the absorber at 100°C. The heat pump uses a waste heat stream as a heat source and cools this stream from 60°C to 55°C, while the refrigerant evaporates at 50°C. Condensing steam at 250°C is used to drive the cycle so that the end desorption temperature reaches 240°C.

The heating capacity delivered by the condenser and absorber is 1000 kW. Use the thermodynamic data of water–lithium bromide (use, for instance, the work of Kim and Infante Ferreira, 2006) where needed. Because, as a salt, lithium bromide has negligible vapor pressure, the produced refrigerant vapor is pure water. Assume that saturation conditions are attained at the outlet of the condenser, evaporator, absorber and desorber.

Calculate the following:

1. The weak and strong solution concentration circulating through the solution loop.

2. The circulation ratio that follows from these concentrations.

3. The enthalpy at relevant parts of the system.

4. The heat and pumping power requirement of the heat pump.

5. The COP of the cycle.

7.2.4.1.1 Analysis   If equilibrium conditions can be assumed, then the operating conditions can be indicated in a $\log p - 1/T$ diagram. Figure 7.24 illustrates the operating conditions in a $\log p - 1/T$ diagram of water–lithium bromide.

The evaporating temperature determines the low-pressure level of the cycle ($p_L = 0.12$ bar), while the condensing pressure determines the intermediate-pressure level of the cycle ($p_H = 1.0$ bar). The high pressure in the cycle is determined by the condensation temperature of the vapor produced in the high-pressure desorber. In this example, it is assumed that 50% of the flow pumped from the absorber enters the intermediate desorber, and that the other 50% enters the high-pressure desorber. From Figure 7.24, it is clear that the concentration of the solution increases from 32% to 38% as the solution passes the absorber. Both desorbers reduce the solution concentration to the original 32% so that after the two flows are rejoined, the concentration remains 32%. Figure 7.25 illustrates how the cycle is arranged. The system consists of a high-pressure desorber, an intermediate-pressure desorber, a condenser, an absorber, an evaporator, two solution heat exchangers, a pump and several expansion valves. This configuration has three pressure levels. The low-pressure level in

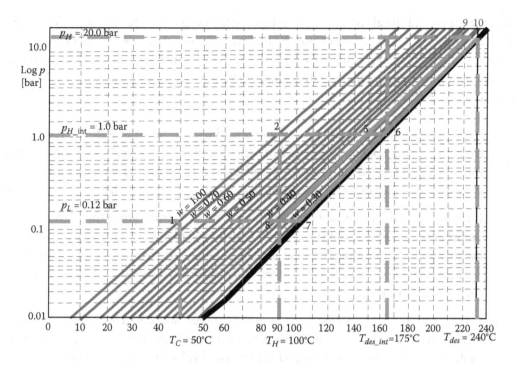

FIGURE 7.24   Double-effect heat pump process in water-lithium bromide log p − $1/T$ diagram.

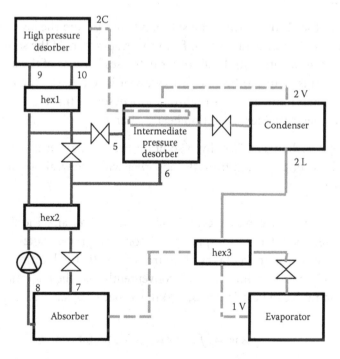

FIGURE 7.25   Schematic of the double-effect absorption heat pump in parallel configuration. The black lines indicate the strong and the weak solution flows. The gray lines indicate the refrigerant flows. The dotted lines indicate the vapor phase, and the solid lines indicate the liquid phase.

the evaporator and the absorber is determined by the evaporation temperature. The intermediate pressure in the condenser and the intermediate-pressure desorber depends on the pressure in the condenser. The high pressure in the high-pressure desorber depends on the condensation pressure (temperature) in the intermediate-pressure generator. The main components operate as follows:

- High-pressure desorber: Heat is supplied to the desorber to desorb the refrigerant from the strong solution coming from the absorber. The superheated water vapor flows to the intermediate-pressure desorber, and the resulting weak solution flows through the high-temperature solution heat exchanger (hex1) to the throttling device.

- Intermediate-pressure desorber: Superheated vapor coming from the high-pressure desorber is condensed, and the rejected heat is used to desorb more refrigerant from the strong solution coming from the absorber. The superheated water vapor flows to the condenser, and the weak solution flows to the absorber through the low-temperature solution heat exchanger (hex2).

- Condenser: The superheated water vapor is cooled and condensed into the liquid state. While the refrigerant is condensed, heat is rejected to the heating fluid. After that, the condensate flows through the expansion valve to the evaporator.

- Evaporator: The refrigerant evaporates while it receives energy from the low-temperature heat source.

- Absorber: In the absorber, the weak solution coming from the low-temperature heat exchanger absorbs refrigerant vapor from the evaporator. At the same time, the concentration of the solution in the absorber increases because of the absorbed refrigerant. Because this is an exothermic process, heat is rejected to the heating fluid. The strong solution is pumped to the high-pressure desorber through the low- (hex2) and the high-temperature (hex1) solution heat exchangers to close the cycle.

- Solution heat exchangers: Usually, a solution heat exchanger is used to preheat, for example, the strong solution from the absorber with the weak solution coming from the desorber.

Figure 7.26 shows the cycle in a concentration-enthalpy diagram. The absorption process is given by line 7-8. The intermediate-pressure desorption process takes place from 5 to 6, and the high-pressure desorption takes place from 9 to 10. The location of these states is also indicated in Figure 7.25, while Table 7.4 conveniently lists the relevant enthalpy values. The circulation ratio can be calculated, making a species balance around the absorber.

$$f \times w_s = (f-1) \times w_w + \dot{m}_{ref} \times 1.00$$

The circulation ratio is again given by Equation 7.14 and is equal to 11.3. Vasilescu and Infante Ferreira (2014) investigated the effect of the distribution ratio of the strong solution

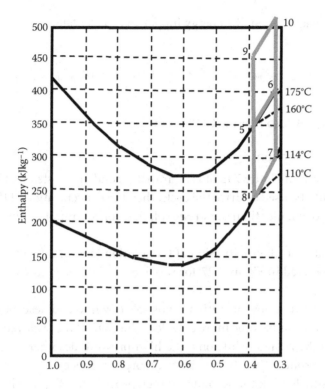

FIGURE 7.26   Double-effect cycle in a water-lithium bromide enthalpy-concentration diagram.

TABLE 7.4   Conditions at Several Locations of the Double-Effect Water–Lithium Bromide Heat Pump Cycle

State	T (°C)	Concentration (%)	Enthalpy (kJ/kg)
Vapor at absorber inlet, 1 V	50	100	2373
Liquid at condenser outlet, 2 L	100	100	200
Solution at desorber inlet, 9	225	38	460
Solution at desorber inlet, 10	240	32	520
Solution at intermediate desorber inlet, 5	160	38	355
Solution at intermediate desorber outlet, 6	175	32	400
Solution at absorber inlet, 7	114	32	300
Solution at absorber outlet, 8	100	38	240
Vapor at outlet desorber, h_{vap}	175	100	2607
Vapor at outlet desorber, h_{vap_hp}	225	100	2886
Liquid at outlet int. desorber, h_{vap_hpL}	175	100	763

flow between the two desorbers. An even distribution in general gives good results, and that is also the ratio assumed here so that the strong solution entering both desorbers is equal to $f/2$. Also, the refrigerant flow leaving each desorber is half of the total refrigerant flow $\dot{m}_{ref}/2$. The mass flow rate of refrigerant can be obtained from

$$\dot{m}_{ref} = \frac{\dot{Q}_{abs} + \dot{Q}_{cond}}{[f \times (h_7 - h_8) + (h_{1V} - h_7)] + \frac{1}{2}[h_{vap} - h_{2L}] + \frac{1}{2}[h_{vap\_hp\_L} - h_{2L}]}$$

so that $\dot{m}_{ref} = 0.346$ kg/s. Then, the energy input can be calculated

$$\dot{Q}_{des\_high\_pressure} = \frac{\dot{m}_{ref}}{2} \times \left[\frac{f}{2} \times (h_{10} - h_9) + (h_{vap\_hp} - h_{10}) \right] \qquad (7.25)$$

$\dot{Q}_{des\_high\_pressure} = 468$ kW, so that the COP is now 2.1.

7.2.5 Multistage Operation

In the example given in Figure 7.15, it is clear that if condensing steam of 180°C or a lower temperature should be used to drive the cycle, then the concentration of the weak solution becomes equal to or smaller than the strong solution concentration so that the cycle cannot operate anymore.

A possible solution is the subdivision of the solution loop into two stages, two separate loops. This is illustrated in Figure 7.27, for which an end desorption temperature of 150°C has been assumed.

Figure 7.27 shows that now operation is possible down to end desorption temperatures of 130°C. The system has become more complex. The COP of the cycle reduces because two desorbers need to be heated, while only the high-pressure desorber produces refrigerant vapor. The vapor produced in the low-stage desorption process, process 5-6, is absorbed at the same pressure level in the second absorber, process 7H-8H, which also delivers heat to the process flow that needs to be heated. Because this vapor flow is directly absorbed,

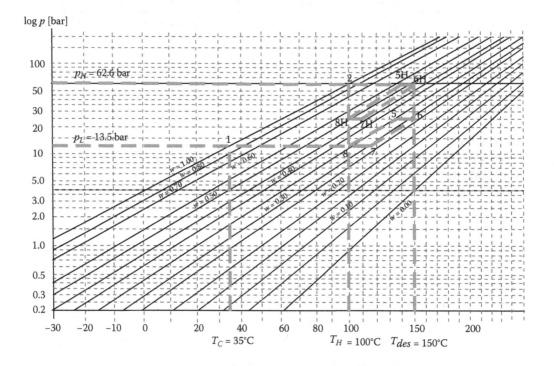

FIGURE 7.27   Two-stage cycle in ammonia-water $\log p - 1/T$ diagram.

TABLE 7.5   List of Large-Capacity Absorption System Vendors

Vendor	Type	Working Fluid	Heating Fluid
Broad	Single effect/double effect	H_2O-LiBr	Hot water/steam
Carrier/Sanyo	Single effect/double effect	H_2O-LiBr	Hot water/steam/direct fired
Colibri	Single effect/double effect	NH_3-H_2O	Hot water/steam/process gas
Ebara	Single effect/double effect	H_2O-LiBr	Hot water/steam
Hitachi	Double effect	H_2O-LiBr	Steam/direct fired
Johnson Controls (York)	Single effect/double effect	H_2O-LiBr	Hot water/steam
Shuangliang	Single effect/double effect	H_2O-LiBr	Hot water/steam/direct fired
Thermax	Single effect/double effect	H_2O-LiBr	Hot water/steam/direct fired
Trane	Single effect/double effect	H_2O-LiBr	Hot water/steam/direct fired

Note: Evaporator capacities in the range 3.5 to 5.0 MW.

there are no concerns about its purity so that rectification is not necessary for the low-stage desorber. The vapor produced in the high-pressure desorber, process 5H-6H, circulates through the refrigerant loop and needs to be purified.

7.2.6 Stage of Development

Absorption heat pumps can be useful when both heating and cooling are necessary, and this is a strong incentive for the development of such systems. A list of vendors of absorption systems with large capacities is shown in Table 7.5. These vendors sell mainly chillers, but the machines can also operate as heat pumps. In Europe, Trane also sells Thermax absorption systems (more details available at http://www.thermaxglobal.com). Also, Carrier (http://www.carrier.com) cooperates with Sanyo for what concerns absorption systems.

7.3 SOLID-VAPOR ADSORPTION HEAT PUMPS

Solid sorption heat pumps have zero ozone-depletion potential (ODP) as well as zero global warming potential (GWP) because they use environmentally friendly refrigerants such as water. When compared with the liquid-vapor absorption heat pump systems, solid-vapor adsorption systems have a wider range of working temperatures and have no problems of corrosion and solution crystallization. Moreover, these systems are simple and reliable and, except for the (automatic) valves, have no moving parts, making them practically maintenance free. Because of poor heat transfer characteristics and limited capacity of an adsorbent, an adsorption heat pump is likely bigger than an absorption heat pump for the same capacity. The low specific cooling power (SCP) significantly limits this technology from its wider use. Also, the COP of solid-vapor sorption systems is usually too low to be economically competitive. Infante Ferreira and Kim (2014) have determined the average second law efficiency of water–silica gel adsorption systems from experimental data reported in literature as 0.25, while liquid-vapor sorption systems generally attain values of 0.50 (see the definition of the reference Carnot efficiency in Equation 7.5). For a solid sorption refrigeration system, the working performance is mainly dependent on the sorption working pair, heat and mass transfer inside sorption beds and the sorption heat pump cycle self.

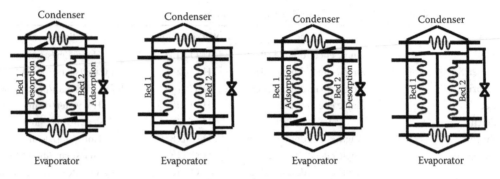

Phase I –adsorption in bed 2   Interphases–valves closed   Phase II –adsorption in bed 1   Interphases–valves closed

FIGURE 7.28 Continuous solid-vapor adsorption cycles require two bed reactors, which switch their function between desorption and adsorption processes.

Figure 7.28 illustrates the working principle of this cycle. The evaporator and condenser are generally separate heat exchangers and have a function similar to their function in liquid-vapor absorption systems. The thermal compressor consists of two beds with the adsorbent through which the external fluids flow. Saturation of the adsorbent bed with refrigerant requires its regeneration, which takes place during the desorption process. The external fluid flow must also switch between the two beds. During desorption, the heating flow goes through the bed, while during adsorption the flow that needs to be heated flows through the bed.

During desorption, the refrigerant vapor is released from the adsorbent bed and the pressure in the desorption volume increases until the pressure is high enough to lift the automatic valve. The vapor is condensed in the condenser so that the pressure further remains constant during condensation. During adsorption, the pressure decreases in the adsorption volume until the pressure is low enough to lift the automatic valve on the evaporator side. The vapor produced in the evaporator is adsorbed in the adsorbent bed. In this way, the evaporator pressure is maintained constant during the adsorption process. Figure 7.29 illustrates how the system operates as a function of time.

Figure 7.29 shows that, as an adsorbent bed starts operating, a large amount of heat must be delivered to the desorber: A large temperature difference between in- and outlet temperature of the heating medium is than attained. As time progresses and the saturation of the bed with refrigerant approaches, this difference becomes smaller and smaller, and at some moment, the function of the two adsorbent beds is reversed. A cycle is, in this example, 8 minutes, of which only 7 minutes are available for capacity delivery.

The sorbents utilized can be divided into physical and chemical materials according to the nature of the forces involved in the solid-gas sorption process. Physical adsorption is different from chemical adsorption in the aspect that physical adsorption does not involve any change in its chemical composition or in its phase in the process. Physical adsorption is based on intermolecular forces, while in chemical adsorption a chemical reaction is the driving principle. Physical and chemical adsorption technologies are separately discussed.

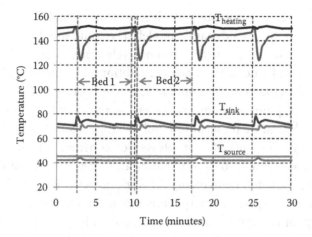

FIGURE 7.29 Temperature change with time of the external flows applied to an adsorption heat pump cycle. Continuous solid-vapor adsorption cycles require two bed reactors, which switch their function between desorption and adsorption processes. As the function of the beds switch, the system stops delivering capacity.

7.3.1 Physical Adsorption

Adsorbents like zeolite, silica gel, activated carbon and alumina are physical adsorbents, having highly porous structures with surface-volume ratios on the order of several hundreds. They have numerous micropores that can selectively catch and hold certain types of substances, including refrigerants. When they are saturated with the refrigerant, they can be regenerated simply by being heated. The fundamental process behind adsorption heat equipment is the storage of heat. Because refrigerant accumulation is limited, each adsorber bed must be loaded and unloaded intermittently, as illustrated in Figure 7.28 and Figure 7.29. Continuous operation is only possible when multiple adsorbers are combined. The refrigerant vapor that cycles through the system is adsorbed and desorbed intermittently. The refrigerant released in the evaporator is adsorbed in the adsorber. When the adsorber is in the desorption phase, the produced refrigerant vapor is condensed in the condenser. In continuous systems using multiple adsorption beds, sorption and desorption take place simultaneously.

Because solid sorption processes are intermittent, the role of dynamic effects is substantial and becomes more important when cycle times become smaller. Sorption bed(s) need to be heated and cooled every cycle. In the heat pump cycle, the bed is heated during regeneration. In this phase, the regeneration (sorption) heat and the bed heating load need to be supplied to the bed. During the adsorption phase, the sensible heat delivered during the regeneration phase and the adsorption heat need to be extracted from the bed to enable high adsorption bed loadings. In most solid sorption operations, two beds are used to enable continuous operation. A part of the bed heating load of one bed can be reused to heat the other bed. During normal continuous operation, however, the sorption beds never reach full equilibrium. The amount of refrigerant that is effectively used in these cases depends on cycle time, secondary flow rates and temperature levels. Predicting actual sorption efficiency for certain operating conditions is not possible with a time-independent approach (prediction requires a dynamic model). Dynamic simulations show that a sorption efficiency of 50% scales well with common cycle times of around 10–20 minutes for most configurations under normal operating conditions. The sorption efficiency is defined as the ratio of actual to theoretical amount of active refrigerant.

Solid sorption heat pump systems are merely thermodynamic systems based on heat exchangers, in which heating (desorption) and cooling (adsorption) processes are always accompanied by large amounts of heat released or consumed during the sorption process. Therefore, a good design to optimize heat and mass transfer in the adsorbent bed heat exchanger is important for enhancing the working performance of solid sorption heat pump systems. All heat inputs and outputs require a driving force: Evaporation, condensation, bed heating for desorption and bed cooling for adsorption are enabled by external flows at different temperatures. Where heat is delivered to the system, the secondary flow requires a temperature higher than the process temperature T_{des}. The two processes at T_H extract heat from the system and require a process flow temperature below the internal heat pump temperature during condensation and during adsorption. In practice, sorption takes place at a gradually increasing or decreasing bed temperature. If the external fluid input temperature is constant, bed temperature and external fluid exit temperature gradually approach the external fluid entrance temperature as illustrated in Figure 7.29. Saturated refrigerant liquid can be assumed at the outlet of the condenser, and saturated refrigerant vapor can be assumed at the outlet of the evaporator. If the condenser and the adsorbing bed are connected to the same external flow (process flow that needs to be heated), the eventual equilibrium temperature will be the same. During normal operation, they both vary in time. Because the condenser in most cases is kept at a constant temperature and the adsorbent bed switches between T_H and T_{des} each cycle, the temperature differences between them can vary significantly. These temperature differences also depend on the system configuration. Just as for the absorption cycles, if the two flows are connected to the same heat sink, they can be connected in series or parallel. Pressure differences between system components are the driving force of the refrigerant flow between these components. Increasing the pressure differences will increase not only the flow rate but also the corresponding flow losses.

In the theoretical cycle, the temperature differences needed for heat transfer are neglected so that the process side temperatures will be equal to the temperature of the external fluid inputs. In the actual cycle, the cycle temperatures must account for the temperature difference required for heat and mass transfer to and from the external fluid (e.g. T_{des} must be lower than the temperature of the external heating flow). Just as for absorption cycles, the assumption that condensation and end adsorption temperature are identical simplifies the modeling of the process.

The equilibrium evaporation and condensation pressure are determined by T_C and T_H, the source and sink temperatures of the heat pump, respectively. The evaporation heat load gives the heat that is subtracted from the heat source at environmental temperature T_C. The condensation heat is part of the heat delivered at heat pump application temperature T_H. A silica gel–water heat pump, as the most commonly used refrigerant–solid sorbent

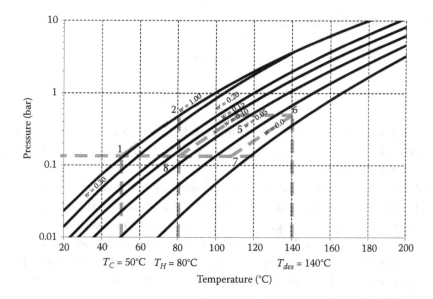

FIGURE 7.30 Adsorption cycle in a silica gel–water log p – $1/T$ diagram. Desorption takes place between states 5 and 6; adsorption takes place between states 7 and 8. State 2 indicates the condensation temperature and pressure (pure water) and state 1 the evaporation temperature and pressure (pure water).

combination, is considered. Figure 7.30 shows the cycle in a log p – $1/T$ diagram of the silica gel–water system.

Evaporator and condenser equilibrium pressures are determined from the saturation line of pure water. The heat released during condensation, per kilogram adsorbent bed, is determined using

$$\Delta H_{cond} = \eta_w \times \Delta w \times (h_{v\_5-6} - h_{2L}) \tag{7.26}$$

with Δw the theoretical amount of active refrigerant that is released and adsorbed each sorption cycle; η_w is the sorption efficiency. Most sorption cycle loads are directly dependent on the amount of active refrigerant. The active amount of refrigerant used in a sorption cycle is defined as the difference between adsorption and desorption equilibrium bed loadings. These bed loadings are determined by the bed temperatures and pressures during the two separate phases. The desorption heat load is the heat required per cycle to regenerate the adsorbent bed at the regeneration temperature T_{des}. Bed heating losses are determined by estimating the bed heating load (divided into refrigerant and adsorbent heating load). With the sorption and liquid/vapor heat loads and the bed heating losses, the system efficiencies can be determined.

Sorption equilibrium pressures are estimated using Clausius-Clapeyron in the form given by Cacciola et al. (1993):

$$\ln p = A(w) + \frac{B(w)}{T} \tag{7.27}$$

in which p is expressed in millibars and

$$A = a_0 + a_1 w + a_2 w^2 + a_3 w^3 \tag{7.28}$$

$$B = b_0 + b_1 w + b_2 w^2 + b_3 w^3 \tag{7.29}$$

The a and b values are fitted to experimental data (see Table 7.6); the w values are in loading percentages (100% = 1 kgkg^{-1}); T is the sorption bed temperature (K) and p is the equilibrium pressure (mbar).

The difference between the adsorption and desorption bed loading represents the active amount of refrigerant based on equilibrium states. This is an indication for the theoretical maximum amount of active refrigerant at a certain operation condition (operating temperatures and pressures). In real adsorption heat pumps, the actual amount of active refrigerant at a certain operation condition is mainly determined by the cycle time. The sorption efficiency gives the ratio between the actual amount of active refrigerant to the theoretical amount of active refrigerant:

$$\eta_w = \frac{\Delta w_{actual}}{\Delta w_{theoretical}} \tag{7.30}$$

where $\Delta w_{theoretical}$ is the theoretical amount of active refrigerant:

$$\Delta w_{theoretical} = w_{ads} - w_{des} \tag{7.31}$$

with w_{ads} the water concentration after the adsorption process (states 8 and 5 in Figure 7.30, charged bed) and w_{des} the water concentration after the desorption process (states 6 and 7 in Figure 7.30, discharged bed). The adsorption heat per kilogram absorbent bed is determined using

$$\Delta H_{ads} = \eta_w \frac{\int_{w_{des}}^{w_{ads}} \Delta h_{ads}\, dw}{M_{ref}} \tag{7.32}$$

TABLE 7.6   Constants in Silica Gel Equilibrium Sorption Equations, Equations 7.27 to 7.29

i	a	b
0	23.308	−7200.8
1	−0.4429	244.8
2	0.023159	−11.52
3	−0.0003396	0.1719

To solve this in discrete steps, the sorption heat is determined using Riemann sums:

$$\Delta H_{ads} = \eta_w \frac{\left(\displaystyle\sum_{i=1}^{\frac{w_{des}}{\delta w}} \delta w \times \Delta h_{ads} - \sum_{i=1}^{\frac{w_{ads}}{\delta w}} \delta w \times \Delta h_{ads} \right)}{M_{ref}} \tag{7.33}$$

where

$$\Delta h_{ads}(w) = -BR \tag{7.34}$$

with B the slope of the equilibrium line, Equation 7.29, and $R = 8.314$ JK^{-1}mol^{-1}. Δh_{ads} is expressed in Joule per mole, δw is the bed loading step size (for instance, 0.0001 kgkg^{-1}); w_{des} and w_{ads} are the discharged and charged bed loadings, respectively.

To determine the heating losses, it is assumed that there are no cyclic losses for the condenser and evaporator because both are held at a constant temperature. It is further assumed that the sorption bed is heated using high-temperature heat. The energy required to heat the bed is therefore added to the energy input, reducing the overall performance. This heating load is divided into heating loads required for the adsorbent, the refrigerant and the extra heat capacity related to the metallic parts of the bed. These assumptions result in the heat load required to heat the adsorbent and the extra heat capacity used for the bed construction:

$$Q_{bed\_loss} = m_{bed} \times (c_{p\_bed} + c_{p\_construction}) \times (T_{des} - T_H) \tag{7.35}$$

The heat loss load is determined by the bed mass m_{bed}. The extra heat capacity is a factor to incorporate the adsorbent bed construction, time effects and, if applicable, the reuse of bed heat.

Besides the pure adsorbent, the refrigerant contains must also be heated to desorption temperature. In practice, desorption will take place during bed heating, reducing the amount of refrigerant to heat. The refrigerant bed loading w_{ads} is used for this first estimation, which is the state in which the bed desorption phase starts. The specific heat of adsorbed water is approximated by the specific heat of water vapor.

$$Q_{bed\_water} = m_{bed} \times w_{ads} \times c_{p\_water\_vapor} \times (T_{des} - T_H) \tag{7.36}$$

where $c_{p\_water\_vapor}$ is the specific heat of water vapor at 25°C, 1864 Jkg^{-1}K^{-1} (Smith et al., 2001).

The total high-temperature energy input is obtained by combining the high-temperature desorption heat with the bed heating loads:

$$Q_{des} = \Delta H_{ads} \times m_{bed} + Q_{bed\_loss} + Q_{bed\_water} \tag{7.37}$$

The COP of the cycle can be calculated using

$$COP = \frac{(\Delta H_{cond} + \Delta H_{ads}) \times m_{bed}}{Q_{des}} \tag{7.38}$$

7.3.2 Chemical Adsorption

In chemical adsorption systems, a second adsorbent bed (reactor) can be used instead of the evaporator and condenser. In this case, two different chemicals (salts) with a different operational range (vapor pressure saturation lines) are used. A low-temperature salt (LTS) will adsorb the refrigerant, while a high-temperature salt (HTS) is desorbing and vice versa (Yu et al., 2008).

Chemical adsorption is characterized by the strong chemical bond between the refrigerant and the adsorbent. Therefore, it is more difficult to reverse and thus requires more energy to remove the adsorbed molecules than in physical adsorption. Adsorption capacity is also much larger. Besides, unlike physical adsorption, chemical adsorption is a monovariant process. That is, once pressure is fixed, the reaction temperature is automatically determined. This is illustrated in Figure 7.31, where the temperature of heat rejection to the process flow needing heating T_H determines the operating pressures and also the temperature of the heat sources T_C and T_{des}.

The calculation of cycle heat loads for the LTS bed (two-salt system) and the combination of condenser and evaporator (one-salt system) are analogous. The main difference in the calculation procedure is the absence of cyclic LTS bed heating for the one-salt system as the evaporator and the condenser are kept at constant temperature. In the description of the chemical sorption system, the deviations of the one-salt system are discussed further in the chapter. To estimate chemical sorption heat pump performance, the LTS and HTS sorption heat loads for each operation stage must be known. The heat loads required to

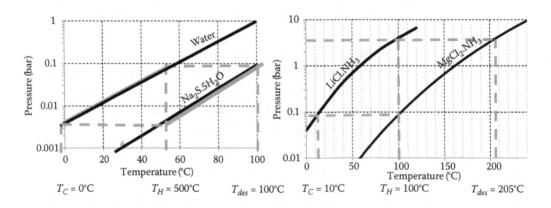

FIGURE 7.31 Adsorption cycle in a Na$_2$S-water log p – $1/T$ diagram (left) and in a LiCl.NH$_3$-MgCl$_2$.NH$_3$ log p – $1/T$ diagram (right, two-salt system). High-temperature desorption takes place at T_{des}; adsorption takes place at T_H. Condensation (left) and high-pressure adsorption (right, two-salt system) take place at T_H. Evaporation (left) and low-pressure desorption (right) take place at T_C.

heat the sorption bed(s) and the refrigerant contained in the system before and during the HTS desorption stage are used to estimate the total heat(ing) losses.

Realistic performance numbers are approached by imposing a sorption efficiency and a relative bed heat capacity to the idealized equilibrium steady-state performance. In the definition of the theoretical chemical sorption cycle, the following temperature levels play a role: T_H, temperature of heat delivery to the process flow that needs to be heated (HTS and LTS adsorption temperature [one salt: condensation temperature]); T_C, temperature of heat uptake; LTS desorption temperature (one salt: evaporation temperature); T_{des}, high-temperature heat input; HTS desorption temperature.

In models, the LTS and HTS sorption heat loads are considered equal for all in- and output temperatures. In the example on the right-hand side of Figure 7.31, the LiCl (LTS) sorption heat is 1285 kJ kg_{HTS}^{-1}, and the $MgCl_2$ (HTS) sorption heat is 1373 kJ kg_{HTS}^{-1}. The variation in performance is only determined by the heat losses. Van der Heuvel (2010) has reported LTS heat losses of about 25%–30% of the total heat input. He also reported maximum HTS bed heat losses of about 45% of the HTS sorption heat.

Sorption beds need to be heated and cooled every cycle. High-temperature HTS desorption takes place until the maximum temperature T_{des}. The LTS bed has to be heated to the heat delivery temperature T_H using LTS adsorption heat. All heat inputs and outputs require a driving force. Evaporation, condensation, bed heating/desorption and bed cooling/adsorption are enabled by the external flows at different temperatures. Where heat is put into the system, the secondary flow requires a temperature higher than the process temperature. Cooling to T_C and heating HTS to T_{des} requires external temperatures below or above these temperatures. In practice, sorption takes place at a gradually increasing or decreasing bed temperature. If the external fluid input temperature is constant, bed temperature and secondary fluid exit temperatures gradually approach the external fluid inlet temperature. Chemical sorption processes take place in discrete steps according to the reaction phases in which the refrigerant is adsorbed. Using $MgCl_2$ as an example, the maximum amount of ammonia adsorbed by $MgCl_2$ is 6 molecules. While the first and second ammonia molecules are adsorbed in separate steps (transitions) that can be recognized as such, the third to the sixth step occur simultaneously and cannot be recognized separately. Anhydrous $MgCl_2$ absorbs up to 6 moles of NH_3 according to the following chemical reactions:

$$MgCl_2(s) + NH_3(g) \rightarrow Mg(NH_3)Cl_2(s) \tag{7.39}$$

$$Mg(NH_3)Cl_2(s) + NH_3(g) \rightarrow Mg(NH_3)_2Cl_2(s) \tag{7.40}$$

$$Mg(NH_3)_2Cl_2(s) + 4NH_3(g) \rightarrow Mg(NH_3)_6Cl_2(s) \tag{7.41}$$

Cycle time dependency of bed loading and bed efficiency are captured in the sorption efficiency previously defined for the physical adsorption processes η_w. This parameter is used to determine the active amount of refrigerant using the amount of refrigerant between the equilibrium adsorption state and the equilibrium desorption state. The heating of the HTS bed during the HTS desorption phase is used as an indication for the heat(ing) losses during a single sorption cycle. The heating losses are divided into the same three terms as for physical sorption. The heat loads required to heat a sorption bed can be approximated using constant specific heats (temperature independent). These loads are assumed to be fed to the system at constant temperature (T_C or T_{des}).

The most commonly used chemical adsorbent has been calcium chloride ($CaCl_2$) in solar cooling applications. Calcium chloride adsorbs ammonia to produce $CaCl_2 \cdot 8NH_3$ and water to produce $CaCl_2 \cdot 6H_2O$ as a product. This adsorbent expands significantly during the process and thus requires room for the expansion in an adsorber design (Wang et al., 2004).

Increase of COP by recycling adsorption heat can be realized between multiple numbers of adsorbers.

7.3.3 Multieffect Operation

Li et al. (2014) reported a significant number of advanced solid-vapor sorption systems developed to improve energy efficiency by reducing the primary energy input. These advanced solid sorption refrigeration cycles mainly include a heat recovery sorption cycle, mass recovery sorption cycle, mass and heat recovery sorption cycle, double/multieffect sorption cycle, combined double-way sorption cycle, double-effect and double-way sorption cycle or two-/multistage sorption cycle. Li et al. indicated that a good design of the reactor–heat exchangers is important for enhancing the working performance of solid-vapor sorption systems and gave an overview of efforts made to reduce the thermodynamic losses associated with heat and mass transfer in these exchangers.

Considering the definition of COP given by Equation 7.38, it is clear that improvement of the COP requires improvement of the sorption efficiency (Equation 7.30) or reduction of the bed heating losses. Li et al. (2014) extensively discussed the efforts to increase the COP reported in literature and categorized the strategies as heat management, mass recovery and combined heat and mass recovery. Multibed regenerative cycles use the adsorption heat of one adsorber to heat another adsorber in the regeneration phase by selectively circulating the heat transfer medium. The regenerative efficiency is usually low, and the system is complicated when the conventional convective heat exchange methods are employed to achieve heat recovery process.

7.3.4 Stage of Development

Adsorption technology has been mostly applied for cooling applications, so in the following discussion refrigeration systems are considered instead of heat pumps. Remember that the COP of a heat pump is equal to one plus the COP of the refrigeration cycle, when operating under similar conditions. Saman et al. (2004) reported that in 2004 there were two major manufacturers of large adsorption chillers with cooling capacities between 70 and 350 kW: Nishiyodo and Mayekawa. Nishiyodo Air Conditioner Company owns a patent for machines based on silica gel–water, while Mayekawa's machines are based on zeolite-water. GBU in Germany offers large silica gel–water adsorption chillers under the brand NAK; these have cooling capacities up to 1 MW. GBU is an agent of a Japanese manufacturer in Europe. Power Partners Incorporated in the United States also offers large silica gel adsorption chillers under the brand ECO-MAX; these have cooling capacities up to 1 MW. Two German companies deliver adsorption chillers with capacities under 20 kW: SorTech (silica gel–water and zeolite-water) and Invensor (silica gel–water). SorTech proposes the aggregation of their systems up to eight units, leading to cooling capacities up to 128 kW. Nishiyodo's machine was already introduced in 1986 (Wang and Oliveira, 2005). According to the manufacturer's specification, ECO-MAX C-20 produces 70-kW cooling from 90.6°C hot water with a cooling COP of 0.50 when heat is rejected at 29.4°C. The operation weight of this model is 8.2 tons, and the dimensions are 2.0 × 2.86 × 2.4 m³. Performance of this adsorption chiller is, as expected, lower than that of a commercially available single-effect LiBr-water absorption chiller. For example, WFC-SC20 from Yazaki produces 70-kW cooling from 88°C hot water with cooling COP of 0.7 when heat is rejected at 31°C. Its operation weight is 1.2 ton, and the dimensions are 2 × 1.1 × 1.3 m³ (Yazaki Energy Systems Inc., 2015). The adsorption chiller is seven times heavier and five times bulkier than the absorption chiller. This comparison clearly shows two major problems associated with adsorption technology: lower COPs and cooling power density than absorption systems. Due to the extra losses associated with the heating of the bed reactor each cycle, Equations 7.35 and 7.36, adsorption chillers show lower performance than absorption chillers. Also, their cooling power densities are much lower.

Table 7.7 gives an overview of the systems currently available on the market. The refrigeration COPs given in the table apply for the conditions given in the manufacturers' documentation, which is not standardized and is difficult to compare. For this reason, a column with the second law efficiency has been added, again based on the operating conditions given in the documentation from the manufacturers. It must be clear that these values are significantly lower than the values that apply for absorption and vapor compression cycles. The Carnot efficiency has been obtained by making use of the definition for thermally driven systems given in Equation 7.5.

Demir et al. (2008) gave an overview of the employed adsorbent-refrigerant pairs, condensation, evaporation and desorption temperature of research studies reported on solid-vapor adsorption. The overview included COP values for these studies and included both theoretical and experimental references.

TABLE 7.7 Adsorption Chillers Available in the Market

Brand/Company	Working Pair	Capacity Range (kW)	COP_{ref} (-)	η_{2nd_law} (COP/COP_{Carnot}) (-)
GBU (European distributor of Japanese brand)	Water/silica gel	71–352	0.60	0.33
ECO-MAX, USA	Water/silica gel	35–1178	0.50	0.26
Mayekawa/Ad-Ref-Noa	Water/zeolite	105–430	N.A.	N.A.
Invensor	Water/zeolite	10–18	0.58	0.22
SorTech	Water/silica gel	16–128	Max 0.65	N.A.
SorTech	Water/zeolite	13–104	Max 0.53	N.A.

Note: N.A., not available.

7.4 EJECTOR-BASED HEAT PUMPS

7.4.1 Technology Description

The ejector heat pump (EHP) cycle makes use of a high-pressure, vapor-driven ejector instead of a mechanical compressor as in a vapor compression heat pump cycle. When closed, the only rotating equipment of an EHP system is a pump. Open systems make use of steam from an external source and have no moving parts. The system has no need for lubrication, has a long operating lifetime, is highly reliable, has low maintenance costs and produces little vibration and noise. Water (an environmentally friendly refrigerant with high latent heat of vaporization) may be used in these systems as the working fluid and will practically always be used in the open systems. The COP of the EHP is smaller than the COP of vapor compression and absorption heat pump systems.

Figure 7.32 shows a schematic view of an EHP. This system includes an evaporator, a condenser, an expansion valve, an ejector, a pump and a boiler. The components boiler, ejector and pump replace the compressor of the vapor compression heat pump cycle.

The motive vapor with high pressure and temperature enters the ejector (1). By converting the thermal energy of vapor into kinetic energy in the converging-diverging nozzle, the flow becomes supersonic with a low-pressure region at the primary nozzle exit plane. This forces the low-pressure refrigerant from the evaporator (2) into the nozzle and creates a lower pressure in the evaporator. This results in the evaporation of refrigerant in the evaporator while heat is absorbed from the heat source. Then, the primary (motive) and secondary (suction) fluids mix, and the resulting mixed fluid leaves the ejector and enters the condenser (3) after that pressure has been recovered in the diffuser. After condensing the refrigerant vapor in the condenser (4), a part of the saturated liquid expands back to the evaporator (5) by passing through the expansion valve, and the rest is directed to the boiler (6) by a pump.

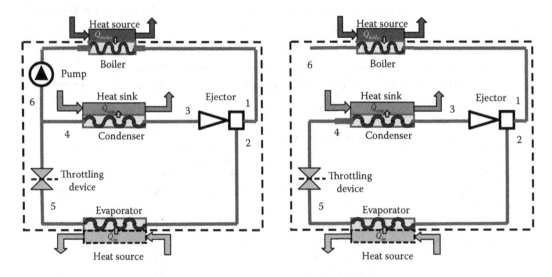

FIGURE 7.32   Ejector heat pump system: closed cycle (left) and open cycle (right).

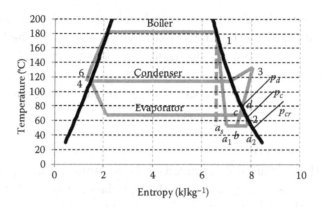

FIGURE 7.33 Ejector heat pump system in *T-s* diagram.

The ejector operation processes are shown in a *T-s* diagram in Figure 7.33. Saturated vapor with pressure p_1 and temperature T_1 enters the converging-diverging nozzle at point 1 and accelerates to a supersonic flow. In the nozzle, the motive fluid expands isentropically to the pressure at which the secondary fluid reaches the sonic velocity and reaches state a_s in the *T-s* diagram. Applying the primary nozzle efficiency η_m, the actual state of motive fluid is at state a_1. The expansion of the primary fluid causes a low-pressure region at the nozzle exit plane. This drives the secondary fluid from the evaporator into the ejector. This saturated vapor with pressure p_2 and temperature T_2 enters the ejector at point 2 and expands isentropically to pressure p_{cr} to reach the sonic velocity. The mixing process starts when the secondary fluid reaches the sonic velocity (at state a_2 in the figure); then, the mixture passes through a constant pressure p_{cr} process with final state at state b. The mixed fluid velocity is always supersonic. This supersonic stream compresses isentropically to p_c (the pressure before the shock wave) in the b_{ec} region (Figure 7.33) with final actual point at c due to existing exit mixing chamber efficiency η_{mix}. The supersonic flow enters the constant-diameter region at point c. At this stage, a normal shock forms in that region due to the ejector backpressure. The strong compression effect of this shock converts the supersonic flow to the subsonic flow and increases the pressure suddenly to p_d (the pressure after the shock wave). Then, the mixed fluid enters the subsonic diffuser and expands isentropically to p_3 at point 3 in the *T-s* diagram (considering η_{dif} as the diffuser efficiency).

Ejector models are generally one dimensional and use lumped parameters to describe the various state properties. The flow inside the ejector is modeled with the conservation equations of mass, momentum and energy. Most models assume that the two phases are in thermodynamic equilibrium. All models divide the ejector in multiple sections to describe the flow. Typically, at least four sections are present in mathematical models: motive nozzle, suction nozzle, mixing section and diffuser.

Sherif et al. (2000) divided the motive nozzle into a converging and a diverging section for the calculation of the flow. Banasiak and Hafner (2011) also divided the mixing section into two zones: a premixing chamber where a 0D double-flow model is used and a constant-area mixing chamber where a 1D double-flow model is applied. Because the flow phenomena inside an ejector are complex, irreversibilities in the flow are typically lumped together in various isentropic ejector component efficiencies. Ersoy and Bilir (2010) gave an overview of ejector component efficiencies used in the literature. Only one study tried to empirically establish ejector component efficiencies, the work of Liu and Groll (2013).

7.4.2 Heat Pump Efficiency

The COP of these heat pumps is determined by the entrainment ratio φ, which is the ratio between the mass flow entering the ejector through the suction nozzle and the mass flow entering the ejector through the motive nozzle. Using the notation of Figure 7.32,

$$\varphi = \frac{\dot{m}_2}{\dot{m}_1} \tag{7.42}$$

In an open cycle, the condensation heat is used to evaporate the vapor from the solution, which needs to be concentrated, so the COP of the heat pump is given by

$$COP = \frac{\dot{m}_1 \times (1+\varphi) \times \Delta h_{cond}}{\dot{m}_1 \times \Delta h_{cond}} = 1 + \varphi \tag{7.43}$$

The temperature difference between condensation and evaporation temperature is in fact the temperature lift delivered by the heat pump. With a given evaporation temperature, the condensation temperature results from the pressure recovery ratio of the ejector p_R, which is the ratio between pressure at the exit of the diffusor of the ejector and the pressure at its suction nozzle (evaporating pressure).

7.4.3 Sizing the Ejector

Weerstra (2014b) extended the ejector model of Liu et al. (2012) to include a diverging section of the motive nozzle. His approach is discussed here. The following assumptions are generally made to analyze ejectors:

- The flow is steady and one dimensional.
- The flow is homogeneous.
- The two phases are in thermodynamic equilibrium.
- The effect of thermal diffusion is neglected.
- The gravitational force effect on the flow is neglected.

The geometric parameters and operation conditions are shown in Figure 7.34.

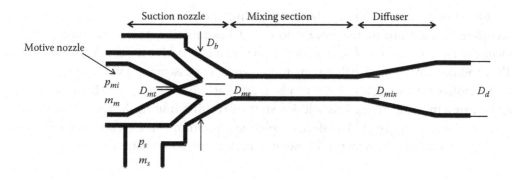

FIGURE 7.34   Geometric parameters of the ejector.

7.4.3.1 Model of Motive Nozzle Flow

Based on the critical flow model, the motive nozzle of the ejector is modeled using the following assumptions:

- The flow inside the motive nozzle is steady and one dimensional.

- The nozzle is a converging-diverging nozzle.

- At the nozzle throat, the flow reaches a sonic condition, which is then further accelerated in the diverging part of the nozzle. At the exit, the flow has a supersonic velocity.

- The inlet flow velocity is neglected.

The governing equations for the *converging section* of the motive nozzle flow are

$$h_{mi} = h(p_{mi}, T_{mi}) \tag{7.44}$$

$$s_{mi} = s(p_{mi}, T_{mi}) \tag{7.45}$$

$$h_{mt\_is} = h(p_{mt}, s_{mi}) \tag{7.46}$$

$$h_{mt} = h_{mi} - \eta_m(h_{mi} - h_{mt\_is}) \tag{7.47}$$

$$V_{mt} = \sqrt{2(h_{mi} - h_{mt})} \tag{7.48}$$

$$c_{mt} = c(p_{mt}, h_{mt}) \tag{7.49}$$

$$\rho_{mt} = \rho(p_{mt}, h_{mt}) \tag{7.50}$$

$$\dot{m}_m = \rho_{mt} A_{mt} V_{mt} \tag{7.51}$$

$$s_{mt} = s(p_{mt}, h_{mt}) \tag{7.52}$$

This set of equations can be solved by equating the two velocities V_{mt} and c_{mt} because, according to the third motive nozzle flow model assumption, these velocities should be identical. c_{mt} is the speed of sound for the refrigerant at motive nozzle throat conditions. The conditions at the motive throat are unknown, but by assuming a value for the pressure in the motive throat p_{mt}, a solution can be found such that $V_{mt} - c_{mt} = 0$ because both V_{mt} and c_{mt} are a function of p_{mt}. This solution strategy requires finding the root of the function $f(p_{mt}) = V_{mt}(p_{mt}) - c_{mt}(p_{mt})$. An effective solution algorithm is the secant method: a root-finding algorithm defined by the following generic recurrence relation:

$$y_n = y_{n-1} - f(y_{n-1})\frac{y_{n-1} - y_{n-2}}{f(y_{n-1}) - f(y_{n-2})} \tag{7.53}$$

As can be seen from the recurrence relation, the secant method requires two initial values, y_0 and y_1, which should ideally be chosen to lie close to the root. To solve the first (converging) section of the motive nozzle, the variable y_n and its function $f(y_n)$ are defined as

$$y_n = p_{mt} \tag{7.54}$$

$$f(y_n) = V_{mt} - c_{mt} \tag{7.55}$$

The governing equations for the *diverging section* of the motive nozzle flow are

$$h_{me\_is} = h(p_{me}, s_{mt}) \tag{7.56}$$

$$h_{me} = h_{mt} - \eta_m(h_{mt} - h_{me\_is}) \tag{7.57}$$

$$V_{me,1} = \sqrt{2(h_{mt} - h_{me}) + V_{mt}^2} \tag{7.58}$$

$$\rho_{me} = \rho(p_{me}, h_{me}) \tag{7.59}$$

$$V_{me,2} = \frac{\dot{m}_m}{\rho_{me} A_{me}} \tag{7.60}$$

The diverging part of the motive nozzle is also solved by the secant method. The parameters for the secant method are

$$y_n = p_{me} \tag{7.61}$$

$$f(y_n) = V_{me,1} - V_{me,2} \tag{7.62}$$

TABLE 7.8   Inputs and Outputs of the Motive Nozzle Model

i	Inputs (9)
A_{mt}	Area nozzle throat
A_{me}	Area nozzle exit
η_m	Isentropic efficiency nozzle
T_{mi}	Temperature motive nozzle inlet
p_{mi}	Pressure motive nozzle inlet
p_{mt1}, p_{mt2}	Initial guesses of nozzle throat pressure
p_{me1}, p_{me2}	Initial guesses of nozzle exit pressure
	Outputs (5)
$\dot{m}_m$	Mass flow motive nozzle
V_{me}	Velocity motive nozzle exit
ρ_{me}	Density motive nozzle exit
p_{me}	Pressure motive nozzle exit
h_{me}	Enthalpy motive nozzle exit

The model of the motive nozzle flow consists of 16 equations and 25 variables. This means that 9 input variables are required to solve the model. The in- and outputs of the model are listed in Table 7.8.

7.4.3.2 Model of Suction Nozzle Flow
The expansion process of the suction nozzle is modeled in the same way as the expansion process of a converging nozzle using the following assumptions:

- The flow is steady and one dimensional.

- The inlet flow velocity is neglected.

- The heat transfer between the fluid and the nozzle wall is neglected.

The governing equations for the suction nozzle flow are as follows:

$$\dot{m}_s = \varphi \times \dot{m}_m \tag{7.63}$$

$$h_s = h(T_{evap}, x_{evap}) \tag{7.64}$$

$$s_s = s(T_{evap}, x_{evap}) \tag{7.65}$$

$$h_{b\_is} = h(p_b, s_s) \tag{7.66}$$

$$h_b = h_s - \eta_s(h_s - h_{b\_is}) \tag{7.67}$$

$$V_{b,1} = \sqrt{2(h_s - h_b)} \tag{7.68}$$

$$\rho_b = \rho(p_b, h_b) \tag{7.69}$$

$$V_{b,2} = \frac{\dot{m}_s}{\rho_b A_b} \tag{7.70}$$

The set of suction nozzle equations is again solved by the secant method. The parameters for the secant method are

$$y_n = p_b \tag{7.71}$$

$$f(y_n) = V_{b,1} - V_{b,2} \tag{7.72}$$

The model of the suction nozzle flow consists of nine equations and 16 variables; hence, 7 input variables are required to solve the model. The in- and outputs of the model are listed in Table 7.9.

7.4.3.3 Model of Mixing Section Flow
The assumptions behind the model of the mixing section flow are as follows:

- At the inlet plane, the motive stream has a velocity V_{me} and a pressure p_{me} and occupies the area A_{me}.

- At the inlet plane, the suction stream has a velocity V_b and a pressure p_b and occupies the area A_b.

- At the outlet plane, the flow becomes uniform and has a velocity V_{mix} and a pressure p_{mix}.

- The heat transfer between the fluid and the mixing section wall is neglected.

TABLE 7.9   Inputs and Outputs of the Suction Nozzle Model

i	Inputs (7)
A_b	Area nozzle exit
φ	Entrainment ratio
η_s	Isentropic efficiency nozzle
$\dot{m}_m$	Mass flow motive nozzle
T_{evap}	Evaporating temperature
p_{mi}	Pressure motive nozzle inlet
p_{b1}, p_{b2}	Initial guesses of nozzle exit pressure
	Outputs (4)
$V_{b,1}$	Velocity suction nozzle exit
ρ_b	Density suction nozzle exit
p_b	Pressure suction nozzle exit
h_b	Enthalpy suction nozzle exit

The governing equations for the mixing section are the mass conservation, momentum conservation and energy conservation equations, respectively:

$$\rho_{me} A_{me} V_{me} + \rho_b A_b V_b = \rho_{mix} A_{mix} V_{mix} \tag{7.73}$$

$$\begin{aligned} p_{me} A_{me} + \eta_{mix} \rho_{me} A_{me} V_{me}^2 + p_b (A_{mix} - A_{me}) \\ + \eta_{mix} \rho_b (A_{mix} - A_{me}) V_b^2 = p_{mix} A_{mix} + \rho_{mix} A_{mix} V_{mix}^2 \end{aligned} \tag{7.74}$$

$$\dot{m}_m \left(h_{me} + \frac{V_{me}^2}{2} \right) + \dot{m}_s \left(h_b + \frac{V_b^2}{2} \right) = (\dot{m}_m + \dot{m}_s) \left(h_{mix} + \frac{V_{mix}^2}{2} \right) \tag{7.75}$$

The model is solved by rearranging the conservation equations in the following forms:

$$C_1 = \frac{\rho_{me} A_{me} V_{me} + \rho_b A_b V_b}{A_{mix}} \tag{7.76}$$

$$C_2 = \frac{p_{me} A_{me} + \eta_{mix} \rho_{me} A_{me} V_{me}^2 + p_b (A_{mix} - A_{me}) + \eta_{mix} \rho_b (A_{mix} - A_{me}) V_b^2}{A_{mix}} \tag{7.77}$$

$$C_3 = \frac{\dot{m}_m \left(h_{me} + \dfrac{V_{me}^2}{2} \right) + \dot{m}_s \left(h_b + \dfrac{V_b^2}{2} \right)}{\dot{m}_m + \dot{m}_s} \tag{7.78}$$

$$p_{mix} = C_2 - C_1 V_{mix} \tag{7.79}$$

$$h_{mix} = C_3 - \frac{V_{mix}^2}{2} \tag{7.80}$$

$$\rho_{mix,1} = \frac{C_1}{V_{mix}} \tag{7.81}$$

$$\rho_{mix,2} = \rho(p_{mix}, h_{mix}) \tag{7.82}$$

$$\rho_{g,mix} = \rho(p_{mix}, x = 1) \tag{7.83}$$

$$\rho_{f,mix} = \rho(p_{mix}, x = 0) \tag{7.84}$$

$$x_{mix} = x(p_{mix}, h_{mix}) \tag{7.85}$$

The final step in solving the mixing section model is using the secant method with the parameters:

$$y_n = V_{mix} \tag{7.86}$$

$$f(y_n) = \rho_{mix,1} - \rho_{mix,2} \tag{7.87}$$

The model of the mixing section flow consists of 10 equations and 26 variables. This means that 16 input variables are required to solve the model. The in- and outputs of the model are listed in Table 7.10.

TABLE 7.10   Inputs and Outputs of the Mixing Section Model

i	Inputs (16)
A_b	Area suction nozzle exit
A_{me}	Area motive nozzle exit
A_{mix}	Area mixing section
η_{mix}	Isentropic efficiency mixing section
$V_{mix,1}, V_{mix,2}$	Initial guesses of mixing velocity
$\dot{m}_m$	Mass flow motive nozzle
$\dot{m}_s$	Mass flow suction nozzle
V_{me}	Velocity motive nozzle exit
ρ_{me}	Density motive nozzle exit
p_{me}	Pressure motive nozzle exit
h_{me}	Enthalpy motive nozzle exit
$V_{b,1}$	Velocity suction nozzle exit
ρ_b	Density suction nozzle exit
p_b	Pressure suction nozzle exit
h_b	Enthalpy suction nozzle exit
	Outputs (7)
V_{mix}	Velocity mixing section
ρ_{mix}	Density mixing section
$\rho_{f,mix}$	Density liquid mixing section
$\rho_{g,mix}$	Density vapor mixing section
p_{mix}	Pressure mixing section
h_{mix}	Enthalpy mixing section
x_{mix}	Quality mixing section

7.4.3.4 Model of the Diffusor Flow

By assuming that the mixed stream at the outlet of the mixing section is in homogeneous equilibrium, the governing equations for the diffuser flow are

$$C_4 = 0.85\,\rho_{mix}\left[1-\left(\frac{A_{mix}}{A_d}\right)^2\right]\left[\frac{x_{mix}^2}{\rho_{g,mix}}+\frac{(1-x_{mix})^2}{\rho_{f,mix}}\right] \tag{7.88}$$

$$p_d = \frac{1}{2}\rho_{mix}V_{mix}^2 C_4 + p_{mix} \tag{7.89}$$

$$h_d = \frac{\dot{m}_m h_{mi} + \dot{m}_s h_s}{\dot{m}_m + \dot{m}_s} \tag{7.90}$$

$$x_d = x(p_d, h_d) \tag{7.91}$$

The model of the diffuser flow consists of four equations and 16 variables. This means that 12 input variables are required to solve the model. The in- and outputs of the model are listed in Table 7.11.

TABLE 7.11 Inputs and Outputs of the Diffusor Section Model

i	Inputs (12)
A_{mix}	Area mixing section
A_d	Area diffuser exit
$\dot{m}_m$	Mass flow motive nozzle
$\dot{m}_s$	Mass flow suction nozzle
h_{mi}	Enthalpy motive nozzle inlet
h_s	Enthalpy suction nozzle inlet
V_{mix}	Velocity mixing section
ρ_{mix}	Density mixing section
$\rho_{f,mix}$	Density liquid mixing section
$\rho_{g,mix}$	Density vapor mixing section
p_{mix}	Pressure mixing section
x_{mix}	Quality mixing section
	Outputs (3)
$\dot{m}_d$	Mass flow diffuser exit
p_d	Pressure diffuser exit
x_d	Quality diffuser exit

7.4.3.5 External Input Parameters

The ejector model described has a total of 12 external input parameters, which can be independently varied. These parameters can be divided in three different categories:

1. Geometric parameters

   - Motive nozzle throat diameter D_{mt}

   - Motive nozzle exit diameter D_{me}

   - Suction nozzle exit diameter D_b

   - Constant-area mixing section diameter D_{mix}

   - Diffuser exit cross-section diameter D_d

2. Ejector efficiencies

   - Isentropic efficiency motive nozzle η_m

   - Isentropic efficiency suction nozzle η_s

   - Isentropic efficiency constant-area mixing section η_{mix}

3. Operating conditions

   - Evaporating temperature/suction nozzle inlet T_{evap}

   - Temperature motive nozzle inlet T_{mi}

   - Pressure of motive vapor p_{mi}

   - Entrainment ratio $\varphi = \dot{m}_s / \dot{m}_m$

The most crucial boundary condition that should be met involves the prescribed ejector outlet pressure p_d. The pressure recovery ratio of an ejector is defined as the ratio of the pressure at the diffuser exit of the ejector to the pressure of the suction flow:

$$p_R = \frac{p_d}{p_{evap}} \tag{7.92}$$

This boundary condition can be expressed by the pressure recovery ratio. A higher diffuser exit pressure is desired in an EHP cycle as it means a higher condensation temperature (pressure).

7.4.4 Stage of Development

Thermal vapor compression is commonly applied in multieffect distillation for production of high-purity water for domestic and industrial use (Alfa Laval, 2015). Boot et al. (1998) reported its application in multistage evaporation processes. The International Energy

TABLE 7.12   Experimental Ejector Coefficients Reported by Weerstra (2014a)

i	Parameter	Minimum	Maximum
η_m	Isentropic efficiency motive nozzle	0.70	0.95
η_s	Isentropic efficiency suction nozzle	0.60	0.90
η_{mix}	Isentropic efficiency motive nozzle	0.80	1.00
φ	Entrainment ratio	0.40	0.60
P_R	Pressure recovery ratio	1.10	1.30

Agency–Industrial Energy Technology and Systems (IEA-IEST, 2014) reported its use in breweries. The technology is frequently applied in practice mostly as an open system.

Weerstra (2014a) reported experimental efficiency values in the ranges listed in Table 7.12. The minimum value of assumed ejector efficiencies is 60%, whereas the maximum value of assumed efficiencies is around 100%. Liu and Groll (2013) showed with experiments that their ejector expansion device had, on average, motive nozzle and mixing section efficiencies of approximately 80% and suction nozzle efficiency of approximately 70%.

Weerstra (2014a) concluded that there is a trade-off between the mass entrainment ratio and the pressure recovery ratio. High-pressure recovery ratios can be obtained at the expense of the mass entrainment ratio and vice versa. On average, his experimental pressure recovery ratio was 1.2 with an entrainment ratio of 0.50.

The geometric parameters of the ejector have a substantial effect on the performance of the system. Altering the geometry of the ejector by adjusting the size of the motive nozzle throat alters the mass entrainment ratio and the pressure recovery ratio. Another way of altering the mass entrainment ratio and pressure recovery ratio is by altering the flow passage area of the suction nozzle throat. The mass entrainment ratio is related to the heating capacity; the pressure recovery ratio is related to the temperature lift.

Figure 7.35 illustrates the impact of the range of entrainment and pressure recovery ratios measured by Weerstra (2014a) on the performance of an open EHP with water as the working fluid. The COP values are slightly lower than for liquid-vapor absorption heat pumps, but its major limitation is the temperature lift that can be attained.

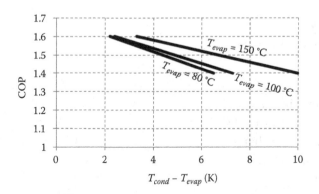

FIGURE 7.35   The COP of a steam ejector heat pump as a function of operating conditions.

7.5 CONCLUDING REMARKS

This chapter presented the alternative thermally driven technologies available for the implementation of heat pumps: liquid-vapor absorption, solid-vapor adsorption and ejector-based heat pumps. The chapter discussed the methods for the prediction of the performance of these cycles when the externally imposed operating conditions are prescribed: heat sink in- and outlet temperatures and corresponding heating capacity and heat source inlet temperature.

LIST OF SYMBOLS

A	Crossflow area	m^2
c	Speed of sound	ms^{-1}
c_p	Specific heat at constant pressure	$kJkg^{-1}K^{-1}$
D	Diameter	m
f	Circulation ratio	–
(g)	Vapor phase	–
h	Enthalpy	$kJkg^{-1}$
m	Mass	kg
$\dot{m}$	Mass flow	Kgs^{-1}
M	Molar mass	$kgkmol^{-1}$
p	Pressure	kPa
p_R	Pressure recovery ratio	–
Q	Heat	kJ
$\dot{Q}$	Heat flow	kW
R	Gas constant	$kJkg^{-1}K^{-1}$
s	Entropy	$kJkg^{-1}K^{-1}$
(s)	Solid phase	–
T	Temperature	K
v	Specific volume	m^3kg^{-1}
V	Velocity	m
w	Concentration of refrigerant in solution	$kgkg^{-1}$
$\dot{W}$	Power	kW
x	Quality	–
y	Parameter in Equation 7.53	–

Greek Symbols

Δh	Enthalpy change	$kJkg^{-1}$
ΔH	Load per kilogram absorbent bed	$kJkg^{-1}$
Δw	Concentration change	$kgkg^{-1}$
φ	Entrainment ratio	–
η	Efficiency	–
ρ	Density	kgm^{-3}

Subscripts

b	Suction nozzle exit
C	Source (cold)
abs	Absorber, absorption
ads	Adsorption
cond	Condenser
d	Diffusor exit
des	Desorber
dif	Diffusor
e	Exit
evap	Evaporator
f	Saturated liquid
g	Saturated vapor
G	Vapor phase
H	Sink (hot), high (pressure side)
hex	Heat exchanger
hp	Heat pump, high pressure
i	Inlet
in	At inlet, into the system
is	Isentropic
L	Low (pressure side), liquid phase
LG	Liquid-vapor phase change
m	Motive nozzle
mix	Mixing section
out	At outlet, out of the system
pump	Pump
rect	Rectifier
ref	Refrigerant
s	Strong solution; suction nozzle; superheating; subcooling
t	Throat
v | *vap*	Vapor
w	Weak solution, sorption related

Abbreviations

COP	Coefficient of performance
EHP	Ejector heat pump
GWP	Global warming potential
hex	Heat exchanger
HP	Heat pump
HTS	High-temperature salt
LTS	Low-temperature salt
N.A.	Not available

ODP Ozone-depletion potential
SCP Specific cooling power

REFERENCES

Alfa Laval, Thermo vapour compression distiller, http://www.alfalaval.com (accessed June 1, 2015).

Arora C.P., *Refrigeration and air conditioning*, 17th reprint, Tata McGraw-Hill, New Delhi, 2006.

Arun M.B., Maiya M.P., Murthy S.S., Performance comparison of double-effect parallel-flow and series flow water–lithium bromide absorption systems, *Applied Thermal Engineering*, 21 (2001), 1273–1279.

Banasiak K., Hafner A., 1D computational model of a two phase R744 ejector for expansion work recovery, *International Journal of Thermal Sciences*, 50 (2011), 2235–2247.

Bogart M., *Ammonia absorption refrigeration in industrial processes*, Gulf, Houston, 1981.

Boot H., Nies J., Verschoor M.J.E., de Wit J.B., *Handbook industriële warmtepompen* (in Dutch), Kluwer bedrijfsinformatie, Blaricum, the Netherlands, 1998.

Cacciola G., Hajji A., Maggio G., Restuccia G., Dynamic simulation of a recuperative adsorption heat pump, *Energy*, 18 (1993), 1125–1137.

Demir H., Mobedi M., Ulku, S., A review on adsorption heat pump: Problems and solutions, *Renewable and Sustainable Energy Reviews*, 12 (2008), 2381–2403.

Ersoy H.K., Bilir N., The influence of ejector component efficiencies on performance of ejector expander refrigeration cycle and exergy analysis, *International Journal of Exergy*, 7 (2010), 425–438.

Garousi Farshi L., Seyed Mahmoudi S.M., Rosen M.A., Analysis of crystallization risk in double effect absorption refrigeration systems, *Applied Thermal Engineering*, 31 (2011), 1712–1717.

Grossman G., Zaltash A., ABSIM – Modular simulation of advanced absorption systems, *International Journal of Refrigeration*, 24 (2001), 531–543.

Herold K.E., Radermacher R., Klein S.A., *Absorption chillers and heat pumps*, CRC Press, Boca Raton, FL, 1996.

Infante Ferreira C., Kim D.-S., Techno-economic review of solar cooling technologies based on location-specific data, *International Journal of Refrigeration*, 39 (2014), 23–37.

International Energy Agency, Application of industrial heat pumps, *IEA industrial energy-related systems and technologies, Annex 13 and IEA heat pump programme Annex 35*, Report No. HPP-AN35-1, ISBN 978-91-88001-92-4, IEA Heat Pump Centre, Borås, Sweden, 2014.

International Institute of Refrigeration (IIR), NH_3-H_2O: *Thermodynamic and physical properties*, IIF-IIR, Paris, 1994.

Kim D.S., Solar absorption cooling, PhD thesis, Delft University of Technology, the Netherlands, 2006.

Kim D.S., Infante Ferreira C.A., A Gibbs energy equation for LiBr aqueous solutions, *International Journal of Refrigeration*, 29 (2006), 36–46.

Lemmon E., Huber M., McLinden M., *NIST Standard reference database 23: Reference fluid thermodynamic and transport properties – REFPROP*, Version 9.1, National Institute of Standards and Technology, Gaithersburg, MD, 2013.

Li T.X., Wang R.Z., Li, H., Progress in the development of solid-gas sorption refrigeration thermodynamic cycle driven by low-grade thermal energy, *Progress in Energy and Combustion Science*, 40 (2014), 1–58.

Liu F., Groll E.A., Study of ejector efficiencies in refrigeration cycles, *Applied Thermal Engineering*, 52 (2013), 360–370.

Liu F., Groll E.A., Li D., Investigation on performance of variable geometry ejectors for CO_2 refrigeration cycles, *Energy*, 45 (2012), 829–839.

Niebergall W., Sorptions-Kältemaschinen, In: *Handbuch der Kältetechnik*, edited by R. Plank, Band 7, Springer-Verlag, Berlin, 1981.

Power R.B., *Steam jet ejectors for the process industries*, McGraw-Hill, New-York, 1993.

Saman W., Krause M., Vajen K., Solar cooling technologies: Current status and recent developments, In *Proceedings of 42nd ANZSES Conference Solar 2004*, Perth, Australia, 2004, 1–10.

Sherif S.A., Lear W.E., Steadham J.M., Hunt P.L., Holladay J.B., Analysis and modeling of a two-phase jet pump of a thermal management system for aerospace applications, *International Journal of Mechanical Sciences*, 42 (2000), 185–198.

Smith J.M., van Ness, H.C., Abbott M.M., *Introduction to chemical engineering thermodynamics*, McGraw-Hill, Singapore, 2001.

Stolk A., Koudetechniek A1 – Koudeopwekking' Lecture notes i77A (in Dutch), Werktuigbouwkunde en Maritieme Techniek, TU Delft, the Netherlands, 1990.

Van der Heuvel C.M., Hybrid adsorption compression heat pump systems, Delft University of Technology, the Netherlands, ME-SPET, 2010.

Vasilescu C., Infante Ferreira C., Solar driven double-effect absorption cycles for sub-zero temperatures, *International Journal of Refrigeration*, 39 (2014), 86–94.

Wang L.W., Wang R.Z., Wu J.Y., Wang K., Compound adsorbent for adsorption ice maker on fishing boats, *International Journal of Refrigeration*, 27 (2004), 401–408.

Wang, R., Wang, L., Wu, J., *Adsorption Refrigeration Technology*, Wiley, New York, 2014.

Wang R.Z., Oliveira R.G., Adsorption refrigeration – An efficient way to make good use of waste heat and solar energy, In *Proceedings of International Sorption Heat Pump Conference*, Denver, USA, 2005.

Weerstra J., Dynamic behavior of CO_2 refrigeration system with integrated ejector, Delft University of Technology, ME-SPET, 2014a.

Weerstra J., Ejector technology for CO_2 refrigeration systems, Delft University of Technology, the Netherlands, ME-SPET, 2014b.

Yazaki Energy Systems Inc., Specifications of water-fired chillers and chiller-heaters, http://www .yazakienergy.com/waterfiredspecifications.htm (accessed June 1, 2015).

Yu Y.Q., Zhang P., Wu J.Y., Wang R.Z., Energy upgrading by solid-gas reaction heat transformer—A critical review, *Renewable and Sustainable Energy Reviews*, 12 (2008), 1302–1324.

CHAPTER 8

Solid-State Heat Pumps

8.1 INTRODUCTION

Although mechanically or thermally driven heat pumps are among the most applied in the chemical process industry, heat pumps based on other driving forces are also available and already used in consumer goods with potential to be used as well in the chemical process industry. This chapter describes the most important available solid-state heat pump technologies: magnetic heat pumps (MHPs), thermoelectric (TE) heat pumps and thermoacoustic (TA) heat pumps (TAHPs). Although the solid-state heat pumps are less efficient as compared to conventional compression cycle systems, they are popular due to the long lifetime, absence of moving parts and lack of potentially hazardous refrigerants.

Magnetic heat pumping is an adiabatic heating method that applies the magnetocaloric effect (MCE). From the point of view of basic physics, it shows an analogy to the conventional gas compression-expansion method. It has been applied for many years in cryogenics to reach very low temperatures. After the discovery of the 'giant' magnetocaloric effect (GMCE) in $Gd_5(Si_2Ge_2)$ in 1997, which increases the MCE, many scientists and industrial representatives of the refrigeration community conceded that this new technology (applying permanent magnets and the GMCE) has good future potential for a remarkable penetration into the refrigeration/heat-pumping market.

Thermoelectric heat pumps are based on the Peltier effect, observable when a direct electric current passes around a circuit incorporating two different metals; one contact area is heated and the other cooled, depending on the direction of current flow. The Peltier effect conveniently provides a means for pumping heat without using moving parts. In a circuit containing two junctions between dissimilar conductors, heat may be transferred from one junction to the other by applying a direct current (DC) voltage. To be effective, the conductors must provide high TE power and low thermal conductivity combined with adequate electrical conductivity. Such a combination of properties is not to be found in metallic conductors, and this principle could not be applied to heat pumping until the advent of semiconductor materials, typically bismuth, antimony, selenium and tellurium alloys.

Thermoacoustics (TA) relates to the physical phenomenon that a temperature difference can create and amplify a sound wave, and vice versa that a sound wave is able to create a temperature difference. A sound wave is associated with changes in pressure, temperature and density of the medium through which the sound wave propagates. In addition, the medium itself is moved around an equilibrium position. An acoustic wave is brought into interaction with a porous structure with a much higher heat capacity compared to the medium through which the sound wave propagates. This porous structure acts as heat storage (regenerator). Within TA, a distinction is made between a TA engine or prime mover (TA engine) and a TAHP. The first relates to a device creating an acoustic wave by a temperature difference, while in the second an acoustic wave is used to create a temperature difference.

8.2 MAGNETIC HEAT PUMPS

Magnetic heat pumping or magnetic refrigeration (MR) is a cooling technology based on the MCE that can be used to attain extremely low temperatures, along with the ranges used in common refrigerators. Compared to conventional gas compression refrigeration, MR is safer and more environmentally friendly, quieter and more compact and has reasonable cooling efficiency. The hot topic of MR has been addressed in several review papers (Yu et al., 2003; Sangkwon et al., 2006; Bjork et al., 2010; Romero Gomez et al., 2013).

8.2.1 Magnetocaloric Effect

The MCE is a magnetothermodynamic phenomenon in which a temperature change of a suitable material (e.g. iron) is caused by exposing the material to a changing magnetic field (Figure 8.1a). In that part of the refrigeration process, a decrease in the strength of an

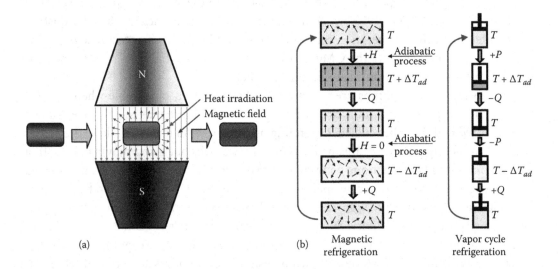

(a) (b) Magnetic refrigeration Vapor cycle refrigeration

FIGURE 8.1   Gadolinium alloy heats up inside the magnetic field and loses thermal energy to the environment, so it exits the field cooler than when it entered (a). Analogy between magnetic refrigeration and vapor cycle or conventional refrigeration: H, externally applied magnetic field; Q, heat quantity; P, pressure; ΔT_{ad}, adiabatic temperature variation (b).

externally applied magnetic field allows the magnetic domains of a magnetocaloric material to become disoriented from the magnetic field by the agitating action of the thermal energy present in the material. If the material is isolated such that no energy transfer is allowed (adiabatic process), then the temperature drops as the domains absorb the thermal energy to perform their reorientation.

The MCE is an intrinsic property of all magnetic materials, and it can be quantified as the reversible change in temperature ΔT_{ad} in the material when the field change takes place in an adiabatic process or the reversible change of magnetic entropy ΔS_m if the change in field is brought about in an isothermal process. The relationship between the two properties can be illustrated by a schematic diagram T-s (Figure 8.2a). This diagram represents the thermal dependency of the entropy of a magnetic system on the applied field. The existence of MCE at temperature T_0 may cause an adiabatic temperature change in the system ($\Delta T_{ad} = T_1 - T_0$) or an isothermal change of magnetic entropy ($\Delta S_m = S_1 - S_0$). The first occurs when the entropy is kept constant, while the second is produced when the temperature is kept constant. Both ΔT_{ad} and ΔS_m are characteristic values of the MCE, according to the initial temperature T_0 and the value of change in the magnetic field. With the increasing value of the external magnetic field change, the ordering of the magnetic spin increases and the magnetic entropy is decreased (Romero Gomez et al., 2013).

Important characteristics of a magnetic material are its total entropy S_T and the magnetic entropy of the system S_m. The entropy can be modified by varying the magnetic field, by the temperature and by other thermodynamic parameters. Entropy is a measure of the order in the magnetothermodynamic system: A high order is related to low entropy and vice versa. Applying a magnetic field in a ferromagnetic material causes a magnetic ordering of spin of the molecules, orienting themselves in the same direction and lowering the entropy of the system (Figure 8.1b). The temperature of

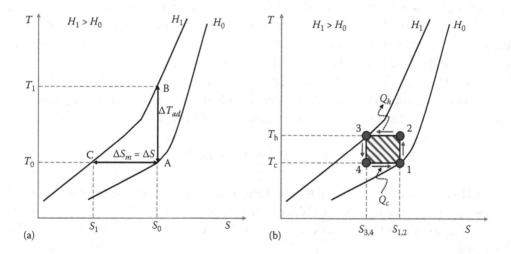

FIGURE 8.2 Thermal dependence of entropy depending on the applied field in a ferromagnetic material (a). T-s diagram of an MR Carnot cycle (b).

the material directly influences the kinetics of the electrons and the vibrations of the molecules. Lowering the temperature (releasing energy from the system) promotes a more orderly system and therefore less entropy. Magnetic entropy S_m and its change are closely related to the MCE value and to the magnetic contribution to heat capacity. The magnetic entropy change is also used to determine the characteristics of magnetic refrigerators, such as the capacity of the coolant as well as other characteristics (Romero Gomez et al., 2013).

8.2.2 Thermodynamics of MCE

The total entropy of a magnetic material can be presented in general at constant pressure as

$$S_T(H, T) = S_m(H, T) + S_r(T) + S_e(T) \tag{8.1}$$

where H is the intensity of the magnetic field (not enthalpy), and the contribution to the total entropy S_T is given by magnetic entropy of the magnetization of the material S_m, lattice entropy caused by the vibrations of crystal lattice S_r and electronic entropy of the material's free electrons S_e. The lattice and electronic entropy can be considered independent from the magnetic field and only depend on temperature. The magnetic entropy, however, is highly dependent on both the magnetic field and the temperature. When applying a magnetic field under adiabatic conditions in a ferromagnetic sample, the total entropy remains constant during the process of magnetization. Thus, when the magnetic entropy is reduced, lattice and electronic entropy increase to compensate because of the spin lattice connections and vibrations. This causes a temperature increase ΔT_{ad}, which depends on the applied magnetic field strength. When the external field is removed, the magnetic spin system returns to its original alignment by capturing energy from the lattice, which decreases thermal entropy, and the sample returns to its original temperature. Individual magnetic moments align with the external field, thereby decreasing the magnetic entropy of the sample and maintaining the S_T. If the application of magnetic field on the sample is isothermal, the total entropy decreases due to a decrease in the magnetic contribution because the lattice and electronic entropy do not vary as a result of keeping the temperature constant (Romero Gomez et al., 2013).

With the joint application of the two principles of thermodynamics on a ferromagnetic sample under a magnetic field and considering only the sample as a thermodynamic system, the change in internal energy can be expressed as follows:

$$dU = TdS - pdV + \mu_0 HdM \tag{8.2}$$

where H is the intensity of the magnetic field, p is the pressure, V is the volume of the sample, μ_0 is the magnetic permeability of the vacuum and M is the magnetic momentum of the sample. If the system's volume is not modified ($dV = 0$), then the simplified form holds:

$$dU = TdS + \mu_0 HdM \tag{8.3}$$

Rewriting this equation, which contains extensive magnitudes, depending on specific values per unit mass, it follows that

$$du = Tds + \mu_0 H d\sigma \tag{8.4}$$

In this way, σ is defined as the specific magnetization (magnetic momentum per unit of mass). The total specific entropy change of the system expressed according to H and T can be represented as

$$ds = \left(\frac{\partial s}{\partial T}\right)_H dT + \left(\frac{\partial s}{\partial H}\right)_T dH \tag{8.5}$$

The specific heat c of a substance under a constant state or parameter χ can be defined as

$$c_\chi = \left(\frac{\partial q}{\partial T}\right)_\chi \tag{8.6}$$

When this equation is combined with the second law ($ds = \partial q/T$), the specific heat of a substance for an isobaric process and constant magnetic field (c_{pH}) can be defined follows:

$$c_{pH} = T\left(\frac{\partial s}{\partial T}\right)_H \tag{8.7}$$

The dependence of the entropy with the magnetic field can be expressed in terms of magnetization through a Maxwell relation:

$$\left(\frac{\partial s}{\partial H}\right)_T = \mu_0 \left(\frac{\partial \sigma}{\partial T}\right)_H \tag{8.8}$$

Introducing the last two equations into the expression for calculating ds, the following expression is obtained for the entropy:

$$ds = \left(\frac{c_{pH}}{T}\right) dT + \mu_0 \left(\frac{\partial \sigma}{\partial T}\right) dH \tag{8.9}$$

The reversible change of temperature (ΔT_{ad}) that the sample undergoes in an adiabatic process of magnetization (process A-B of Figure 8.2a) is carried out satisfying this equation under the condition of $ds = 0$. Thus, the MCE can be quantified when the field variation takes place in an adiabatic process, according to

$$MCE_{ad} = \Delta T_{ad} = -\mu_0 \int_{H_0}^{H_1} \left(\frac{T}{c_{pH}} \right) \left(\frac{\partial \sigma}{\partial T} \right)_H dH \qquad (8.10)$$

It must be taken into account that this equation is not as trivial as it appears as the temperature itself is an implicit function of H given that the temperature will change due to adiabatic temperature change, such as when altering the magnetic field. This should be included when carrying out integration. The adiabatic temperature change can, of course, also be determined by direct measurements of the sample's temperature (Romero Gomez et al., 2013).

When the sample is subjected to a variation of the magnetic field in an isothermal process ($dT = 0$, process A-C of Figure 8.2a), the reversible change in entropy ΔS is equal to the magnetic entropy change ΔS_m. The specific entropy change Δs can be determined establishing $dT = 0$, resulting in the following expression that quantifies MCE if the field variation is performed under an isothermal process:

$$MCE_{isot} = \Delta s = \Delta s_m = -\mu_0 \int_{H_0}^{H_1} \left(\frac{\partial \sigma}{\partial T} \right) dH \qquad (8.11)$$

In a process in which the applied field increases ($\Delta H > 0$), the MCE sign is given by the sign of $(\partial\sigma/\partial H)_H$, distinguishing

- Direct MCE when $(\partial\sigma/\partial H)_H < 0$, resulting in $\Delta s_m < 0$ and $\Delta T_{ad} > 0$
- Inverse MCE in the opposite case, when $(\partial\sigma/\partial H)_H > 0$, where $\Delta s_m > 0$ and $\Delta T_{ad} < 0$

For most magnetic materials, there is a decrease of magnetization with temperature; therefore, $(\partial\sigma/\partial H)_H$ is negative. So, it is seen that $\Delta s_m < 0$ and $\Delta T_{ad} > 0$ for positive field changes. It can be concluded that the MCE will be large if the magnetic field variation is large, if magnetization changes rapidly with temperature, or if the material has a low specific heat (Romero Gomez et al., 2013).

The change in magnetization with temperature and specific heat is intrinsic to the material, while the change in magnetic field can be controlled externally. In absolute terms, the variation of magnetization with temperature is large around the phase transition involving a change of magnetization in the material. Transition or phase change is the transformation of a system from one phase to another. The main characteristic is an abrupt change in one or more physical properties. With regard to the phase transition in magnetic systems, two transition modes are given: first-order magnetostructural phase transitions and the continuous-phase or second-order transitions (more details provided in the review papers of Yu et al., 2003; Romero Gomez et al., 2013).

8.2.3 Thermodynamic Cycle

A magnetic refrigerator comprises a magnetic working material, a (de)magnetizing system, hot heat exchangers (HHXs), cold heat exchangers (CHXs) and a heat transfer system with a thermal fluid. The heat transfer fluid is responsible for pumping the heat between the working magnetic material and the HHXs and CHXs. Depending on the working temperature, the transfer fluid may be a gas or a liquid. The general working principle of a magnetic refrigerator is similar to a gas compression process (Figure 8.1b), wherein compression is replaced by magnetization, and expansion by adiabatic demagnetization: (1) magnetization (temperature increase), (2) cooling (extraction of heat), (3) demagnetization (temperature decrease) and (4) heating (injection of heat). Basically, the working material (refrigerant) absorbs heat from the load at a low temperature (CHX) and transfers it to the high-temperature source (hot end heat exchanger). As a result of the cyclical repetition of this process, the load is cooled. In magnetic refrigerators, the working material is a magnetic material, which changes its temperature and entropy under the effect of a magnetic field (Romero Gomez et al., 2013). The combination of thermodynamic processes of isothermal magnetization (where the refrigerant is magnetized as the temperature is kept constant and during this process the MCE manifests itself as a change in entropy), adiabatic magnetization (where the coolant temperature increases due to an adiabatic temperature change) and processes at a constant field allows the achievement of magnetic refrigerators with different thermodynamic cycles.

8.2.3.1 Carnot Cycle

In MR cycles, the Carnot cycle can be considered as the reference cycle as it allows the direct study of manifestations of the MCE. The cycle consists of two adiabatic and two isothermal processes and can be illustrated in a T-s diagram between two lines of constant field (Figure 8.2b). The magnetic refrigerant is partially magnetized (process 1-2), increasing its temperature adiabatically from T_c to T_h. Then, the intensity of the applied magnetic field is increased to complete magnetization isothermally, making it necessary to remove the refrigerant material's thermal insulation to allow the heat exchange with the heat transfer fluid (process 2-3). In this process, the coolant keeps its temperature constant, while the fluid absorbs the heat generated in the refrigerant due to the magnetization. In process 3-4, the applied magnetic field decreases, and this lowers the temperature of the magnetic refrigerant adiabatically, from T_h to T_c. Finally, the cycle is completed when the material is completely demagnetized during process 4-1. In this process, the magnetic refrigerant absorbs heat from the fluid, recovering the energy lost during demagnetization. To ensure an efficient performance of the system, the fluid rejects the absorbed energy of the magnetized coolant (process 2-3) to the hot sink and absorbs energy from the cold source (system to be cooled) that is transferred to the magnetic refrigerant in demagnetizing process 4-1 (Romero Gomez et al., 2013).

For practical Carnot cycle–based refrigeration, the temperature interval between the hot and cold source is restricted by the adiabatic temperature change of the magnetocaloric material in the processes of (de)magnetization. Therefore, the working temperature of the hot and cold source cannot be chosen freely. When the temperature rises, the specific heat and lattice entropy increase (associated vibrations of atoms), causing the decrease of adiabatic temperature change. When the lattice entropy is very large compared with that of magnetic entropy, the material's adiabatic temperature change is insignificant and requires very large magnetic fields to reduce vibration of the atoms. Furthermore, the Carnot cycle uses a different variable magnetic field in each of the four working points. This requires an electromagnet or a superconducting magnet where the field can be manipulated, which is inefficient with respect to energy consumption and makes the Carnot cycle unsuitable for normal refrigeration. This limits its application to temperatures lifts lower than 20 K. By adding a regenerator to the MR system, the heat expelled by the lattice system in one stage of the cycle is restored and returned to the lattice system in another stage. Thus, the capacity used for system cooling load can be effectively used for the increase of effective entropy change and temperature span (Romero Gomez et al., 2013).

8.2.3.2 Brayton Cycle

Heat transfer in the Brayton cycle occurs differently from that of the Carnot cycle. Heat transfer is performed in processes where the magnetic intensity remains constant, thus obtaining higher temperature ranges and consequently higher heat transfer between the magnetocaloric material and the fluid. An MR Brayton cycle with regeneration is shown in Figure 8.3a. The cycle consists of four processes: two adiabatic and two where the intensity of the applied magnetic field remains constant. Considering the cycle operating processes shown in Figure 8.3a from point 1, when the working magnetic material is at temperature T_1, it undergoes a temperature rise to T_2 (process 1-2) caused by the MCE in the adiabatic magnetization. Thus, the material (in the presence of a constant magnetic field) transfers heat to the hot source Q_h, lowering its temperature to T_{2a}. The additional cooling of T_{2a} to T_3 is achieved with the aid of the regenerator (process 4a-1). At point 3, the material undergoes an adiabatic demagnetization (process 3-4); thus, its temperature drops to T_4. The magnetic material at temperature T_4 exchanges heat (process 4-4a)

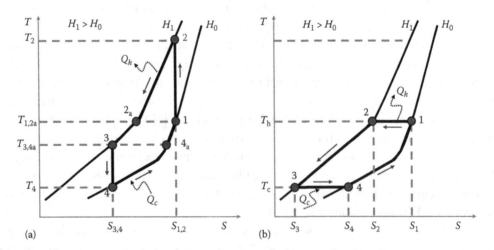

FIGURE 8.3   *T-s* diagram of an MR Brayton cycle with regeneration (a) and of an Ericsson cycle with regeneration (b).

with the cold source exchanger Q_c, absorbing heat so that its temperature increases to T_{4a}. Finally, the regenerator transfers heat to the magnetic material (process 4a-1), thereby completing the cycle.

The Brayton cycle and Ericsson cycle (discussed hereafter) are ideal for working with regeneration, in such a way that a magnetic refrigerator working temperature span is achieved independently from the working cycle. This enables room temperature MR. The theoretical Brayton cycle is characterized by the lower cooling capacity and greater heat rejection compared with the theoretical Ericsson cycle. Nevertheless, the differences between the actual Ericsson and Brayton cycles are small due to the deviation of the true isothermal and adiabatic magnetization in real processes.

8.2.3.3 Ericsson Cycle

The Ericsson cycle is a regeneration cycle similar to the Brayton cycle (Romero Gomez et al., 2013). The only difference is that isothermal (de)magnetization is used instead of adiabatic, as shown in Figure 8.3b. In process 1-2 of isothermal magnetization, the magnetic material rejects heat Q_h to the hot source at temperature T_h, and in process 3-4 absorbs Q_c from the cold source at temperature T_c. The regeneration corresponds to 2-3 and 4-1 heat exchange processes.

Regeneration, both in the Ericsson cycle and in the Brayton, are only possible with the existence of a difference in temperature that ensures heat transfer. This represents the existence of irreversible processes and therefore a decrease in the efficiency of the device.

8.2.3.4 Cascade Magnetic Cycles

Magnetocaloric materials have a temperature at which the MCE is maximum, corresponding to the Curie temperature. As we move away from such a temperature, the MCE decreases. In the case of refrigeration processes with large temperature differences (large span), there is reduced efficiency due to the decrease of MCE because of the deviation of Curie temperature. One solution to this problem is to implement a cascade system of cycles, wherein each cycle has a different material with a Curie temperature in proximity of its application, so that its working domain and operating temperature range is optimum. A major advantage of an MR cascade system over that of a conventional one is that the MR machine does not require heat exchangers between cycles. This is due to the fact that magnetic cooling (MC) material is solid so that the same fluid can be transferred to both hot and cold loops.

8.2.3.5 Active Magnetic Regenerator Cycle

In the active magnetic regenerator (AMR) cycle, the magnetic material serves not only as a refrigerant providing the temperature change as a result of (de)magnetization but also as a regenerator for the flow of heat transfer fluid. With the exception of the Carnot cycle, the AMR is the most efficient cycle of MR for room temperature. The conventional AMR cycle consists of adiabatic (de)magnetization and two processes where the intensity of the applied magnetic field remains constant. Due to the nature of the refrigerant (solid), the AMR cycle includes a heat transfer fluid that associates the refrigerant with the cold and hot source heat exchangers. The MC regenerator material is immersed into the heat transfer fluid flow, and by means of pistons or pumps, the transfer fluid can move through the regenerator. The AMR cycle cannot be illustrated by a *T-s* diagram as each part of the regenerator executes single thermodynamic cycles, which bind to one another through the heat transfer fluid.

The AMR refrigerator comprises a magnet, a regenerator with magnetocaloric material, cold and hot source heat exchangers and a device to allow the flow of heat transfer fluid through the active regenerator. Some authors assume that the AMR cycle can be considered as a group or a series of cascade cycle refrigerators carried out by the AMR material. However, the cascade approach is erroneous as each solid element does not pump heat from the regenerator directly to the next neighboring solid element, but all elements accept or reject heat to the heat transfer fluid simultaneously and are indirectly coupled through the fluid. This makes the difference between the two cycles clear (Romero Gomez et al., 2013).

The AMRs are porous to facilitate heat exchange with the transfer fluid and can be formed by thin sheets arranged in parallel, perforated sheets or small spheres of magnetocaloric material. The MC material of the regenerator can be a single material or can consist of several materials with different scaled Curie temperatures. With several materials, it is possible to increase the working range of the magnetic refrigerator. This is an apparently intuitive idea. However, issues such as the number of materials used and the relative amount of each are being investigated to achieve optimal performance. Research groups have developed several prototypes of refrigerators with AMR cycles. These can be classified according to the type of magnetic source (permanent magnets or electromagnets), the type of magnetocaloric material and the relative movement of the active elements of the device. With regard to heat transfer fluid used by the investigators, depending on the temperature range, these can be natural water, glycol water, distilled water, gases (helium) or coolants, among others.

8.2.4 Working Materials

The MCE is an intrinsic property of a magnetic solid. This thermal response of a solid to the application or removal of magnetic fields is maximized when the solid is near its magnetic ordering temperature. Thus, the materials considered for MR devices should be magnetic materials with a magnetic phase transition temperature near the temperature region of interest (Smith et al., 2012). For refrigerators that could be used in the home, this temperature is room temperature. The temperature change can be further increased when the order parameter of the phase transition changes strongly within the temperature range of interest (Bruck, 2005).

A magnetocaloric material may provide three contributions to the total entropy: a magnetic, an electronic and a lattice contribution. The entropy is a measure of order in the magnetothermodynamic system: A high order is related to low entropy and vice versa. Dipoles (i.e. electronic spins) may show different orientations. If in a paramagnet, ferromagnet or diamagnet these entities are oriented in the same direction, the order and the magnetization are high. Applying a magnetic field aligns electronic spins, and lowering the temperature (by releasing energy from the system) also leads to a more ordered system. Therefore, the external magnetic field yields the stress parameter and the magnetization the order parameter of such magnetic materials.

The magnitudes of the magnetic entropy and adiabatic temperature changes are strongly dependent on the magnetic ordering process. The magnitude is generally small in antiferromagnets, ferrimagnets and spin glass systems but can be much larger for ferromagnets that undergo a magnetic phase transition. First-order phase transitions are characterized by a discontinuity in the magnetization changes with temperature, resulting in a latent heat, while second-order phase transitions do not have this latent heat associated with the phase transition (Smith et al., 2012).

From the previous entropy analysis of magnetic materials, only magnetic entropy S_m is changeable with the magnetic field change. In the range of room temperature, the influence of lattice entropy S_r is too remarkable to neglect. Therefore, part of the cooling capacity of the magnetic system is consumed for cooling the lattice system for the entropy flow from the lattice system, although the temperature decreases to some extent during adiabatic demagnetization. Thus, the gross cooling capacity is less than that of the condition of $(S_r + S_e) \approx 0$.

To apply the MCE with high performance, optimal properties of magnets and magnetocaloric materials are required. For this, the different families, which show a large GMCE, have to be taken into consideration. The properties of the best magnets and designs are described in the literature (Yu et al., 2003; Jeong et al., 2006; Bjork et al., 2010). As pointed out in many papers, the magnetic material must fulfill a series of properties and characteristics to be used as cooling material (Yu et al., 2003; Gschneidner et al., 2005; Romero Gomez et al., 2013):

- Low Debye temperature values

- Curie temperature near working temperature

- Large temperature difference ΔT_{ad} in the vicinity of phase transition

- No thermal or magnetic hysteresis to enable high operating frequency and, consequently, large cooling power

- Low specific heat and high thermal conductivity, thereby allowing large changes in temperature and facilitating the processes of heat transfer and increasing efficiency

- Large electric resistance to avoid the eddy current loss

- High electrical resistance to avoid Foucault currents in the processes of rapid change in magnetic field

- Essentially zero magnetic hysteresis

- Good mechanical properties

- Resistance to corrosion

- Lack of toxicity and low environmental impact

- Low manufacturing costs necessary for commercial viability

Among the available materials, pure gadolinium may be regarded as the ideal substance for MR, just as the ideal gas is for conventional refrigeration. Gadolinium's temperature increases when it enters certain magnetic fields; similarly, the temperature drops when it leaves the magnetic field. The effect is stronger for the gadolinium alloy $Gd_5Si_2Ge_2$ (Gschneidner and Gibson, 2001). Remarkably, praseodymium alloyed with nickel ($PrNi_5$) has such a strong MCE that it allows approaching within one thousandth of a degree (mK) of absolute zero (Emsley, 2001). Because the MCE occurs below room temperature, these materials would not be suitable for refrigerators operating at room temperature. Also, just as conventional systems are usually not operated with ideal gases; magnetic refrigerators perform better with specially designed alloys. One advantage of pure gadolinium is that its physical properties may be described by basic physical laws. This allows the numerical calculation of magnetothermodynamic charts of high resolution. Producing such charts for MC alloys would demand a tremendous amount of high-quality experimental data, which usually are not available.

Gschneidner et al. (2005) published the following list of promising categories of magneto-caloric materials for application in magnetic refrigerators: binary and ternary intermetallic compounds; gadolinium-silicon-germanium compounds; manganites; lanthanum-iron–based compounds; manganese-antimony arsenide and iron-manganese-arsenic phosphides. At present, a number of toxic substances in such compounds are being replaced by more acceptable elements. A discussion of the different types of materials with their distinct properties is found in extended reviews (Bruck, 2005; Gschneidner et al., 2005). Currently, the total entropies and the related refrigeration capacity, the adiabatic temperature change and the costs of the materials are under investigation. Bruck (2005) stated that other properties, such as corrosion resistance, mechanical properties, heat conductivity, electrical resistivity and the environmental impact are also important.

8.2.5 Stage of Development

The MCE-based technique can be used to attain extremely low temperatures, along with the typical ranges used in refrigerators. Some commercial ventures to implement this technology are under way, claiming to reduce energy usage by 40% compared to current domestic refrigerators.

Application of the GMCE calls for a magnetic field change in a magnetocaloric material. This can be performed using different MR principles:

- Alternatively changing magnetic fields in static blocks of magnetocaloric material by application of electromagnets

- Rectilinear motion of magnetocaloric material with static permanent magnet assemblies

- Rectilinear motion of permanent magnet assemblies with static MC material blocks

- Rotary motion of magnetocaloric material with static permanent magnet assemblies

- Rotary motion of permanent magnet assemblies with static magnetocaloric material blocks

At least 28 prototypes have been reported so far, and some of their characteristics were listed in the literature (Gschneidner and Pecharsky, 2008). The best-known prototypes have been built by Astronautics Corporation, USA (a rotary type of magnetic refrigerator operated with a frequency of up to 4 Hz, filled with gadolinium spheres, having a magnetic field induction of 1.5 T and a cooling capacity of 95 W with a maximum temperature span of 20 K); Material Science Institute in Barcelona; Chubu Electric/Toshiba in Yokohama; University of Victoria in British Columbia (prototype that applies the layered bed technique with two different materials, such as by choosing different alloys at different positions in the refrigerator, the performance of the refrigerator is increased); Sichuan Institute of Technology/Nanjing University in Nanjing, China (first prototype that applied a material with the GMCE exceeding the adiabatic temperature difference of gadolinium); Laboratoire d'Electronique Grenoble in Grenoble (France) and Cooltech Applications (France).

Concerning the performance of MHPs, note that single-stage systems allow only small temperature lifts (e.g. only 7.8 K for magnetic fields up to 1.5 T). Large lifts will require a large number of stages with associated heat transfer losses between stages. Another drawback is the operating frequency of the cycle in combination with the specific capacity of the cycle. One of the most successful prototypes shows a maximum frequency of 4 Hz. Among the best capacity that has been reported is 1,500 J/kg, that is, about 6 kW can be taken by the cycle per kilogram of magnetocaloric material.

Cooltech Applications (France) – a company focused on MR systems – developed Cold Power units characterized by 200–700 W, 0.3 Hz, temperature of 2°C–5°C in the cabinet and about 32°C ambient, and positive return on investment (ROI) of 3 years (more details are available at http://www.cooltech-applications.com).

The MCE cannot be used directly for cooling; instead, specialized cooling cycles with regeneration are required for MR to be used for cooling at room temperature at an effective cost. All described MCE-based refrigeration cycles cannot be applied in refrigeration technology at room temperature. In this way, the Carnot cycle can only be effective in cryogenic applications. Nevertheless, the regenerative Brayton and Ericsson cycles are well suited to real applications with little differences between them. The most used MR cycle at room temperature is the AMR cycle due to its greater performance in comparison with the rest of the cited cycles. Therefore, almost all of the implemented prototypes so far are based on the AMR cycle (Romero Gomez et al., 2013).

8.3 THERMOELECTRIC HEAT PUMPS

Thermoelectric (TE) heat pumps use the TE effect and have improved over time to the point at which they are useful for certain refrigeration tasks. TE heat pumps and coolers are mostly based on functional units that are called Peltier modules or thermoelectric cooling (TEC) elements. TE (Peltier) heat pumps are generally only around 10%–15% as efficient as the ideal heat pump (Carnot cycle), so much lower than the range of 40%–60% achieved by conventional compression cycle systems (reverse Rankine systems using compression/expansion). However, this area of technology is currently the subject of active research in materials science (Semenyuk, 2010; Yang et al., 2013). A major reason for its popularity is the long lifetime, as there are no moving parts, and when renewable energy is used, then it would be better for the environment.

8.3.1 Thermoelectric Effect

The TE effect is the direct conversion of temperature differences to electric voltage and the other way around. A TE device creates a voltage when there is a difference in temperature; conversely, it creates a temperature difference when a voltage is applied to it. Figure 8.4 shows the schematic of a Peltier device: Insulating heat spreaders are used on top and bottom, assumed to be connected to heat and cold reservoirs, while p-type and n-type TE materials are stacked electrically in series and thermally in parallel. This can work as a Peltier element or a TE generator, depending on whether a current source or a resistive device is connected to the electrical connections. The applied temperature gradient causes charge carriers in the material to diffuse from the hot to the cold side. The TE effect is used to generate electricity, as well as measure or change the temperature of objects. The term *TE effect* includes separately identified effects: Seebeck, Peltier and Thomson effect. Joule heating – the heat generated when a current is passed through a resistive material – is related, but it is not generally termed a TE effect. In contrast to the Peltier, Seebeck and Thomson effects, which are reversible thermodynamically, Joule heating is not. Figure 8.5 shows the application of these effects to a TE generator and cooler.

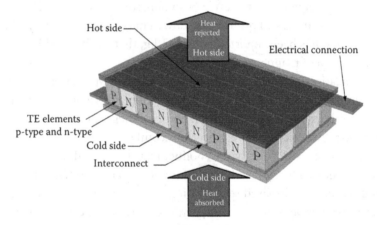

FIGURE 8.4 Peltier element schematic: Thermoelectric legs are thermally in parallel and electrically in series.

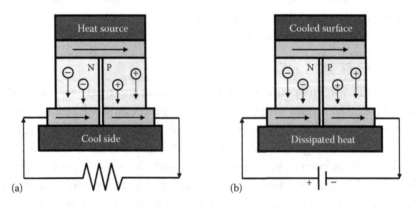

FIGURE 8.5 Thermoelectric circuit made of materials of different Seebeck coefficient, configured as a thermoelectric generator (a). Seebeck circuit configured as a thermoelectric cooler (b).

8.3.1.1 Seebeck Effect

The Seebeck effect (i.e. conversion of temperature differences into electricity) is named after physicist Thomas Johann Seebeck, who discovered in 1821 that a compass needle would be deflected by a closed loop formed by different metals (joined in two places) with a temperature difference between the junctions. This effect was due to the metals responding to the temperature difference in different ways, thus creating a current loop and a magnetic field. The Seebeck effect is a classic example of an electromotive force (EMF) and leads to measurable currents or voltages in the same way as any EMF.

The Seebeck voltage appears to be proportional to the temperature difference between the junction and the connections to the meter. Widely used as temperature measurement devices, thermocouples make use of this principle. The Seebeck effect creates a back-EMF opposing the externally applied voltage in reaction to this temperature difference, thereby reducing the effective potential difference over the plates. Consequently, it also reduces the heat flow from the cold to the hot side as induced by the Peltier effect, thus making the Seebeck effect an important loss factor for modules used as coolers or heaters (Elderson, 2015).

The voltage generated by the Seebeck effect is the result of moving charge carriers. To illustrate this, let us consider a slice of semiconductor rod with infinitesimal length in the flow direction of the electrons. An incremental temperature change corresponds to a change in the electrons' kinetic energy. That is, electrons coming from the hotter side exhibit a higher velocity than those at the colder side, which results in a net flow of negative charge towards the cold side. In turn, this leads to the buildup of a net positive charge at the hot side and a negative charge at the cold side. The electric field created in this manner slows the hot electrons moving towards the cold side, while speeding up the cold electrons going up. The potential difference (Seebeck voltage) associated with the electric field created per unit of temperature is material specific and described by the Seebeck coefficient. The Seebeck coefficient α determines the EMF ε or induced Seebeck voltage V_S appearing in reaction to a temperature gradient ∇T over a piece of semiconductor:

$$\varepsilon = \nabla V_S = - \alpha(T)\nabla T \tag{8.12}$$

The Seebeck coefficient is also strongly dependent on the local temperature. By convention, α represents the potential of the hot side with respect to the cold side of the material piece per unit of temperature difference. The Seebeck coefficient is largest for those materials that also exhibit large Peltier coefficients. Moreover, as the Seebeck and Peltier effects are interdependent in operation, these two coefficients are also related through the Thomson effect, discussed further in the chapter.

8.3.1.2 Peltier Effect

The Peltier effect – named after physicist Jean Charles Athanase Peltier, who discovered it in 1834 – is the presence of heating or cooling at an electrified junction of two different conductors. Heat may be generated or removed at the junction when a current flows through a junction between two conductors. The total heat generated at the junction is not determined by the Peltier effect alone, as it may also be influenced by Joule heating and thermal gradient effects.

The Peltier effect at the junction of an n-type (negative) and p-type (positive) semiconductor rod is caused by the energy difference of moving charge carriers in the rods as created by the presence of a band gap in their electron energy spectra: Passing charge carriers between rods with opposite doping leads to energy being released within the connector plate at the side where charge carriers are moving towards, and taken up from, the plate from which charge carriers move away.

The magnitude of the heat flow through the rods per unit of current is a material-specific property that depends on its energy density spectrum and is referred to as its Peltier coefficient. The values for this coefficient are generally largest for doped semiconductors. As the difference between the heat flowing in and out of the plate determines the energy released to (or extracted from) the connector plates, the difference between the Peltier coefficients of the n-type and p-type materials used correspondingly determines the temperature difference created between the hot and cold plate by means of the Peltier effect. Because the Peltier coefficients of either type of semiconductor have opposite signs (positive for the n type and negative for the p type), combining both types into P/N couples within Peltier devices appears to be the most effective configuration to maximize the net heat flow. The Peltier coefficient π represents the amount of Peltier heat displaced per unit of time resulting from applying one unit of electrical current – by convention, the positive direction of heat transfer is that of increasing potential. For semiconductors, the Peltier coefficient is strongly dependent on temperature. The Peltier heat $\dot{Q}_p$ is proportional to the current I and Peltier coefficient π:

$$\dot{Q}_p(T) = I\pi(T) \tag{8.13}$$

The working principle of Peltier refrigeration devices is based on the Peltier effect in reaction to the external voltage supplied. Depending on the external conditions and properties of the material used, a TE equilibrium is reached for P/N couples – a state of temperatures and gradients, heat flows, electrical potentials and currents as a result of the three TE effects occurring in parallel, together with the Joule heating and Fourier conductive effects.

8.3.1.3 Thomson Effect

The reversible heating or cooling caused by a flow of electric current and a temperature gradient being present at the same time in a homogeneous conductor is known as the Thomson effect, predicted and observed in 1851 by W. Thomson, who later became better known as Lord Kelvin. By applying thermodynamic theory to the physical characteristics of the Seebeck and Peltier effect, Thomson recognized the interdependency of the two effects, which would imply the existence of this third TE effect. Thomson heating or cooling arises from the work done by electrons of a current that passes along or against the direction of electric intensity.

Depending on the direction of both the current and the temperature gradient, heat is either released into or taken up from the conductor. In a conductive rod present in a Peltier element in operation, energy is released into the lattice as the electric current exerts work along the electrical field. As opposed to the Peltier and Seebeck effects, for which only the net effect of two different materials can be measured, the heat released through the Thomson effect is directly measurable for a single material. A third TE material property named the Thomson coefficient is thus introduced to describe the net energy accumulation due to the Thomson effect per unit of volume, temperature gradient and electric current. The Thomson coefficient κ represents the amount of heat absorbed or evolved as one unit of charge passes one unit of temperature difference, and its definition is known as the first Thomson relation, given by

$$\kappa(T) = \frac{d\pi(T)}{dT} - \alpha(T)$$

(8.14)

In practical Peltier modules, the so-called Kelvin heat resulting from this effect is almost negligible. However, the value of the Thomson coefficients of the conductor materials used in modules is of most particular value as it describes the relationship between the Peltier and Seebeck coefficients; the other two TE coefficients follow directly from the Thomson coefficient. The second Thomson relation expresses the Peltier coefficient as the product of the Seebeck coefficient and the local temperature, as follows:

$$\pi(T) = \alpha(T)T$$

(8.15)

This way, the Thomson effect defines the continuity of the Peltier effect through the definition of its coefficient, by describing the part of the Peltier heat flow that is the result of the interaction of current with changes in temperature. Although the Seebeck and Peltier effects occur at junctions between dissimilar conductors, which might suggest that these are actually interfacial phenomena, the Thomson effect shows that all three TE coefficients are interrelated bulk properties of the conductor materials, which determine to a large extent the behavior of Peltier modules.

8.3.1.4 Joule Heating

Joule heating involves the increase in temperature of the lattice atoms in the semiconductor material caused by the passing of charge carriers. The movement of electrons and holes through the material as induced by an electric potential difference (and thus the presence of an electric field), results in collisions of these charge carriers with the lattice atoms. These collisions cause the transfer of both energy and momentum from the charge carriers to these atoms (i.e. the relaxation of energy from electrons to phonons in the lattice) due to which the material heats up. Therefore, unlike the heat transfer by TE effects, the heat transferred by Joule heating is independent of the flow direction of the electrons as heat carriers are distributed randomly into all directions. Its magnitude depends on the electric resistivity and the magnitude of the charge carrier flow. The Joule heat $\dot{Q}_J$ produced per unit of length L of TE material is inversely related to its temperature-dependent electrical conductivity γ and the cross-sectional area A and proportional to the square of the electrical current I running through it, as follows:

$$\dot{Q}_J = \frac{I^2}{\gamma(T)} \frac{L}{A} \tag{8.16}$$

Whereas this phenomenon is exploited in electric resistance heaters and in food processing, in TE devices Joule heating results in an efficiency decrease as the semiconductor material heats up over its whole length, resulting in decreasing the heat flow from the heat source into the P/N couples. As the current also passes the plates connecting the semiconductor rods, Joule heating also takes place in the connecting plates. The dependency of the electrical resistivity of the semiconductors on temperature makes Joule heating become larger at higher temperatures. The heat released into the lattice by both the Thomson and Joule effects is conducted through the material in reaction to the temperature gradient according to Fourier's law.

8.3.1.5 Fourier Heat Conduction

According to Fourier's law, heat conduction occurs in the opposite direction of a thermal gradient within a material, from a hot to a cold location. The phenomenon is described as a consequence of diffusion of heat carriers from a location with a high density of these carriers to a location of low density. The heating energy transported in this manner then depends on the relaxation time and the velocity in the flow direction of these carriers, which determine the distance it travels before it loses its excess energy due to scattering. Along with this mean free path, the specific heat of the carriers (or the average amount of excess energy each heat carrier can lose on scattering) determines the magnitude of this heat flow. This material-specific behavior of heat transfer through the atom lattice is reflected in the thermal conductivity of the material. Fourier's law correlates the magnitude of heat flow to the temperature difference over the increment in the flow direction through the thermal conductivity of the material in that direction. The heat conduction through any material by the Fourier mechanism as caused by a temperature gradient is described by means of its thermal conductivity λ, as dependent on the ratio of cross-sectional area A and the length L. As the heat flows in the opposite direction of the temperature gradient, its magnitude $\dot{Q}_F$ is described by

$$\dot{Q}_F = -\lambda(T) \frac{A}{L} \nabla T \tag{8.17}$$

In Peltier modules, Fourier heat conduction results in heat flow from the hot to the cold side of the element, in the opposite direction of the desired heat flow. This results in a decrease of both the cooling and the heating power, such that the efficiency drops as the conductivity of the rod materials is higher. The Fourier heat flow increases with increasing temperature difference between the hot and cold side of the module. In the temperature range of operation of modules used in refrigeration devices, generally the conductivity increases with temperature, so that the losses due to Fourier conduction are higher at higher operating temperatures. As opposed to other heat transfer mechanisms discussed previously, conduction takes place in all the parts of the module: P/N couples, connector plates, ceramic end plates and the junctions of these different parts.

Other effects that can influence the performance of Peltier modules are discussed in the literature. Among these are the drag of phonons, which is found in semiconductors and contributes positively to the performance but is known not to lead to significant changes in the conditions of modules. The same applies to temperature- and time-dependent variations in the Fermi levels of materials, which therefore are not further discussed. Next to the conduction through the rods themselves, a small portion of heat is also transferred to the air cavities in between them as an additional source of loss, along with the conduction through this air. Furthermore, radiation mainly from the hot to the cold plate might also magnify the heat transfer in this direction. Both phenomena were found to exert a small, but not negligible, influence on the performance (Elderson, 2015).

A more influential, but hardly understood, phenomenon that significantly reduces the cooling power of modules is the internal electrical resistance by interface layers, which not only leads to a temperature-dependent increase of the total impedance of the array of semiconductor rods but also the generation of additional Joule heat. Even though its influence was proven experimentally, this effect has not yet resulted in readily usable empirical correlations to quantify the resulting energy flow.

8.3.2 Figure of Merit and Performance

The comparison between the functionality of different TE materials for use inside a Peltier module is generally accomplished by means of the figure of merit ZT. A material with a high figure of merit would show a high performance level in such a device. Recalling the heat transfer mechanisms as described previously, this would require high electrical conductivity and a large Seebeck coefficient, combined with a low thermal conductivity. The dimensionless and temperature-dependent number ZT is defined as

$$ZT = \frac{\gamma \alpha^2}{\kappa} T \tag{8.18}$$

A better value for ZT leads to a higher efficiency of conversion from electrical to thermal power in modules. Because the Thomson effect is not taken into account in its definition, the ZT value is less meaningful for devices in which this phenomenon plays a more important role. The maximum coefficient of performance (COP) achievable based on a material with a certain figure of merit and known temperatures for the heat source T_{source} and sink T_{sink} is given by

$$COP_{c,max} = \frac{T_{source}}{T_{sink} - T_{source}} \cdot \frac{\sqrt{1+Z\bar{T}} - \dfrac{T_{sink}}{T_{source}}}{\sqrt{1+Z\bar{T}} + 1} \quad \text{where } \bar{T} = \frac{1}{2}(T_{sink} + T_{source}) \tag{8.19}$$

$$COP_{h,max} = \frac{T_{sink}}{T_{sink} - T_{source}} \cdot \frac{\sqrt{1+Z\bar{T}} - \dfrac{T_{source}}{T_{sink}}}{\sqrt{1+Z\bar{T}} - 1} \qquad (8.20)$$

The temperature in the ZT value is taken here as the average between the source and the sink. When applied to a thermocouple, the ZT value of both semiconductor types is assumed to be equal. The first factor in the equations corresponds to the Carnot efficiency of cooling and heating – the theoretical maximum efficiency of any conversion process operating between the given temperatures. The second factor then gives the fraction of this efficiency that can theoretically be achieved under optimal conditions of the material. According to the formula given for the optimum COP values, a ZT of at least 2 would be required to make Peltier-based refrigeration competitive with vapor compression cooling systems.

Riffat and Ma (2003) showed that the COP of a TE refrigeration cycle drops off substantially when ΔT becomes larger. While at $\Delta T = 10$ K the COP exceeds 4, for temperature differences between heat rejection and heat intake of $\Delta T > 30$ K, the obtainable COP (heating) is in the range of 1.9 and less.

Table 8.1 lists the ZT values of state-of-the-art TE materials along with the calculated maximum COP values for applications around 300 K. Given its high performance when compared to other materials in the practical range of temperature for heat pumping and refrigeration devices, Bi_2Te_3 alloys are currently the most frequently applied in TE modules.

8.3.3 Stage of Development

In practice, two unique semiconductors (n type and p type) are used as they need to have different electron densities. The semiconductors are placed thermally in parallel to each other and electrically in series and then joined with a thermally conducting plate on each side. When a voltage is applied to the free ends of the two semiconductors, there is a flow of DC current across the junction of the semiconductors, which causes a temperature difference. The side with the cooling plate absorbs heat, which is then moved to the other side

TABLE 8.1 Figure-of-Merit Values for State-of-the-Art Semiconductor Materials at 300 K Average Temperature and Calculated Resulting Maximum COP Values

Material	$\bar{ZT}$ at 300 K	$COP_{c,max}$ (at $\Delta T_{lift} = 10$ K)	$COP_{h,max}$ (at $\Delta T_{lift} = 10$ K)	η_{2nd_law} (−)
Bi_2Te_3-based alloys	0.95	4.5	5.5	0.145
$CoSb_3$	0.15	0.54	1.5	0.017
$CoSb_3$ alloys (projected optimum)	0.68	3.4	4.4	0.109
Zn_4Sb_3	0.30	1.5	2.5	0.048
Zn_4Sb_3 alloys	0.38	1.9	2.9	0.061
$CeFe_{4-x}Co_xSb_{12}$	0.08	0.07	1.1	0.002
PbTe-based alloys	0.05	−0.13	0.87	−0.004

Source: Adapted from Elderson (2015).

of the device where the heat sink is. TEC units are typically connected side by side and sandwiched between two ceramic plates. The cooling ability of the total unit is then proportional to the number of TEC units in it.

A single-stage TEC typically produces a maximum temperature difference of 70 K between its hot and cold sides. The more heat moved using a TEC, the less efficient it becomes as it needs to dissipate both the heat being moved and the heat it generates itself from its own power consumption. The amount of energy that can be absorbed W is proportional with the Peltier coefficient π, which is dependent on temperature and materials used, the current I and time t:

$$W = \pi I t \tag{8.21}$$

Thermoelectric junctions are about four times less efficient in refrigeration applications than conventional means, so due to this lower efficiency, TEC is generally used only in environments where the solid-state nature (no moving parts, low maintenance, compact size and orientation insensitivity) outweighs the drawback of low efficiency. TEC performance is a function of ambient temperature, hot and cold side heat exchanger (heat sink) performance, thermal load, Peltier module (thermopile) geometry and Peltier electrical parameters. Common TE materials used as semiconductors include bismuth telluride (most commonly used), lead telluride, silicon germanium and bismuth-antimony alloys. New high-performance materials for TEC are being actively researched.

The following requirements are typical for TE materials: narrow band-gap semiconductors (due to room temperature operation); heavy elements (high mobility and low thermal conductivity); large unit cell; highly anisotropic or highly symmetric properties; complex compositions. The main benefits of using a TEC include no moving parts (less maintenance); no greenhouse gases (GHGs) (e.g. chlorofluorocarbon [CFC]) used as working fluids; tight temperature control (within fractions of a degree); flexible shape (form factor) and small size; usable in more severe environments; long lifetime (mean time between failures [MTBF] > 100,000 h); controllable via changing the input voltage/current. Nonetheless, there are also some disadvantages of using a TEC unit, such as limited amount of heat flux able to be dissipated; relegation to applications with low heat flux; not as efficient (lower COP) as vapor compression systems.

The TECs are used for applications that require heat removal ranging from milliwatts to several kilowatts, and they can be made for small applications (e.g. beverage cooler) or large applications (e.g. submarine). Note that TEC units have a limited high-performance lifetime, and their health strength can be measured by the change of their alternating current resistance (ACR): When a TEC is worn out, the ACR will increase.

Peltier elements are commonly used in consumer products, as for example in camping equipment, portable coolers, cooling electronic components and small instruments, water extraction from the air in dehumidifiers, climate-controlled jackets, heat sinks for microprocessors, wine coolers, scientific devices, thermal cyclers, fiber-optic applications and military electronic equipment.

8.4 THERMOACOUSTIC HEAT PUMPS

Thermoacoustics relates to the physical phenomenon that a temperature difference can create and amplify a sound wave and vice versa that a sound wave is able to create a temperature difference. Acoustic work (sound) can be used to generate temperature differences that allow the transport of heat from a low-temperature reservoir to an ambient at higher temperature, thus forming a TA refrigeration system (Herman and Travnicek, 2006). TAHPs can work also as heaters and share the same advantages of TA coolers (Bassem et al., 2011).

Note that a sound wave is associated with changes in pressure, temperature and density of the medium through which the sound wave propagates. In addition, the medium itself is moved around an equilibrium position. An acoustic wave is brought into interaction with a porous structure with a much higher heat capacity compared to the propagation. This porous structure acts as heat storage–regenerator (Bruinsma and Spoelstra, 2010; van de Bor and Infante Ferreira, 2011). The material used for the regenerator on the properties has an important influence on a TAHP (Kruk, 2013).

8.4.1 Thermoacoustic Effect

The TA effect was discovered in the second half of the eighteenth century, but it did not receive much attention as a potential technology for energy conversion until near the middle of the twentieth century. Glassblowers were the first to observe the TA effect; they noticed that the heated pipes for blowing glass made sounds. In 1777, Byron Higgins observed that placing the flame at both ends of an open tube resulted in the appearance of an audible sound. Another described example of this phenomenon was mentioned in combustion, in so-called singing pipes, described by Putnam and Dennis (1956), who explained the experimental findings of Higgins in 1777. The TA effect was first described by Lord Rayleigh (1894 and reprinted in 1945) in his famous book *The Theory of Sound*. He stated that 'if heat be given to the air at the moment of greatest condensation, or be taken from it at the moment of greatest rarefaction, the vibration is encouraged'. The effect was observed in helium cylinders by Kramers (1949), who discovered sound wave excitation in the liquid and gas phases inside a column at 4 K and 300 K. At that stage, the TA phenomenon received attention as a problem to be overcome, and research was directed at suppressing these oscillations. However, in the second half of the twentieth century, researchers studied the potential engineering applications of the effect, developing standing-wave devices that reached a maximum efficiency close to 20% Carnot. Later, it was realized that looped-tube travelling-wave devices were more promising and offered higher performance than standing-wave devices. Successful travelling-wave devices began to be built only in the last decade of the twentieth century (Kruk, 2013).

Within TA, a distinction is made between a TA engine or prime mover (a device creating an acoustic wave by a temperature difference) and a TAHP (where an acoustic wave is used to create a temperature difference). Note that a TA engine converts thermal energy to acoustic energy via its core, comprising a porous material that is sandwiched between two heat exchangers (hot and ambient). The two heat exchangers impose a temperature gradient along the porous material, which is required to sustain and amplify the acoustic wave that is generated. TAHPs or refrigerators utilize the acoustic energy received within a porous medium that is sandwiched between two heat exchangers (ambient and cold) to transport heat from the cold to the ambient heat exchanger (i.e. in the reverse direction to that in a heat engine) (Spoelstra, 2007, 2008; van de Bor and Infante Ferreira, 2011; Abduljalil, 2012).

The porous material is generally referred to as a *stack* in standing-wave TA devices (where the phase difference between pressure and velocity is close to 90°) and as a *regenerator* in travelling-wave devices (where the phase difference between pressure and velocity is close to 0°). In the absence of solid material within the acoustic field, the acoustic oscillations are essentially adiabatic. However, the presence of a solid surface will cause viscous and thermal boundary layers to form in the acoustic field. In these boundary layers, heat transfer takes place between the oscillating gas parcels and the solid surface, resulting in a positive transfer of heat from one end of the solid surface to the other, whose direction depends on whether the device is a heat engine or a refrigerator. The phenomenon of nonzero net heat transport that occurs in an acoustic field is designated as the *thermoacoustic effect*.

Figure 8.6 illustrates the main common arrangements for TA devices that are under investigation by researchers, as reported in the literature (Abduljalil, 2012). A standing-wave device can be a simple resonance tube containing the core, which is placed at an optimized location based on highest efficiency or highest power. Travelling-wave devices often consist of a looped tube that contains the TA core. In such a system, a secondary ambient heat exchanger is commonly used to cool the tube – known as a thermal buffer tube (TBT) – behind the HHX. The TA Stirling heat engine (TASHE), where the gas within

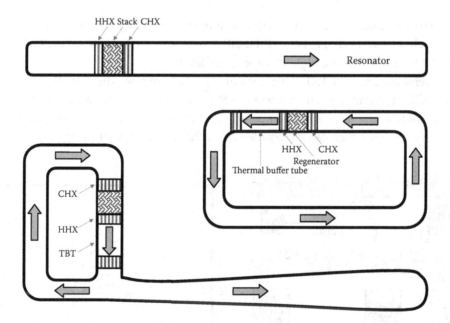

FIGURE 8.6 Standing-wave TA device (top), travelling-wave looped-tube TA device (center) and TA Stirling heat engine (bottom). The stack/regenerator is sandwiched between the hot and cold heat exchangers; the arrows indicate the acoustic power flow.

the regenerator undergoes a thermodynamic cycle similar to that of a conventional Stirling engine, comprises a linear tube connected to a looped tube, which contains the core of the system with the secondary CHX. A secondary CHX may be used to cool behind the TBT.

8.4.2 Thermodynamic Cycle

When a temperature gradient is imposed across a regenerator (porous structure) by, for example, a CHX and an HHX, an acoustic wave passes by from the cold side and an acoustic cycle takes place with a parcel of gas (see Figure 8.7). To be able to create or move heat, work must be done, and the acoustic power provides this work. The processes that take place in a TAHP are as follows:

- *Compression.* The gas is being compressed by the passing pressure wave. Because the gas is in close thermal contact with the regenerator, the temperature stays the same locally.

- *Cooling.* The gas parcel is moved back to its original position, and it is still hotter than the structure (regenerator), resulting in heat transfer from the gas to the structure.

- *Expansion.* The pressure wave that first compressed the gas parcel is now expanding it. Again, the gas is not cooled here due to the close thermal contact with the regenerator.

- *Heating.* Successively, the gas parcel is moved to a hotter part of the regenerator. Because the temperature over there is higher than the gas parcel, the gas is heated.

The thermodynamic cycle of a TAHP is run in the reverse direction of the well-known Stirling cycle (Tijani and Spoelstra, 2011, 2013), meaning that acoustic energy is used to pump heat from a lower to a higher temperature level.

The reverse process of a TAHP happens in a TA heat engine. The acoustic wave has the function of both pistons normally present in a Stirling engine. In this way, it is possible to create and amplify a sound wave by a temperature difference. The thermal energy is converted into acoustic energy, which can be regarded as mechanical energy (Spoelstra, 2007, 2008).

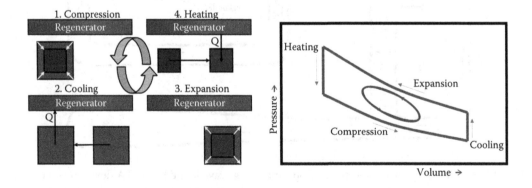

FIGURE 8.7   Working principle of the thermoacoustic cycle.

8.4.3 Thermoacoustic Technology

A *thermoacoustic heat pump* is the front runner in using a different mechanism for the work input. Although it is a relatively new technology, the proof-of-principle stage has been successfully completed, with scaling up currently undergoing intense efforts (Spoelstra, 2007, 2008; Bruinsma and Spoelstra, 2010; Tijani et al., 2011). The reason is that the TAHP features a wide applicability range, much larger than the heat pumps previously mentioned. Basically, the TAHP is a TA device that uses high-amplitude sound waves to pump heat from one place to another.

Figure 8.8 illustrates the TAHP application to a distillation column as well as the working principle (Kiss et al., 2012). The TA device consists of heat exchangers, a resonator and a regenerator (on traveling-wave devices) or stack (on standing-wave devices). The regenerator is made of a material with a much higher heat capacity compared to the gas medium through which the sound wave propagates. The function of regenerators in TA systems is an exchange and temporary storage of heat. A TA layout can be generally divided into two segments (Kruk, 2013): (1) the part containing a thermodynamic regenerator, two heat exchangers and a thermal buffer column and (2) the acoustic system that provides adequate conditions for the acoustic wave (a resonator and sound source). Depending on the type of engine, a driver or loudspeaker might be used as well to generate sound waves. To limit the space used, an electric driver (linear motor) generates the acoustic power cased inside a resonator, the temperature lifts being determined by the size of resonator, as well as the properties and pressure of the acoustic medium (Gardner and Howard, 2009). The resonator, housing the TA engine and the TAHP, determines the operating frequency, acts as a pressure vessel and transports the acoustic power between the components. Note that there is no concern about the noise levels because these are similar to the current industrial standards (e.g. below 85 dB), about 140–160 dB internal but only 55 dB external.

Because the medium used is a gas (air, noble gas [e.g. helium]), the system complies with safety and environmental concerns, while virtually having no limitations in its applications – it can be applied down to cryogenic temperatures, with temperature lifts of 100 K or more. The reliability of a TAHP is considered high, as it has basically no moving parts for the thermodynamic cycle when a waste heat–driven system is used, although there is a moving

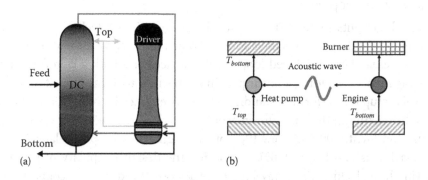

FIGURE 8.8   TA heat pump: mechanically/electrically driven (a) and thermally driven (b).

part in case of a linear motor. Other moving parts deal with the supply and removal of the heat to and from the system.

A TAHP is flexible with respect to changes in temperatures and powers. If the amount of waste heat is reduced, all powers within the TA system will drop accordingly and vice versa, at the cost of slightly decreased efficiency of the system. Changes in temperature are accommodated in a similar way. The acoustic response time of the system is fast; therefore, the overall system response will be determined mainly by the thermal inertia. This system can be rather easily started because it already uses high-temperature heat to drive the system. The cost estimates are based on material quantities that are necessary to realize the system, mainly stainless steel for the resonator, the heat exchangers and the regenerator. Alternative materials such as aluminum or copper could be used for the heat exchangers, although the manufacturing costs will increase.

Thermoacoustic systems can be scaled by using dimensionless numbers. Based on these numbers, scaling rules can be obtained that relate systems of different sizes, using different working media and working pressure. Note that heat transfer and heat losses will not scale according to these rules.

Although the working principle of TA technology is rather complex, the practical implementation is simple. This offers great advantages with respect to the economic feasibility of this technology, as well as additional benefits, such as the following:

- No moving parts for the thermodynamic cycle, so it is reliable and has a long life span.

- It has an environmentally friendly working medium, such as air or noble gas.

- The use of air or noble gas as a working medium offers a large window of applications because there are no phase transitions.

- Simple materials with no special requirements, which are commercially available in large quantities and therefore relatively inexpensive, are used.

- A large variety of applications can be covered on the same technology base.

8.4.4 Stage of Development

This section highlights some of the recent achievements in TA technology, which is still under development despite implementation in some places. Near the end of the twentieth century, the researchers developed looped-tube devices and took a step further towards higher performance by adding a linear part in connection with the looped tube, discovering that this setup outperformed standing-wave TA engines by more than 50%. Such an engine had a maximum thermal efficiency of 30% – equivalent to 41% of Carnot efficiency (Backhaus and Swift, 2000). A new type of TA engine was proposed later by the same authors (Backhaus and Swift, 2002), combining the design simplicity of standing-wave devices with the high efficiency of travelling-wave devices. The idea was to connect a standing-wave engine to a travelling-wave engine in series, such that the output acoustic power of the standing-wave core was fed to the cold (ambient) side of the regenerator of the travelling-wave engine, which would then be amplified and used for the desired application.

Praxair Incorporated developed – with support from the US Los Alamos National Laboratory – TASHEs and refrigerators for the liquefaction of natural gas (Wollan et al., 2002). The combination of TA engines with orifice pulse tube refrigerators (OPTRs) was capable of producing significant cryogenic refrigeration with no moving parts. A prototype powered by a natural gas burner was built and tested, at a natural gas liquefaction capacity of about 500 gal/d. The unit liquefied 350 gal/d, with a projected production efficiency of 70% liquefaction and 30% combustion of an incoming gas stream. A larger system (liquefaction capacity of 20,000 gal/d and an efficiency of 80%–85% liquefaction) had undergone preliminary design. In the 500-gal/d system, the combustion-powered TASHE drove three OPTRs to generate refrigeration at methane liquefaction temperatures. Each refrigerator was designed to produce over 2 kW of refrigeration.

Sakamato and Watanabe (2004) worked on the construction and testing of a prototype looped-tube TA cooling system with a length of 3.3 m and 100-Hz fundamental frequency. In addition to air, a mixture of air and helium was examined; two identical ceramic honeycomb structures, each with a cell density of 1200 per square inch, were used to form the 50 mm long stacks installed within the looped tube. The system was tested in a vertical orientation in which the prime mover stack was placed with its cold end up. The two stacks were situated symmetrically in the loop, and the input power to the heater was 120 W. The system was only tested at a mean gas pressure equal to atmospheric pressure. The waveforms, the pressure amplitude distribution and the temperature variation at the cold end of the refrigeration stack were measured. The results showed significant advantages of the helium-air mixture over air alone: The system reached onset condition much faster, while the cooling power and the COP were significantly higher for the gas mixture. According to the pressure distribution data, the highest-pressure amplitude was detected somewhere near the refrigeration stack. While observing the frequency spectra, it was noticed that higher harmonics started to form, especially along the so-called steady-state stage.

Tijani and Spoelstra (2011) reported that they were able to run a TASHE with a thermal efficiency of 32%, corresponding to 49% of Carnot efficiency. A few years later, Tijani and Spoelstra (2013) reported that a hot air–driven TA Stirling engine was designed, built and tested. The engine produced about 300 W of acoustic power with a performance of 41% of the Carnot performance at a hot air temperature of 620°C. The performance of the engine can be improved by suppressing the heat leak down the TBT and improving the HHX. A better solution might be to use a hot head instead of the HHX. The hot head with fins can be directly heated by the hot gases as usually done in conventional Stirling engines. TA technology has been further developed by this group at the Energy Center of the Netherlands (ECN).

Yang et al. (2014) reported the use of travelling-wave TA high-temperature heat pumps for industrial waste heat recovery. A novel travelling-wave TAHP was presented, aimed to solve the problems occurring in a conventional vapor compression heat pump, such as high discharge temperatures, high-pressure ratio and low efficiency. This system comprises three linear pressure wave generators, which are coupled with three heat pumps into one single closed loop. Theoretically, this system is able to complete the TA conversion with much higher efficiency. Considering various waste heat temperatures (40°C–70°C) and different hot-end temperatures (120°C–150°C), the results showed that this new heat pump system had a high relative Carnot efficiency of about 50%–60%. In using a reliable linear compressor and a TAHP with no moving parts, this technology has an inherent potential for high reliability.

Ghorbanian and Karimi (2014) examined the possible performance enhancement of small gas turbine power plants through the application of TA systems powered only by the waste heat of the gas turbine. Two configurations were considered: a TA refrigerator-assisted gas turbine (TRG) and a combined TAHP and refrigeration-assisted gas turbine (CTHRG). The exergy, rational efficiency (exergy efficiency) and relative power gain (RPG) of these configurations were compared with those from the recuperated gas turbine engine. The reported results indicate that the integration of a TA system to a simple gas turbine cycle not only enhances the energy/exergy efficiency for the overall plant but also increases the output power of the cycle. A cycle improvement, and hence the proper selection of the TA configuration, depends on the application requirements of the gas turbine engine.

Given the advantages of TA technology, such devices are increasingly attracting the attention of research groups and academic institutions in the United States, the Netherlands, China, France, Japan, United Kingdom and other countries. Research work is performed in parallel to study the physics of TA phenomena and to produce practical, economically feasible and highly efficient TA systems that can be commercialized for domestic and industrial use. However, there are a number of engineering issues that need further investigation to overcome the difficulties associated with producing a TA system that satisfies the criteria of practicality, economic feasibility and efficiency. One of the most costly parts of the device is its core, comprising the stack (or regenerator) and the two adjacent heat exchangers. Current research seeks to optimize these parts from various points of view, including design, cost and performance. Special attention should be given to the stack (or regenerator), as it is the part of the system where the energy conversion takes place, in a complex simultaneous oscillation of the compressible fluid parcels and the solid surface of the stack material through the boundary layer. This complicated process may involve highly nonlinear fluid dynamic behavior around the solid boundary, such as turbulence, higher-order harmonics and streaming. While there has been some successful modeling of TA standing-wave devices based on what is known as linear theory, it seems that when the pressure amplitude of the excited acoustic wave is high, this theory does not work sufficiently well to predict accurately the behavior of the system. In travelling-wave devices particularly, this issue has become a growing concern, as they are operated at higher power densities, which are associated with high-pressure amplitudes, in addition to the discovery of new nonlinear effects that do not exist in standing-wave devices (Abduljalil, 2012).

8.5 CONCLUDING REMARKS

Although mechanically or thermally driven heat pumps remain the most applied in the chemical process industry, solid-state heat pumps are also available – and already applied in consumer goods or currently under development – with high applicability potential in the chemical process industry. The ones that stand out and show the most potential for the chemical industry were described in this chapter: magnetic heat pumps, TE heat pumps and TAHPs.

Magnetic refrigeration provides an effective alternative to refrigeration methods based on vapor compression due to the benefits of using the MCE. Still, the development of this magnetic heat pump technology is material dependent, so the research focuses on significantly improved materials that are cheap and abundant and exhibit much larger MCEs over a larger range of temperatures. Such materials need to show significant temperature changes under a field of 2 T or less, so that permanent magnets can be used for the production of the magnetic field.

A thermoelectric type of cooling has the advantage of having no moving parts and no high pressure or toxic fluids or gases. Just a small electric DC is needed to power the system with a cold junction heat exchanger (usually a finned heat exchanger) attached directly to the application. But, usually more power is required to produce the same heating capacity as a vapor compression cycle, even with modern advancement in semiconductor technology.

Thermoacoustic devices show much promise due to their design simplicity and the possibility to manufacture them from standard materials that are available in commercial quantities, thus reducing their initial cost. Another key benefit is that the TA systems operate with almost no mechanical moving parts, thus reducing the cost of maintenance to trivial levels and substantially increasing their operational lifetime. Theoretically, the only moving parts are the linear motors that are used to excite the acoustic power in TA refrigerators or to extract it in the case of TA engines, but these parts can be well designed for an infinite fatigue life without the need for lubrication. The working medium in these devices can be air, nitrogen, a noble gas or a mixture of these – all of them environmentally friendly fluids, hence with a lower impact in case of leakage. Also, TA devices can utilize thermal energy supplied from any source as input energy (e.g. renewable sources or the heat of combustion from power plants). So far, TA devices offer only up to two-thirds of the efficiency of conventional devices, but the research continues to further enhance their efficiency.

LIST OF SYMBOLS

A	Area	m^2
c	Specific heat	kJ/kgK
C	Heat capacity	kJ/K
H	Magnetic field intensity	T
I	Current intensity	A
L	Length	m
$\dot{q}$	Heat flux	J/sm^2
M	Magnetic moment	J/T
P	Pressure	kPa
q	Specific heat	kJ/kg
Q	Heat	kJ
s	Specific entropy	kJ/kgK
S	Entropy	kJ/K
t	Time	s
T	Temperature	K
u	Specific internal energy	kJ/kg

U	Internal energy	kJ
V	Volume	m³
V_S	Voltage	V
W	Work (energy)	J/mol

Greek Symbols

α	Seebeck coefficient	–
γ	Electrical conductivity	S/m
ε	Electromotive force	V
η	Efficiency	–
κ	Thomson coefficient	–
λ	Thermal conductivity	W/mK
μ	Magnetic permeability	N/A²
π	Peltier coefficient	–
σ	Specific magnetization	J/kgT

Superscripts

| 0 | Reference/standard condition |

Subscripts

ad	Adiabatic
c	Cold/refrigeration
e	Electronic
F	Fourier
h	Hot/heat pump
H	At constant magnetic field
isot	Isothermal
J	Joule
m	Magnetic
p	At constant pressure
P	Peltier
r	Crystal lattice
S	Seebeck
T	Total
x	State

Abbreviations

AC	Alternate current
ACR	AC resistance
AMR	Active magnetic refrigeration
CFC	Chlorofluorocarbon
CHX	Cold heat exchanger
COP	Coefficient of performance

CTHRG	Combined TA heat pump and refrigeration-assisted gas turbine
DC	Direct current
ECN	Energy Center of the Netherlands
EMF	Electromotive force
GHG	Greenhouse gas
GMCE	Giant magnetocaloric effect
HHX	Hot heat exchanger
HP	Heat pump
MC	Magnetic cooling
MCE	Magnetocaloric effect
MHP	Magnetic heat pump
MR	Magnetic refrigeration
MTBF	Mean time between failures
OPTR	Orifice pulse tube refrigerators
RPG	Relative power gain
TA	Thermoacoustic
TAHP	Thermoacoustic heat pump
TASHE	Thermoacoustic Stirling heat engine
TBT	Thermal buffer tube
TE	Thermoelectric
TEC	Thermoelectric cooling
TRG	TA refrigerator–assisted gas turbine
ZT	Figure of merit for TE heat pumps

REFERENCES

Abduljalil A. S. A., Investigation of thermoacoustic processes in a travelling-wave looped-tube thermoacoustic engine, PhD thesis, University of Manchester, United Kingdom, 2012.

Backhaus S., Swift G. W., A thermoacoustic-Stirling heat engine: Detailed study, *Journal of the Acoustical Society of America*, 107 (2000), 3148–3166.

Backhaus S., Swift G. W., New varieties of thermoacoustic engines, In *Proceedings of the 9th International Congress on Sound and Vibration*, Orlando, FL, July 8–11, 2002, Paper 502.

Bassem M. M., Ueda Y., Akisawa A., Thermoacoustic stirling heat pump working as a heater, *Applied Physics Express*, 4 (2011), Article Number 107301.

Bjork R., Bahl C. R. H., Smith A., Pryds N., Review and comparison of magnet designs for magnetic refrigeration, *International Journal of Refrigeration*, 33 (2010), 437–448.

Bruck E., Developments in magnetocaloric refrigeration, *Journal of Physics D: Applied Physics*, 38 (2005), R381.

Bruinsma D., Spoelstra S., Heat pumps in distillaton, In *Proceedings from Distillation & Absorption*, 21–28, Eindhoven, the Netherlands, September 12–15, 2010, 21–28.

Elderson M., Testing and modelling a Peltier heat pump for the Green Village buildings, MSc thesis, Delft University of Technology, the Netherlands, 2015.

Emsley J., *Nature's building blocks*, Oxford University Press, Oxford, UK, 2001.

Gardner D. L., Howard C. Q., Waste-heat-driven thermo-acoustic engine and refrigerator, Acoustics Seminar, November 23–25, 2009, Adelaide, Australia.

Ghorbanian K., Karimi M., Thermodynamic analysis of a hybrid gas turbine/thermoacoustic heat pump/refrigeration engine, *International Journal of Exergy*, 15 (2014), 152–170.

Gschneidner K. A. Jr., Gibson K., *Magnetic refrigerator successfully tested*, Ames Laboratory News Release, Ames Laboratory, Ames, IA, December 7, 2001.

Gschneidner K. A. Jr., Pecharsky V. K, Tsokol A. O., Recent developments in magnetocaloric materials, *Reports on Progress in Physics*, 68 (2005), 1479–1539.

Gschneidner K. A. Jr., Pecharsky V. K, Thirty years of near room temperature magnetic cooling: Where we are today and future prospects, *International Journal of Refrigeration*, 31 (2008), 945–961.

Herman C., Travnicek Z., Cool sound: The future of refrigeration? Thermodynamic and heat transfer issues in thermoacoustic refrigeration, *Heat and Mass Transfer*, 42 (2006), 492–500.

Jeong S., Numazawa T., Rowe A., A review of magnetic refrigeration technology, *Journal of the Korea Institute of Applied Superconductivity and Cryogenics*, 8:2 (2006), 1–10.

Kiss A. A., Flores Landaeta S. J., Infante Ferreira C. A., Towards energy efficient distillation technologies – Making the right choice, *Energy*, 47 (2012), 531–542.

Kramers H. A., Vibration of gas column, *Physica*, 15 (1949), 971–984.

Kruk B., Influence of material used for the regenerator on the properties of a thermoacoustic heat pump, *Archives of Acoustics*, 38 (2013), 565–570.

Putnam A. A., Dennis W. R., Organ-pipe oscillations in burner with deep ports, *The Journal of the Acoustical Society of America*, 28 (1956), 260–269.

Riffat S. B., Ma X., Thermoelectrics: A review of present and potential applications, *Applied Thermal Engineering*, 23 (2003), 913–935.

Romero Gomez J., Ferreiro Garcia R., de Miguel Catoira A., Romero Gomez M., Magnetocaloric effect: A review of the thermodynamic cycles in magnetic refrigeration, *Renewable & Sustainable Energy Reviews*, 17 (2013), 74–82.

Sakamoto S., Watanabe Y., The experimental studies of thermoacoustic cooler, *Ultrasonics*, 42 (2004), 53–56.

Sangkwon J., Numazawa T., Rowe A., A review of magnetic refrigeration technology, *Progress in Superconductivity and Cryogenics*, 8 (2006), 1–10.

Semenyuk V., Thermoelectric heat pump as a thermal cycler, *Journal of Electronic Materials*, 39 (2010), 1510–1515.

Smith A., Bahl C. R. H., Bjørk R., Engelbrecht K., Nielsen K. K., Pryds N., Materials challenges for high performance magnetocaloric refrigeration devices, *Advanced Energy Materials*, 2 (2012), 1288–1318.

Spoelstra S., Innovative heat pump technology, *NPT Procestechnologie*, 14e jaargang (2007), 14–15.

Spoelstra S., *Saving energy in distillation with thermoacoustic heat pumps*, ECN Brochure B-07-008, Energy Center of the Netherlands, Petten, The Netherlands, 2008.

Tijani M. E. H., Spoelstra S., A high performance thermoacoustic engine, *Journal of Applied Physics*, 110 (2011), Article 093519.

Tijani M. E. H., Spoelstra S., A hot air driven thermoacoustic-Stirling engine, *Applied Thermal Engineering*, 61 (2013), 866–870.

Tijani M. E. H., Vanapalli S., Spoelstra S., Lycklama à Nijeholt J. A., Electrically driven thermoacoustic heat pump, presented at the 10th IEA Heat Pump Conference, Tokyo, June 27–August 31, 2011.

van de Bor D. M., Infante Ferreira C. A., Quick selection of heat pump types and optimization of loss mechanisms, presented at the 10th IEA Heat Pump Conference, Tokyo, May 16–19, 2011.

Wollan J. J., Swift G. W., Backhaus S., Gardner D. L., Development of a thermoacoustic natural gas liquefier, In *Proceedings of AIChE Meeting*, New Orleans, LA, March 11–14, 2002, 1–8.

Yang J. H., Yip H. L., Jen A. K. Y., Rational design of advanced thermoelectric materials, *Advanced Energy Materials*, 3 (2013), 549–565.

Yang Z., Zhuo Y., Ercang L., Yuan Z., Travelling-wave thermoacoustic high-temperature heat pump for industrial waste heat recovery, *Energy*, 77 (2014), 397–402.

Yu B. F., Gao Q., Zhang B., Meng X. Z., Chen Z., Review on research of room temperature magnetic refrigeration, *International Journal of Refrigeration*, 26 (2003), 622–636.

CHAPTER 9

Industrial Applications of Heat Pumps

9.1 INTRODUCTION

Securing a reliable, economic and sustainable energy supply as well as environmental and climate protection are important global challenges of this century. Renewable energy and improving energy efficiency are key steps to achieve these goals of energy policy. While impressive efficiency gains were achieved in the past two decades, energy use and CO_2 emissions in manufacturing industries could be reduced further if best-available technologies were to be applied worldwide. Thermal energy represents a large part of the global energy usage, and about 43% of this energy is used for industrial applications. Large amounts are lost via exhaust gases, liquid streams and cooling water; the share of low-temperature waste heat is the largest. Heat pumps upgrading waste heat to process heat and cooling, as well as power cycles converting waste heat to electricity, can make a strong impact in the related industries. This chapter illustrates the major industrial applications of heat pumps, with a focus on the chemical process industry (CPI); information about the use of heat pumps in heating, ventilation and air conditioning (HVAC) applications is available elsewhere in specialized books as described in Chapter 1.

9.2 APPLICATION OF INDUSTRIAL HEAT PUMPS

Industrial heat pumps (IHPs) are active heat recovery devices that increase the temperature of waste heat in an industrial process to a higher temperature to be used in the same process or another adjacent process or heat demand. A report from the International Energy Agency (IEA, 2014) focused on the application of industrial heat pumps, covering such important topics as heat pump energy situations, energy usage, market overview, barriers for application of heat pumps, modeling calculation and economic models, research and development (R&D) projects, case studies, communication and dissemination. The IEA Heat Pump Programme–Industrial Energy Technology and Systems (HPP-IETS) Annex 35/13, *Application of Industrial Heat Pumps*, a joint venture of the IEA, IETS and HPP,

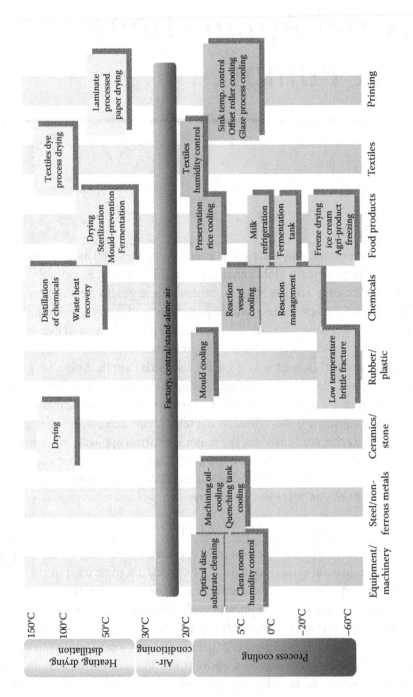

FIGURE 9.1   Applications of industrial heat pumps.

has been initiated to actively contribute to the reduction of energy consumption and emissions of greenhouse gases by the increased implementation of heat pumps in industry. The main objectives of the project include market overviews (country reports), systems aspects and opportunities, apparatus technologies (R&D projects) and system technologies (case studies). The country reports show that the industrial energy consumption in the participating countries varies within 17%–58%, with great differences of the manufacturing sectors: pulp and paper (Austria 20%, Canada 27.6%, Sweden 52.1%); wood (3%–8% of the energy in Austria, Canada, Denmark and Sweden); metal production (Germany 10%–36%); chemical and petrol industry (Netherlands 8.3%–58.8%) and food industry (Denmark 1.4%–25.7%). Figure 9.1 illustrates the applications of industrial heat pumps in various industry sectors (IEA, 2014).

The IEA reports a total of 33 R&D projects (Table 9.1), 115 case studies and 76 applications (Table 9.2) of heat pumps in industry – mostly using waste process heat as the heat source – that have been presented by the participating countries (Austria, Canada, Denmark, France, Germany, Japan, the Netherlands, South Korea and Sweden). These industrial examples show that successful integration of heat pumps is possible. Payback periods lower than 1.5 years are also possible in some examples, while CO_2 emissions and energy costs can be reduced by more than 80% in some cases.

The IEA report showed that in many companies (especially in small- and medium-size enterprises [SME]) only few and aggregated data on the actual thermal energy consumption are available, while disaggregated data such as the consumption of individual (sub)processes have been estimated or determined by costly and time-consuming measurements. The exploitation of existing heat recovery potentials often requires the integration of several processes at different temperature levels and with different operating time schedules. Different technologies available for heat supply have to be combined to obtain optimum solutions. However, the economics of an installation depends on how the heat pump is applied in the process. Identification of feasible installation alternatives for the heat pump is crucial, and considerations of fundamental criteria taking into account both heat pump and process characteristics are useful. The initial procedure should identify a few possible installation alternatives, so the detailed project calculations can concentrate on a limited number of options. The commercially available heat pump types each have different operating characteristics and different possible operating temperature ranges. For a particular application, several possible heat pump types often exist; thus, technical, economic, ecological and practical process criteria determine the best-suited type. For all types, the payback period is directly proportional to installation costs, so it is important to investigate possibilities for decreasing these costs for any heat pump installation.

The IEA report presented some good examples of heat pump technology and its application in industrial processes, field tests and commercial applications, along with an analysis of operating data, in accordance with the annex definition of industrial heat pumps, used for heating, ventilation, air conditioning, hot water supply, heating, drying, dehumidification and other purposes. Although heat pumps for industrial use became available on the market in participating countries in recent years, just a few applications that have been

TABLE 9.1   Selected List of R&D Projects Related to Industrial Heat Pumps in Various Countries

Project (Country)	System	Status	Heating Capacity	Supply Temperature	Refrigerants
Waste heat upgrade for process supply (AT)	Hybrid (absorption/ compression)	Prototype	25 kW	85°C	NH_3-$LiNO_3$ and other working pairs
Thermally driven ejector (CA)	Vapor compression	Laboratory test bench	9 kW (cooling)	10°C	R-134a
Ammonia heat pump (CA)	Compression, single stage	Laboratory prototype	~48 kW	85°C	NH_3
High-temperature drying (CA)	Compression	Industrial-scale prototype (2 units)	65 kW	n/a	R-236fa
Energy-efficient drying with a novel turbo compressor–based high-temperature heat pump (DK)	Rotrex turbo compressor for water vapor compression	Industrial installation	~2.2 MW	100°C	Steam
EDF/JCI (FR)	HT-HP	Experimental tests	700 kW	100°C	R-134a and R-245fa
PACO-Project (FR)	VHT HP–water centrifugal compressor	Experimental tests	700 kW	140°C	Water
Various projects (DE)	High-pressure compression	Projects in industry: greenhouse, paper mill, etc.	Up to 14 MW	Up to 90°C	NH_3
ThermeCO$_2$ (DE)	Transcritical CO_2 heat pump	Installed in industry	Up to 1 MW	90°C	CO_2
Waste heat recovery HP water heater (JP)	Two-stage centrifugal compression	Installed in industry	376–547 kW	90°C	R-134a
HP steam supplier, water to water (JP)	One-/two-stage compression	Installed in industry	370/660 kW	120/165°C	R-245fa/ R-134a and R-245fa
Hot water HP with waste heat (KR)	Hybrid compression/ Absorption	Prototype	30 kW	Over 90°C	NH_3-water

Note: HP, heat pump; HT, heat transfer; VHT, very high temperature.

carried out can be found. Several application barriers were identified as part of the survey in Task 1 (IEA, 2014):

- *Lack of knowledge*: The integration of heat pumps into industrial processes requires knowledge of the capabilities of heat pumps and knowledge about the process itself. Only a few installers and decision makers in the industry have this combined knowledge, which enables them to integrate a heat pump in the most suitable way.

TABLE 9.2   Selected List of Realized Heat Pump Projects and Fact Sheets in Various Countries

Industry (Country)	System	Thermal Capacity	Supply Temperature	Effects
Biomass cogeneration plant (AT)	Absorption	~7.5 MW	95°C	Reduction CO_2 emissions 6 ktpy
Wood drying, high temperature (CA)	Compression	2 × 65 kW	Up to 100°C	Reduction total energy usage up to 50%
Drying air for milk powder (DK)	Hybrid NH_3/ CO_2	1.25 MW	Up to 85°C	Payback period 1.5 years
Food industry: 8 projects (FR)	Closed compression with NH_3	Up to 1.2 MW	65°C	Payback period ~4 years
Coating powder production (DE)	Compression	240 kW	45°C	Payback period ~5 years
District heating (DE)	Compression high-pressure NH_3	13 MW	90°C	Reduction CO_2 emission 12.7 ktpy
Slaughterhouse (DE)	CO_2 transcritical	800 kW	90°C	Reduction of CO_2 emission 0.51 ktpy
Automotive: painting process (JP)	Compression	3.75 MW	n/a	Reduction of CO_2 emissions by 48% and energy cost by 25%
Pharmaceutical production (JP)	Compression	247 kW	45°C	Reduction of CO_2 emissions and primary energy by 24%
Whisky and material alcohol (JP)	Vapor recompression (MVR, TVR)	4.2 t/h	n/a	Reduction of primary energy requirements by 43%
Chemicals: Distillation of propane-propylene splitter (NL)	Mechanical vapor compression	5.8 MW	n/a	Payback period 2 years Savings of 1.2 PJ/year
Margarine production (NL)	Add-on compression	1.4 MW	65°C	Payback period 4 years

Note: ktpy, kiloton per year; PJ, peta Joule.

- *Low awareness of heat consumption in companies*: In most companies, knowledge about the heating and cooling demands of their processes is rare, and this requires expensive and time-consuming measurements to find an integration opportunity for an industrial heat pump.

- *Long payback periods*: Heat pumps have high investment costs, but companies expect low payback time (PBT) of less than 2–3 years. Some companies are willing to accept PBTs up to 5 years when it comes to investments into their energy infrastructure. To meet these expectations, heat pumps need to have long running periods and good coefficients of performance (COPs) to become economically feasible.

- *High-temperature application*: From a technical point of view, one barrier is the temperature limits of most commercially available heat pumps. Many applications are limited to heat sink temperatures below 65°C, but the theoretical potential for the application range of industrial heat pumps increases significantly by developing energy-efficient heat pumps, including refrigerants for heat sink temperatures up to and higher than 100°C.

9.3 WASTE HEAT RECOVERY USING HEAT PUMPS

From the total energy usage, thermal energy (heat) amounts 47% of worldwide energy consumption and 37% of the energy used in Organization for Economic Cooperation and Development (OECD) countries. About 43% of this amount is used for industrial applications, so it comes as no surprise that most large industrial companies are setting ambitious energy-savings programs. McKenna (2009) estimated the annual market potential for surplus heat from industrial processes in the United Kingdom as 36–72 PJ, while Markides (2013) investigated the role of heat pumps, combined heat and power (CHP) schemes and options for the recovery and conversion of waste heat into useful work, towards the creation of a high-efficiency sustainable energy future. The main industrial sectors accounted for in the study of McKenna (2009) were aluminum, cement, ceramics, chemicals, food and drink, glass, gypsum, iron and steel, lime, pulp and paper. Figure 9.2 has been adapted from data reported by Ammar et al. (2012) to illustrate the thermal energy rejected to the environment from some major industrial sources in the United Kingdom (van de Bor et al., 2015). A significant amount of low-grade waste heat is available as water from cooling towers, with temperatures in the range of 45°C–60°C. In particular, distillation alone is responsible for 40% of the thermal energy used in the CPI (Kiss, 2013). Distillation has low thermodynamic efficiency, requiring the input of high-quality energy (e.g. steam) in the reboiler – while rejecting a similar amount of heat at lower temperature, in the condenser, to the cooling water. Several heat pump concepts have been proposed to upgrade that thermal energy and reduce the consumption of valuable utilities, and under certain conditions, the energy savings of heat pump–assisted distillation is usually around 20%–50% (Kiss et al., 2012); this topic is covered in a further section of this chapter.

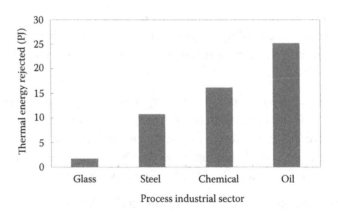

FIGURE 9.2   Low-grade thermal energy rejected from some major UK industrial sources.

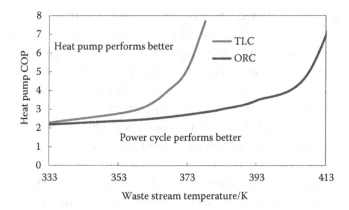

FIGURE 9.3   Performance comparison between heat pumps and power cycle heat engines. ORC, organic Rankine cycle; TLC, trilateral cycle.

In Europe alone, 1,142 PJ of low-temperature heat (below 100°C) and 829 PJ of medium-temperature heat (at levels of 100°C–400°C) are required yearly for the CPI (Pardo et al., 2012). As much as 20%–50% of the energy used is ultimately lost via waste heat contained in streams of hot exhaust gases and liquids. Large amounts of heat contained in spent cooling water (at levels of 45°C–60°C) are rejected to the environment in industrial plants (e.g. to air in cooling towers or as waste water discharged in sea). Presently, the share of waste heat recovery contribution to the total energy usage is still negligible, in spite of the large impact potential.

The potential of several alternative technologies, either for the upgrading of low-temperature waste heat, such as compression-resorption heat pumps (CRHPs), vapor compression (VC) heat pumps and transcritical heat pumps, or for the conversion of this waste heat by using organic Rankine, Kalina and trilateral cycle engines, was investigated by van de Bor et al. (2015) with regards to energetic and economic performance by making use of thermodynamic models. The study focused on temperature levels of 45°C–60°C as at this temperature range large amounts of heat are rejected to the environment; also investigated were the temperature levels for which power cycles become competitive. The heat pumps deliver 2.5–11 times more energy value than the power cycles in this low-temperature range at equal waste heat input. However, the heat engines become competitive with heat pumps at waste heat temperatures at 100°C and above. Figure 9.3 illustrates the comparison between heat pumps and power cycle heat engines (van de Bor et al., 2015). If the COP of the heat pump at a specific waste stream temperature is higher than the line for a power cycle heat engine, the heat pump will show larger economic revenue; otherwise, the power cycle will show larger economic revenue.

9.4 HEAT PUMP–ASSISTED DISTILLATION

In spite of the many well-known benefits of distillation and its widespread use, one major drawback is the significant energy requirements because distillation can generate more than 50% of plant operating costs. To solve this problem, several technologies were proposed to reduce the energy requirements of distillation, with potential energy savings, typically in the range of 20%–50% (Kiss, 2013). Distillation has a relatively low thermodynamic efficiency (Araujo et al., 2007), requiring the input of high-quality energy in the reboiler to perform the separation task. At the same time, a similar amount of heat at lower temperature is rejected in the condenser. Alternatives for heat recovery from a distillation column were proposed in the early 1950s (Freshwater, 1951). Subsequently, the oil crisis of the last decades motivated extensive research on energy efficiency in distillation. Several proposals were made to reduce the energy requirements of distillation, including the application of different heat pump technologies and configurations, as well as heat and thermal integration (Mah et al., 1977; Linnhoff et al., 1983; Omideyi et al., 1984).

Several heat pump concepts have been proposed to upgrade the low-quality energy in the condenser to drive the reboiler of the column, thus reducing the consumption of valuable utilities. Vapor compression (VC) and thermal and mechanical vapor recompression (TVR and MVR, respectively) technologies are used to upgrade the heat by compressing the vapor distillate or a working fluid (Annakou and Mizsey, 1995; Fonyo and Benko, 1998; McMullan, 2003). CRHPs and absorption heat pumps (AHPs) increase the energy efficiency by means of absorption equilibrium (Mučić, 1989). Due to the higher temperature lifts possible, the thermoacoustic heat pump (TAHP) is claimed to have a broader applicability range (Bruinsma and Spoelstra, 2010), while the internally heat-integrated distillation column (HIDiC) enhances both heat and mass transfer, especially when structured packing is used also for exchanging heat (Bruinsma et al., 2012; Kiss and Olujic, 2014).

A large number of alternatives and configurations led to comparison studies with a single case study, the one described by Meszaros and Meili (1994) for butane/isobutene separation. A first selection scheme for heat pumps was developed earlier by Omideyi et al. (1984). Subsequent research simplified the selection guide for multiple heat pump technologies, including also MVR, AHP and TVR (Fonyo and Mizsey, 1994). van de Bor and Infante Ferreira (2013) evaluated the performance of selected heat pumps as a function of the required temperature lift to provide guidelines for their selection in any application. The temperature lift can be related to the temperature difference between the heat sources and sinks (namely, the condenser and reboiler in the case of distillation), which in turn is determined by the product cuts that are separated between the top and bottom of the column.

Kiss et al. (2012) presented applications of the available heat pump-assisted distillation technologies, as well as a quick selection scheme based on an extensive literature survey, taking into account the most promising energy-efficient distillation technologies. Only the key aspects of the overall efficiency were analyzed: boiling point differences ΔT_b or the temperature lift ($\Delta T_{lift} = \Delta T_b + \Delta T_{df}$) to upgrade heat from the source accounting for the driving force (Wallas, 1990), the nature of the components involved, operating pressure of the system, product distribution and purity specifications, reboiler temperatures T_{reb} and duties Q_{reb}, as well as the relative volatility of the components α_{ij}. The efficiency indicators include the (ideal) COP for heat pumps, which relate to the total operating cost (TOC), total investment cost (TIC), as well as the total annual cost (TAC) and the PBT.

9.4.1 Feasibility of Heat Pump–Assisted Distillation

A distillation column can also be considered as a heat engine that produces separation (or entropy reduction) instead of work. In this case, the heat is provided in the reboiler (heat source) and removed at a lower temperature level in the condenser (heat sink). While Kiss et al. (2012) provided a useful scheme for the quick selection of appropriate heat pumps, Plesu et al. (2014) introduced a simple criterion depending on the Carnot power cycle efficiency to decide whether a heat pump is worth considering. This can be simplified to the following form:

$$\frac{Q}{W} = \frac{1}{\eta_{Carnot}} = \frac{T_c}{T_r - T_c} > 10 \tag{9.1}$$

where Q is the reboiler duty, W is the work provided, while η_{Carnot} is the Carnot efficiency, T_r is the reboiler temperature and T_c is the condenser temperature. When the Q/W ratio exceeds 10, then a heat pump should be considered, while when the Q/W ratio is lower than 5 using a heat pump will not bring any benefits (Plesu et al., 2014). Figure 9.4 (Luo et al., 2015) illustrates the dependence of the Q/W ratio on the condenser temperature and the temperature span across the column ($T_r - T_c$).

Note that heat pump–assisted distillation can be applied not only to classic distillation columns (Kiss et al., 2012) but also dividing-wall columns (DWCs) to a lesser extent (Chew et al., 2014; Liu et al., 2015). However, in the case of DWC systems, the temperature span across the column is larger; thus, the temperature lift required by the heat pump is also larger and hence less efficient, so the feasibility is questionable (Kiss, 2013).

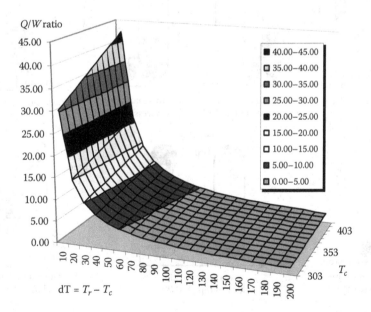

FIGURE 9.4 Dependence of the Q/W ratio on the condenser temperature T_c and the temperature difference between reboiler and condenser ($T_r - T_c$) of the distillation column (DC).

9.4.2 Heat Pump–Assisted Distillation Configurations

Figure 9.5 shows the main configurations used in heat pump–assisted distillation (Kiss, 2014). Most of them were described generically in Chapter 1. In addition, Chapter 6 provides more details about VC, vapor recompression technologies and CRHP, while Chapter 7 covers the AHP and TVR, and Chapter 8 describes the TAHP. One particular configuration is the so-called HIDiC, which for convenience is explained in more detail next.

Heat-integrated distillation column (HIDiC) is the most radical approach of heat pump design, making use of internal heat integration (Kiss, 2014). Instead of using a single-point heat source and sink, the whole rectifying section of a distillation column becomes the heat source, while the stripping part of the distillation column acts as a heat sink, providing a higher potential for energy savings (Kiss, 2013, 2014). The internal heat integration widely enhances the reachable COP because the required temperature difference for heat transfer is kept low with gliding temperatures across both parts. The work input is provided by a compressor installed at the top outlet of the stripper section, while the heat pump cycle is closed by the valve flashing the liquid bottom outlet of the rectifier section. Note that the success of HIDiC technology relies actually on good hardware performance for both heat and mass transfer tasks at the same time.

Although HIDiC claims among the highest energy savings possible in distillation, the capital investment costs are higher compared to conventional distillation, mainly due to the use of an additional compressor and a more complex configuration required for the enhanced heat transfer. Both thermodynamics and bench-scale experimental evaluations

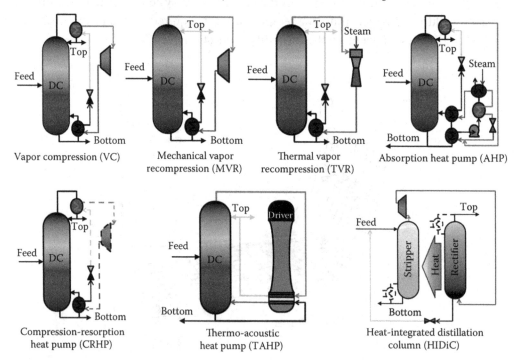

FIGURE 9.5   Heat pump assisted distillation (HPAD) – main configurations applied to a distillation column (DC).

proved that HIDiC holds much higher energy efficiency than conventional distillation columns for close-boiling component separations. Several simulation studies and experimental evaluations have further confirmed that the process can be operated smoothly, with no control difficulties (Kiss and Olujic, 2014).

A number of important issues must be considered during HIDiC design, such as flexibility to changes in the operating conditions, influence of an impurity or a third component and process dynamics and operation. These impose strict constraints on the energy efficiency that can be potentially achieved by a HIDiC system. Therefore, trade-off between process design economics and process operation appears to be important, and it has to be carried out with great caution. It is worth noting that among the design and the operating variables, the compression ratio and the heat transfer coefficient and area are the ones that significantly affect most the performance of a HIDiC. Consequently, an important research task during recent years was the development of more appropriate configurations, such as multitube and multishell, plate-type or even structured-packing HIDiC. The current directions of development are fractionating heat exchangers (shell and tube, plate-fin) or heat-exchanging distillation columns (e.g. concentric columns, parallel columns, partitioning-wall columns).

Although there are several pilot-scale installations, the only design of HIDiC that has reached commercialization stage to date is basically a two-pressure single-shell column introduced recently by a Japanese engineering contractor (Kiss and Olujic, 2014). Toyo Engineering Corporation (http://www.toyo-eng.com/jp/en/) was successful in overcoming difficulties and managed to develop in collaboration with the National Institute of Advanced Industrial Science and Technology (http://www.aist.go.jp/index_en.html) – a main driver behind all similar projects in Japan – a practical and scalable HIDiC configuration named SuperHIDiC. Toyo Engineering Corporation plans to use the novel SuperHIDiC technology in petroleum refining, as well as petrochemical and fine chemical production plants.

Compared to other complex separation systems, HIDiC technology has the key advantage of larger energy savings (up to 70%), although the main problem remains the use of a compressor that adds significantly to the total equipment cost. Therefore, the main challenge of HIDiC technology is to reduce further the investment costs and thus PBTs. Anyhow, it is to be expected that with increasing energy prices, the heat pump–assisted distillation in general and HIDiC in particular will become more economically competitive in the foreseeable future. The first commercialized configuration so far (i.e. SuperHIDiC) is a serious candidate to make a breakthrough in this respect.

9.4.3 Applications of Heat Pump–Assisted Distillation

Each and every energy-efficient distillation technology yields its maximum savings only at given specific conditions (Meszaros and Meili, 1994; van de Bor and Infante Ferreira, 2013). Table 9.3 shows the technologies used and the most important studies collected in the literature survey (Kiss et al., 2012). Mainly, two classes of binary mixtures have been reported – (relatively) small hydrocarbons and water-alcohols – but these cover a large range of industrial processes. The technology selection criteria and the limiting values

TABLE 9.3 Overview of Case Studies, Experiments and Industrial Implementation of Energy-Efficient Distillation Technology for Binary Separations

Technology	Condition of Separation Task	Performance Information	Application	Reference
VC, MVR, bottom flashing, column integration	C_4 splitters: P ~ 5 bar; ΔT_b ~ 10 K; feed upgrade from 19% to 99.1% 1-butene	MVR allowed higher energy and cost savings (~50% lower TAC) than column-integrated schemes, but at almost doubled CapEx. VC savings were similar to MVR with higher energy costs. 36% energy savings from column integration.	Case study	Meszaros and Meili, 1994
VC	Ethanol recovery from fertilization product; key components: ethanol-water	VC with R114 working fluid yielded $3 \leq COP \leq 4$ at $50\ K \leq \Delta T_{lift} \leq 64\ K$. Changing to ethanol as working fluid, total energy savings were up to 50%.	Case study	Omideyi et al., 1985
VC	Theoretically constructed cases	VC with R114 working fluid yielded COP = 3.2 at ΔT_b = 67 K.	Case study	Wallin et al., 1990
MVR	Propane-propylene separation (P between 4 and 16 bar)	Single compressor – optimal P: 9 bar. Double compressor – optimal P: 10 bar. Both featured ~ 37% annual cost savings and ~ 60% less CO_2 emissions. Double compressor had better operability.	Case study	Annakou and Mizsey, 1995
MVR	Separation of N,N-dimethylacetamide (DMAC)/water mixture	47% reduced TACs for double-effect distillation with double MVR-HP and 32% reduced TAC for top MVR-HP.	Case study	Gao et al., 2015
MVR	Bioethanol concentration and dehydration (ΔT_b ~ 22 K)	MVR integrated on extractive DWC. Energy savings of over 40% and reduced total annual cost by 24%.	Case study	Luo et al., 2015
MVR AHP	1-Butane-isobutane separation (ΔT_b ~ 15 K) 2-Acetic acid dehydration (ΔT_b ~ 55 K)	1. MVR yielded COP ~ 6.44 and 29% savings in utilities requirements in case 1. AHP yielded COP ~ 1.87 and 41% utilities savings. AHP lowest PBT. 2. MVR yielded COP ~ 2.63 and AHP yielded COP ~ 1.67. None gave utilities saving. Payback time was negative for all.	Case study	Fonyo and Benko, 1998
MVR AHP	n-Butane/i-butane separation: ΔT_b ~ 10 K at P = 710 kPa	MVR yielded 33% energy savings and 9% less TAC. AHP incorrectly implemented. Both compared to conventional distillation.	Case study	Diez et al., 2009

(Continued)

TABLE 9.3 (CONTINUED) Overview of Case Studies, Experiments and Industrial Implementation of Energy-Efficient Distillation Technology for Binary Separations

Technology	Condition of Separation Task	Performance Information	Application	Reference
TVR	Evaporators, steam stripping, peroxide distillation and other aqueous separations	Usual COP ~ 1.1 with ΔT_{lift} < 40 K. Lowest investment cost per kilowatt output.	Industrial implementation	Feng and Berntsson, 1997; Perry and Green, 1999
AHT	Single- or multieffect distiller combined with an open absorption heat transformer (OAHT)	Waste heat at 70°C elevated to 125°C to produce steam at 120°C in the absorber that drove a four-effect distiller. COP ~ 1.02, performance ratio = 2.19 (single effect) and 5.72 (four effect).	Case study	Zhang et al, 2014
CRHP	Heat stream upgrading	The system delivered 1,000 kW with ΔT_b = 20 K and ΔT_{lift} = 25 K obtained COP ~ 9.1.	Pilot plant implementation	Mučić, 1989
CRHP, MVR	Benzene-toluene distillation	Diabatic CRHP distillation design yielded COP ~ 8 with ΔT_{lift} max 40 K. Conventional MVR yielded COP ~ 7.	Case study	Taboada and Infante Ferreira, 2008
TAHP	Heat stream upgrading	5-kW burner-driven TAHP with argon yielded ΔT_{lift} = 40 K, delivering 200 W acoustic power at 100°C.	Pilot plant implementation	Spoelstra, 2008
TAHP	Benzene-toluene distillation	Electrically driven TAHP yielded COP ~ 2.	Case study	Tijani et al, 2011
HIDiC	Fractionation of C4-C6 mixture: top: n-pentane; bottom: i-pentane P = 190/101 bar	HIDiC achieved 26% energy savings compared with conventional existing column, while requiring ~ 22% more energy than minimum reflux distillation.	Precommercial Implementation	Huang et al., 2008
HIDiC	Cyclohexane N-heptane equimolar feed, 95% product purity; P = 1.1 bar; ΔT_b ~ 17°C; Pressure ratios from 1.4 to 1.7	Primary energy savings ranged from 81% at 1.7 P ratio to 69.7% at 1.4, compared to conventional distillation process. Research also included performance from proof-of-principle experimental study.	100 ktpa case study	Bruinsma et al., 2012
HIDiC	Separation of C5 mixture	Energy savings of 86.7%. Total annual cost reduced by 7.4%.	Case study	Zhu et al., 2015
HIDiC	Methanol-water separation Feed concentration: ~70% Industrial column: 59 rectifying stages, 26 stripping stages	1. VRC, 3.1% lower TAC. 2. External HIDiC, 27.2% lower TAC. 3. Double compressor intensified HIDiC-1, 24.4% lower TAC; HIDiC-2, 34.2% lower TAC; 70.4% energy savings; payback period of 3.3 yr.	Case study	Shahandeh et al., 2015

(Continued)

TABLE 9.3 (CONTINUED) Overview of Case Studies, Experiments and Industrial Implementation of Energy-Efficient Distillation Technology for Binary Separations

Technology	Condition of Separation Task	Performance Information	Application	Reference
HIDiC, MVR	Propane-propylene separation; conventional column: P = 18 bar; MVR: P = 10 bar; HIDiC: P = 18/13 or 18/15 bar	MVR industrial implementation gave COP ~ 7.4 and 37% TAC savings against conventional column (ΔT_{lift} ~ 24.9 K). HIDiC case study yielded COP ~ 10 and 25% TAC savings against MVR with ΔT_b ~ 10.9 K.	Case study/industrial implementation	Olujić et al., 2003, 2006
HIDiC MVR	1-Methanol-water separation (P = 2.6/1.3 bar for HIDiC); 2-Ethyl-benzene-styrene separation (P = 0.24 bar for MVR)	1. Top HIDiC design yielded 84% energy savings, but only 9% lower TAC. 2. Bottom HIDiC design better. MVR pressure ratio: 2.8; while HIDiC 2.2. MVR obtained 49% energy savings, while HIDiC obtained 53%. MVR savings in TAC were 35%, while HIDiC were 28%. All compared to classic distillation.	Case study	Gadalla et al., 2007; Gadalla, 2009
HIDiC MVR	Acetic acid dehydration; vacuum operating pressure	82% energy savings (MVR) and 23% (HIDiC) vs. heat-integrated columns.	Case study	Campbell et al., 2008
MED	Methanol-water (HP = 5–8 bar; LP = 1.4 bar)	A five-column multieffect distillation scheme achieved 33.6% energy savings compared to the existing four-column multieffect distillation.	Case study/industrial implementation	Zhang et al., 2010
MED	CH_3SiCl_3-$(CH_3)_2SiCl_2$ (HP = 2 bar; LP = 1 bar)	Energy savings of 43.7% of multieffect distillation compared to conventional distillation process.	Case study	Wang and Wang, 2012

Note: HP, high pressure; LP, low pressure; P, pressure.

between similar situations were drawn from Table 9.3. In some cases, particularly at low ΔT_b, only the top three technologies providing the maximum savings were selected.

For convenience, Table 9.4 presents a more detailed comparison of the efficiency indicators for various distillation technologies (Kiss et al., 2012): COP and energy savings (%) that directly relate to the TOC as well as to the CO_2 emissions, reduction of the total annual cost ΔTAC and the PBT.

9.4.4 Selection Scheme for Heat Pump–Assisted Distillation

The technology selection scheme proposed by Kiss et al. (2012) aims to provide design guidelines, allowing any process engineer to narrow significantly the number of suitable energy-efficient options for a given separation task at early design stages. However, before approaching the selection schemes, one must ensure that distillation is indeed the most suitable separation method for the newly designed processes. Moreover, heat integration possibilities within the process, plant or with other distillation columns should also be considered in parallel (Linnhoff et al., 1983).

In principle, the selection scheme can be used for new designs or retrofit applications, but the suitable proposals vary for each case. For binary distillation, the scheme offers at least two technology options for each concluding condition, one of them being also suited for retrofits. Those solutions that make use of components external to the distillation process (e.g. VC, MVR, TVR, AHP, CRHP and TAHP) can be used in either new designs or retrofits. Note that the tray-integrated versions of the AHP and CRHP can only be applied in new designs. The solutions that make use of components internal to the distillation process (e.g. HIDiC) are used mainly in new designs.

The maze of choices for binary distillation technologies is illustrated in Figure 9.6 (Kiss et al., 2012). Most technologies are grouped as heat pump–assisted distillation. Consequently, the selection criteria for binary distillation include operating pressure, nature of the components with respect to corrosiveness and fouling, boiling point differences ΔT_b, temperature lift ΔT_{lift}, reboiler duty Q_{reb} and temperature level T_{reb}, as well as the relative volatility between components α_{ij}. The reviewed data (Kiss et al., 2012) includes over 70 technology applications and case studies. The technology should provide about 20%–50% savings to be considered for future application in the conditions reported by the literature. After the initial screening, the data were organized considering the relevant details of the application or case study and the energy and economic savings obtained in each particular case.

The most important criteria for selection are the difference in boiling points ΔT_b or the temperature lift ΔT_{lift}. Basically, ΔT_{lift} adds to ΔT_b the required driving force for heat transfer between the working fluids ΔT_{df}: $\Delta T_{lift} = \Delta T_b + \Delta T_{df}$. According to common practices, the driving force is $\Delta T_{df} = 5$–20 K (Wallas, 1990): less than 5 K for HIDIC, 5–10 K for MVR or TVR, 10 K or lower for CHRP because glide is used or 10–20 K for others, depending on design.

Thus, considering their values, ΔT_{lift} and ΔT_b can be considered more or less the same criterion. The importance of this criterion lies in two key facts. On one side, it is well known that distillation energy requirements are higher when the separation involves close boiling products (Wallin et al., 1990; Fonyo and Benko, 1998; Bruinsma and Spoelstra,

TABLE 9.4   Comparison of Efficiency Indicators for Various Energy-Efficient HPAD Technologies

Technology	Separation Task	ΔT_b (K)	ΔT_{lift} (K)	COP	PES (%)	ΔTAC (%)	Remarks	References
AHP	Butane-isobutane	15	25	1.9	47	−42	PBT = 0.9	Fonyo and Benko, 1998
AHP	Water-acetic acid	55	65	1.7	40	−12	PBT = 7.1	Fonyo and Benko, 1998
CHRP	Steam production	10	30	9.1	78	–	–	Mučić, 1989
CRHP	Benzene-toluene	30	<30	8.3	76	–	–	Taboada and Infante Ferreira, 2008
HIDIC	n-Pentane-cyclopentane	9	20	10.3	71	–	–	Huang et al., 2008
HIDIC	Cyclohexane-n-heptane	17	33	9.6	75	−62	PBT = 1.8	Bruinsma et al., 2012
HIDIC	Propane-propylene	7.7	10.9	10.0	80	−72	PBT = 1.2	Olujić et al., 2006
HIDIC	Methanol-water	35	55	>6.8	41		–	Gadalla, 2009
HIDIC	Benzene-toluene	30		4.0	46	−46	PBT = 0.9	Gadalla et al., 2007
HIDIC	Acetic acid dehydration	35	55	3.2	27	–	–	Campbell et al., 2008
MVR	1-Butene-n-butane	10		5.5	63	−60	PBT = 0.6	Meszaros and Meili, 1994
MVR	Propane-propylene	7.6	11.8	–			−60% CO₂	Annakou and Mizsey, 1995
MVR	Benzene-toluene	30	35	6.7	70	–	–	Taboada and Infante Ferreira, 2008
MVR	Butane-isobutane	15	20	6.4	69	−63	PBT = 1.1	Fonyo and Benko, 1998
MVR	Water-acetic acid	55	60	2.6	23	+65	PBT = 31.8	Fonyo and Benko, 1998
MVR	Butane-isobutane	12.9	35.8	7.2	72	−64	PBT = 0.3	Díez et al., 2009
MVR	Propane-propylene	7.7	24.9	7.5	73	−65	PBT = 1.4	Olujić et al., 2006
MVR	Acetic acid dehydration	35	55	8.4	50	–	–	Campbell et al., 2008
MVR	Methanol-water	35	40	6.6	69	−67	PBT = 0.7	Feng and Berntsson, 1997
TAHP	Methanol-water	35	45	2.0	0	–	–	Tijani et al., 2011
TAHP	Methanol-water	35	55	2.3	14	–	–	Tijani et al., 2011
TVR	Methanol-water	35	40	1.1	23	−19	PBT = 1.5	Feng and Berntsson, 1997
VC	Methanol-water	35	45	3.0	33	−29	PBT = 2.5	Feng and Berntsson, 1997
VC	1-Butene-n-butane	10	20	6.2	67	−57	PBT = 1.6	Meszaros and Meili, 1994
VC	Methanol-water	35	45	4.1	50	–	–	Omideyi et al., 1985
VC	Methanol-water	35	55	3.1	34	–	–	Omideyi et al., 1985

(*Continued*)

TABLE 9.4 (CONTINUED) Comparison of Efficiency Indicators for Various Energy-Efficient HPAD
Technologies

Technology	Separation Task	ΔT_b (K)	ΔT_{lift} (K)	COP	PES (%)	ΔTAC (%)	Remarks	References
VC	Methanol-water	35	43	5.0	60	−57	PBT = 0.9	Wallin et al., 1990
VC	Water-acetic acid	55	65	1.8	−10	+99	PBT = negative	Fonyo and Benko, 1998

Note: Primary energy savings (PES) are the yearly savings compared to the primary energy requirements of a conventional column; PBT is the simple payback time (assuming a plant lifetime of 10 years; primary energy cost = €30/MWh; electrical energy cost = €65/MWh; boiler efficiency = 0.85; efficiency of electricity conversion = 0.42; investments as proposed in papers); negative ΔTAC means savings; ΔT_b is given at the pressure reported in the reference; ΔT_b is boiling points difference, ΔT_{lift} is the temperature lift.

2010). On the other side, the efficiency of the selected technologies greatly depends on the ΔT_b, particularly for heat pumps (Fonyo and Mizsey, 1994; Feng and Berntsson, 1997; de Rijke, 2007). Thus, at lower ΔT_b, technology efficiency and conventional energy requirements are higher. Consequently, the largest number of options are tailored to lower ΔT_b. Similarly, most distillation processes in industry fall in the middle range of ΔT_b. Thus, significant developments in that field are expected from the massive implementation of energy-efficient solutions. Last but not least, high ΔT_b involves huge amounts of work for upgrading heat at industrial scale, as well as relatively high T_{reb} ($\geq$250°C), so this case is less interesting. Solutions with high investment costs or even the introduction of mass separating agents are preferred options for reducing the huge energy requirements of conventional distillation.

Another criterion with high impact is the operating pressure. As shown in Table 9.3 (Kiss et al., 2012), the performance of some technologies changes drastically when the operating pressure drops below the atmospheric pressure. The effect is especially important when the top product is used as working fluid (strong connection between the distillation column and the heat pump) and the work input is provided through a compressor. Moreover, the internals of radical designs such as HIDiC impose certain pressure drop levels that are unsuitable for vacuum applications. When the required products are corrosive or fouling, it is economically wise to avoid contact of these substances with the rotating equipment. Moreover, the equipment internal design for any technology considered should withstand such treatments. Thus, the heat pumps that use working fluids/pairs as heat transfer media with minimal changes in the conventional column design are preferred. Consequently, it is required to evaluate the reboiler temperature to select the most efficient option.

Finally, there are certain technologies that require additional selection criteria. Those were treated as niche applications, as beneficial performance was particularly reported on limited conditions. An example is the limitation on the reboiler duty Q_{reb} imposed by the steam ejector efficiency on TVR (Fonyo and Benko, 1998). Similarly, the strong dependence of VC on the heat transfer fluid properties limits the ranges of T_{reb} suitable for operation. VC is recommended only for T_{reb} matched by the working fluid meeting the strong environmental, safety and performance requirements. Last, the nonideal behavior of certain

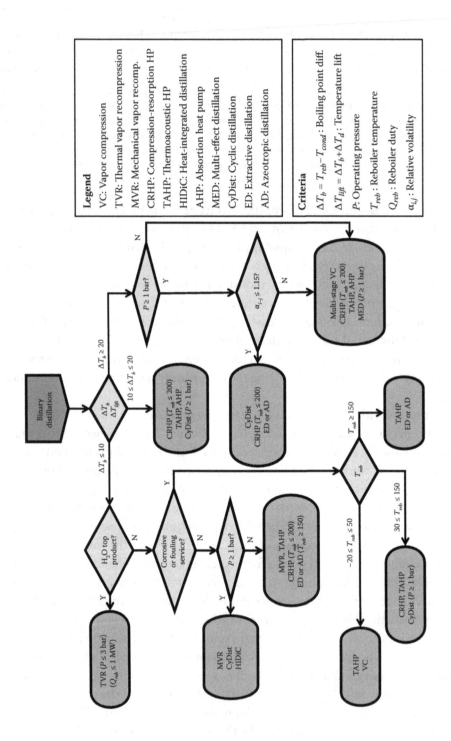

Legend

VC: Vapor compression
TVR: Thermal vapor recompression
MVR: Mechanical vapor recomp.
CRHP: Compression-resorption HP
TAHP: Thermoacoustic HP
HIDiC: Heat-integrated distillation
AHP: Absorption heat pump
MED: Multi-effect distillation
CyDist: Cyclic distillation
ED: Extractive distillation
AD: Azeotropic distillation

Criteria

$\Delta T_b = T_{reb} - T_{cond}$: Boiling point diff.
$\Delta T_{lift} = \Delta T_b + \Delta T_d$: Temperature lift
P: Operating pressure
T_{reb}: Reboiler temperature
Q_{reb}: Reboiler duty
$\alpha_{i,j}$: Relative volatility

FIGURE 9.6   Selection scheme for energy-efficient binary distillation technologies (temperature in degrees centigrade).

components is taken into account in the scheme. In such cases, the relative volatility α_{ij} changes drastically with the composition profile across the distillation column. Thus, flexible technologies capable of handling difficult separations are preferred (e.g. azeotropic distillation [AD] or extractive distillation [ED]). Keep in mind that the selection scheme only narrows down the options for achieving an energy-efficient process, so the final choice should be made based on rigorous simulations.

The physical properties of the components involved should always be available before using the scheme. This is usually not a problem because those data are also required on other stages of process research and design. Although the first variable checked in the scheme is the difference in boiling points, it is advisable first to verify other operational constraints, as the boiling points depend on the chosen operating pressure. Sometimes, the selection of the operating pressure relies on the particular properties of the components involved, such as corrosiveness, fouling and so on. However, in most cases, the use of common utilities (steam, cooling water, etc.) determines the selection of the operating pressure. As mentioned, the use of noncommon utilities can seriously jeopardize the economic operation of the distillation column. Therefore, it is advisable to gather the property data prior to approaching the scheme. After the data-gathering process, the technology selection scheme can be used. Some possible situations are illustrated in the paper of Kiss et al. (2012).

9.4.4.1 Methanol-Water Separation

Methanol and water components are neither corrosive nor fouling. The boiling point for each component can be calculated at atmospheric pressure as a first estimate to check whether common utilities can be used. The boiling point difference at atmospheric pressure is about 35 K, with the top temperature of 65°C (methanol) and the bottom temperature of 100°C (water). When applying the selection scheme, the first criterion verified is the difference in boiling points, which in this particular case fits in the area of wide boiling components ($\Delta T_b \geq 20$ K). The next check is the operating pressure, the preferred value being the atmospheric pressure. The final verification is the presence of nonideal vapor-liquid equilibria (VLE) behavior. As the thermodynamic behavior reveals no azeotropes, the preferred options are VC, CRHP, TAHP and AHP. Comparing the result of the scheme with further in-depth analysis of the literature in Table 9.4, the largest reported primary energy savings (PES) for methanol-water separation so far reported is for CRHP (74%, obtained by slight extrapolation of existing cases), MVR (69%), VC (46%, average of values from different sources) and AHP (43%, interpolated between existing cases). The data available for the TAHP concern a laboratory prototype and show low PES (14%). The reduction in TAC ranks the best solutions as MVR (67%), VC (43%) and AHP (27%). There are no published data available about the CHRP and TAHP. Concerning the PBT, the shortest is reported for MVR (0.7 years), closely followed by VC (0.9 years) and with a significant difference by TVR (1.5 years). The PBT of AHP will be much larger (4.0 years). PBTs of 3.0 years for CHRP, 4.4 years for VCHP and 5.6 years for AHP are obtained using data from van de Bor and Infante Ferreira (2013). These PBTs are based on installed costs and cannot be directly be compared with the PBTs from the other sources based on equipment costs only.

9.4.4.2 Propane-Propylene Splitting

Another example for a revamp of binary separations could be the separation of propane-propylene. These gases are separated in existing plants by distillation at about 10 atm without major concerns about corrosiveness or fouling problems. The boiling point of each component is calculated to verify which energy-efficient technologies could reduce the current energy requirements. At 10 atm, propane boils at about 27°C, while propylene boils at about 20°C. This results in a low ΔT_b (≤10°C). Applying the separation scheme with the complete information, and considering that water is not the top distillate product, the TVR option can be ruled out. Following the selected path, the operating pressure is higher than atmospheric; hence, three viable options are suggested. Of these options, the state-of-the-art system for propane-propylene distillation is MVR, but interesting results were obtained in simulation studies for HIDiC (Olujić et al., 2003, 2006), while the potential of CyDist is now ready to unfold in follow-up studies (Kiss, 2013, 2014).

9.5 CONCLUDING REMARKS

The IEA reported a total of 33 R&D projects, 115 case studies and 76 applications of heat pumps in industry – mostly using waste process heat as the heat source. These industrial examples show that successful integration of heat pumps is possible. Payback periods lower than 1.5 years are also possible, while CO_2 emissions and energy costs can be reduced by more than 80% in some cases.

The potential of several alternative technologies, either for the upgrading of low-temperature (45°C–60°C) waste heat or for the conversion of this waste heat to power, was investigated with regards to energetic and economic performance by making use of thermodynamic models. The heat pumps deliver 2.5–11 times more energy value than the power cycles in this low-temperature range at equal waste heat input. However, the heat engines become competitive with heat pumps at waste heat temperatures of 100°C and above.

Heat pump–assisted distillation technologies were proposed and developed to upgrade the low-level energy and reuse it in the process, thus reducing the consumption of valuable utilities. Temperature lifts of up to 30°C–100°C are feasible, leading to significant PES of up to 80% (Kiss et al., 2012). Moreover, the technology selection scheme described here provides valuable guidelines in the design of energy-efficient distillation processes. Using the selection scheme, process designers can effectively narrow down the number of technology alternatives that can deliver significant energy savings considering the particular process conditions. The selection scheme features several key criteria for the conditions required to successfully apply each technology.

- VC, MVR and TVR usually allow higher savings for low T_{reb} and low ΔT_{lift} (<20 K).

- TAHP and CRHP can be applied from low to middle ΔT_{lift} (in the range of 20–50 K).

- Technologies such as HIDiC and cyclic distillation are applicable only at higher pressures, mainly for noncorrosive compounds, low ΔT_{lift} and relative volatility close to unity.

LIST OF SYMBOLS

P	Pressure	bar
Q	Heat	W
T	Temperature	K
W	Work	W

Greek Symbols

α	Relative volatility	–
η	Efficiency	–

Superscripts

0	Reference/standard condition

Subscripts

b	Boiling
c	Cold/condenser
df	Driving force
h	Hot
r or reb	Reboiler

Abbreviations

AD	Azeotropic distillation
AHP	Absorption heat pump
AHT	Absorption heat transformer
CHP	Combined heat and power
COP	Coefficient of performance
CPI	Chemical process industry
CRHP	Compression-resorption heat pump
CyDist	Cyclic distillation
DWC	Dividing-wall column
ED	Extractive distillation
HIDiC	Heat-integrated distillation column
HP	High pressure/Heat pump
HPAD	Heat pump–assisted distillation
HPP	Heat Pump Programme
HT	High temperature
HVAC	Heating, ventilation and air conditioning
IEA	International Energy Agency
IETS	Industrial Energy Technology and Systems
IHP	Industrial heat pump
ktpy	kiloton per year
LP	Low pressure

MED	Multieffect distillation
MVR	Mechanical vapor recompression
OECD	Organization for Economic Cooperation and Development
ORC	Organic Rankine cycle
PBT	Payback time
PER	Primary energy requirements
PES	Primary energy savings
SME	Small- and medium-size enterprises
TAC	Total annual cost
TAHP	Thermoacoustic heat pump
TIC	Total investment cost
TLC	Trilateral cycle
TOC	Total operating cost
TVR	Thermal vapor recompression
VC	Vapor compression
VHT	Very high temperature
VLE	Vapor-liquid equilibria

REFERENCES

Ammar Y., Joyce S., Norman R., Wang Y., Roskilly A. P., Low grade thermal energy sources and uses from the process industry in the UK, *Applied Energy*, 89 (2012), 3–20.

Annakou O., Mizsey P., Rigorous investigation of heat pump assisted distillation, *Heat Recovery Systems & CHP*, 15 (1995), 3, 241–247.

Araujo A. B., Brito R. P., Vasconcelos L. S., Exergetic analysis of distillation processes – Case study, *Energy*, 32 (2007), 1185–1193.

Bruinsma D., Spoelstra S., Heat pumps in distillation, In *Proceedings from Distillation and Absorption*, September 12–15, the Netherlands, 2010, 21–28.

Bruinsma O. S. L., Krikken T., Cot J., Sarić M., Tromp S. A., Olujić Ž., Stankiewicz A. I., The structured heat integrated distillation column, *Chemical Engineering Research and Design*, 90 (2012), 458–470.

Campbell J. C., Wigala K. R., van Brunt V., Kline R. S., Comparison of energy usage for the vacuum separation of acetic acid/acetic anhydride using an internally heat integrated distillation column (HIDiC), *Separation Science and Technology*, 43 (2008), 9, 2269–2297.

Chew J. M., Reddy C. C. S., Rangaiah G. P., Improving energy efficiency of dividing-wall columns using heat pumps, Organic Rankine Cycle and Kalina Cycle, *Chemical Engineering and Processing*, 76 (2014), 45–59.

de Rijke A. de, Development of a concentric internally heat integrated distillation column (HIDiC), PhD thesis, TU Delft, the Netherlands, 2007.

Díez E., Langston P., Ovejero G., Dolores Romero M., Economic feasibility of heat pumps in distillation to reduce energy use, *Applied Thermal Engineering*, 29 (2009), 1216–1223.

Feng X., Berntsson T., Critical COP for an economically feasible industrial heat-pump application, *Applied Thermal Energy*, 17 (1997), 1, 93–101.

Fonyo Z., Benko N., Comparison of various heat pump assisted distillation configurations, *IChemE Transactions*, 76, A (1998), 348–360.

Fonyo Z., Mizsey P., Economic applications of heat pumps in integrated distillation systems, *Heat Recovery Systems & CHP*, 14 (1994), 3, 249–263.

Freshwater D. C., Thermal economy in distillation, *Transactions of the Institution of Chemical Engineers*, 29 (1951), 149–160.

Gadalla M. A., Internal heat integrated distillation columns (iHIDiCs) – New systematic design methodology, *Chemical Engineering Research and Design*, 87 (2009), 1658–1666.

Gadalla M., Jiménez L., Olujić Ž., Jansens P. J., A thermo-hydraulic approach to conceptual design of an internally heat-integrated distillation column (i-HIDiC), *Computers and Chemical Engineering*, 31 (2007), 1346–1354.

Gao X., Chen J., Tan J., Wang Y., Ma Z., Yang L., Application of mechanical vapor recompression heat pump to double-effect distillation for separating N,N-dimethylacetamide/water mixture, *Industrial and Engineering Chemistry Research*, 54 (2015), 3200–3204.

Huang K., Iwakabe K., Nakaiwa M., Matsuda K., Horiuchi K., Nakanishi T., Consider heat integration to improve separation performance, *Hydrocarbon Processing*, March (2008), 101–108.

International Energy Agency, *Application of industrial heat pumps*, IEA Industrial Energy-Related Systems and Technologies Annex 13 & IEA Heat Pump Programme Annex 35, Report No. HPP-AN35-1, ISBN 978-91-88001-92-4, IEA Heat Pump Centre, Borås, Sweden, 2014.

Kiss A. A., *Advanced distillation technologies – Design, control and applications*, Wiley, New York, 2013.

Kiss A. A., Distillation technology – Still young and full of breakthrough opportunities, *Journal of Chemical Technology and Biotechnology*, 89 (2014), 479–498.

Kiss A. A., Flores Landaeta S. J., Infante Ferreira C. A., Towards energy efficient distillation technologies – Making the right choice, *Energy*, 47 (2012), 531–542.

Kiss A. A., Olujic Z., A review on process intensification in internally heat-integrated distillation columns, *Chemical Engineering and Processing*, 86 (2014), 125–144.

Linnhoff B., Dunford H., Smith R., Heat integration of distillation columns into overall processes, *Chemical Engineering Science*, 38 (1983), 8, 1175–1183.

Liu Y., Zhai J., Li L., Sun L., Zhai C., Heat pump assisted reactive and azeotropic distillations in dividing wall columns, *Chemical Engineering and Processing*, 95 (2015), 289–301.

Luo H., Bildea C. S., Kiss A. A., Novel heat-pump-assisted extractive distillation for bioethanol purification, *Industrial & Engineering Chemistry Research*, 54 (2015), 2208–2213.

Mah R. S. H., Nicholas Jr. J. J., Wodnik R. B., Distillation with secondary reflux and vaporization: A comparative evaluation, *AIChE Journal*, 23 (1977), 651–658.

Markides C. N., The role of pumped and waste heat technologies in a high-efficiency sustainable energy future for the UK, *Applied Thermal Engineering*, 53 (2013), 197–209.

McKenna R. C., Industrial energy efficiency interdisciplinary perspectives on the thermodynamic, technical and economic constraints, PhD Thesis, University of Bath, United Kingdom, 2009.

McMullan A., Industrial heat pumps for steam and fuel savings (DOE/GO-102003–1735), US Department of Energy, 2003, http://www1.eere.energy.gov/industry/bestpractices/pdfs/heatpump .pdf (accessed January 24, 2013).

Meszaros I., Meili A., 1-Butene separation processes with heat pump assisted distillation, *Heat Recovery Systems & CHP*, 14 (1994), 3, 315–322.

Mučić V., Resorption compression heat pump with solution circuit for steam generation using waste heat of industry as heat source, *Heat Pump Centre Newsletter*, 7/1 (1989), 14–15.

Olujić Ž., Fakhri F., de Rijke A., de Graauw J., Jansens P. J., Internal heat integration – The key to an energy-conserving distillation column, *Journal of Chemical Technology and Biotechnology*, 78 (2003), 241–248.

Olujić Ž., Sun L., de Rijke A., Jansens P. J., Conceptual design of an internally heat integrated propylene-propane splitter, *Energy*, 31 (2006), 3083–3096.

Omideyi T. O., Kasprzycki J., Watson F. A., The economics of heat pump assisted distillation systems-I. A design and economic model, *Heat Recovery Systems*, 4 (1984), 3, 187–200.

Omideyi T. O., Parande M. G., Supranto S., Kasprzycki J., Devotta S., The economics of heat pump assisted distillation systems, *Heat Recovery Systems & CHP*, 5 (1985), 511–518.

Pardo N., Vatopoulos K., Krook-Riekkola A., Moya J. A., Perez A., *Heat and cooling demand and market perspective*, JCR Scientific and Policy Reports, JRC Scientific and Policy Reports, ISBN 978-92-79-25311-9, Publications Office of the European Union, Luxembourg, 2012.

Perry R. H., Green D. W. (Eds.), *Perry's chemical engineering handbook*, 7th edition, McGraw-Hill, New York, 1999.

Plesu V., Bonet Ruiz A. E., Bonet J., Llorens J., Simple equation for suitability of heat pump use in distillation, *Computer Aided Chemical Engineering*, 33 (2014), 1327–1332.

Shahandeh H., Jafari M., Kasiri N., Ivakpour J., Economic optimization of heat pump-assisted distillation columns in methanol-water separation, *Energy*, 80 (2015), 496–508.

Spoelstra S., Saving energy in distillation with thermoacoustic heat pumps, ECN Brochure B-07-008, Energy Center of the Netherlands, Petten, The Netherlands, 2008.

Taboada R., Infante Ferreira C. A., Compression resorption cycles in distillation columns, presented at the International Refrigeration and Air Conditioning Conference at Purdue, July 14–17, 2008, West Lafayette, IN, Paper 912.

Tijani M. E. H., Vanapalli S., Spoelstra S., Lycklama à Nijeholt J. A., Electrically driven thermoacoustic heat pump, presented at the 10th IEA Heat Pump Conference, June 27–August 31, 2011, Tokyo.

van de Bor D. M., Infante Ferreira C. A., Quick selection of industrial heat pump types including the impact of thermodynamic losses, *Energy*, 53 (2013), 312–322.

van de Bor D. M., Infante Ferreira C. A., Kiss A. A., Low grade waste heat recovery using heat pumps and power cycles, *Energy*, 89 (2015), 864–873.

Wallas S., *Chemical process equipment selection and design*, Butterworth-Heinemann, Boston, 1990.

Wallin E., Franck P.A., Berntsson T., Heat pumps in industrial processes – An optimization methodology, *Heat Recovery Systems & CHP*, 10 (1990), 4, 437–446.

Wang W., Wang W., Energy saving in methyl-chloro-silane distillation, presented at the International Conference on Computer Distributed Control and Intelligent Environmental Monitoring, CDCIEM 2012, March 5–6, 2012, Zhangjiajie, Hunan, China.

Zhang J., Shengrong L., Feng X., A novel multi-effect methanol distillation process, *Chemical Engineering and Processing*, 49 (2010), 1031–1037.

Zhang X., Hu D., Li Z., Performance analysis on a new multi-effect distillation combined with an open absorption heat transformer driven by waste heat, *Applied Thermal Engineering*, 62 (2014), 239–244.

Zhu C-M., Li Y-A., Zhou W-Y., Shi X-L., Simulation study of heat-integrated distillation column for separation of C5 mixture, *Journal of East China University of Science and Technology*, 41 (2015), 300–307.

CHAPTER 10

Case Studies

10.1 INTRODUCTION

The previous chapters addressed several topics related to the design and selection of heat pumps for industrial processes. The economic feasibility of the application of heat pumps depends on how the heat pump is integrated with the process. Identification of feasible solutions for the heat pump is of major importance. Simultaneous consideration of the heat pump and the process in which it should be integrated is essential for successful integration. The initial selection should identify a few possible installation alternatives (most promising) that should be investigated and compared later in more detail.

In this chapter, five cases in which different heat pump technologies have been applied to industrial processes are described, allowing an overview of the strong and weak points of the different technologies when applied to specific applications. Boot et al. (1998) dedicated one of their chapters to applications of heat pumps in industry. They discussed processes for the concentration of flows by evaporation, drying processes and distillation processes, giving some possible arrangements between processes and heat pumps. An extensive overview of heat pump application cases is given in Task 4 of Annexes 13 and 35 of the International Energy Agency (IEA) on the application of industrial heat pumps (IEA, 2014). The organization of this report with application cases per country makes it less accessible, but the summary helps provide an overview of the reported cases. Most of these applications have supply temperatures below 90°C, indicating that higher temperatures are still more difficult to attain. Except for the ejector heat pump (EHP) case discussed in this chapter, all cases discussed have supply temperatures above 110°C and in the range 110°C to 165°C, illustrating that such supply temperatures should not be a problem for industrial heat pumps.

10.2 VAPOR RECOMPRESSION HEAT PUMP

This case study deals with bioethanol dehydration by extractive distillation (ED) in an integrated process combining vapor recompression (VRC) with dividing-wall column (DWC) technology. Bioethanol is perhaps the most promising renewable fuel, with an important advantage over other alternatives, namely, that it can be directly integrated in existing fuel systems, typically as a 5%–85% mixture with gasoline that does not need any modification of the current engines (Balat et al., 2008). The industrial production of bioethanol relies on various routes: corn to ethanol, sugar cane to ethanol, basic and integrated lignocellulosic biomass to ethanol (Balat et al., 2008). In all cases, the raw materials undergo several pretreatment steps before entering the fermentation stage where bioethanol is actually produced. A common feature of these technologies is the production of diluted bioethanol (typically 5%–12%wt ethanol) that is further concentrated to reach the requirements of the international bioethanol standards (Vane, 2008; Frolkova and Raeva, 2010; Kiss and Ignat, 2012). Depending on the standard requirements, the maximum water content in ethanol is 0.2%vol (EN 15376, EU), 0.4%vol (ANP no. 36/2005, BR) or 1.0%vol (ASTM D 4806, USA).

To reach the purity targets, an energy-demanding separation is needed in practice to overcome the presence of the binary azeotrope ethanol-water (95.63%wt ethanol). The separation is typically carried out by distillation, the first step being a preconcentration distillation column (PDC) that increases the ethanol content from 5%–12% up to 91%–94%wt (Vane, 2008; Frolkova and Raeva, 2010; Kiss and Ignat, 2013). The second step consists of ethanol dehydration, up to concentrations exceeding the azeotropic composition. Quite a number of separation alternatives are available as described in the literature: pervaporation, adsorption, pressure-swing distillation, ED, azeotropic distillation (AD) and hybrid methods combining these options (Vane, 2008; Frolkova and Raeva, 2010). Among them, ED is still the option of choice in the case of large-scale production of bioethanol fuel (Frolkova and Raeva, 2010; Kiss and Ignat, 2012). Typically, ED is performed in a sequence of two columns, one being the ED column (EDC) that separates ethanol, while the other one is the solvent recovery column (SRC) that recovers the mass separating agent (MSA), which is recycled back in the process. Further improvements to the distillation process were proposed, with the aim to increase the energy efficiency of bioethanol purification, such as by ED process optimization (Kiss and Ignat, 2013); thermally coupled distillation columns (Hernandez, 2008); azeotropic and extractive DWCs (Kiss and Ignat, 2012; Kiss and Suszwalak, 2012).

This case study presents a novel heat pump–assisted ED process that efficiently combines VRC with DWC technology and allows a significant reduction of the energy requirements for bioethanol purification (Luo et al., 2015).

A 100-ktpy plant is considered here, processing a feed with 10%wt ethanol by ED using ethylene glycol (EG) as solvent. Rigorous simulations were carried out in Aspen Plus, and for a fair process comparison, the same approach was used as in previous studies (Kiss, 2013a,b). For all process alternatives described hereafter, Aspen Plus simulations were performed using the rigorous RADFRAC unit. The NRTL (nonrandom two-liquid) property model was used as a suitable model due to the presence of a nonideal mixture containing polar components (Luo et al., 2015).

10.2.1 Conventional Extractive Distillation Process

Figure 10.1 presents the process flowsheet, including the mass and energy balance along with the key parameters of this classic bioethanol purification sequence based on ED (Luo et al., 2015), with the optimization of this process described in details in our earlier work

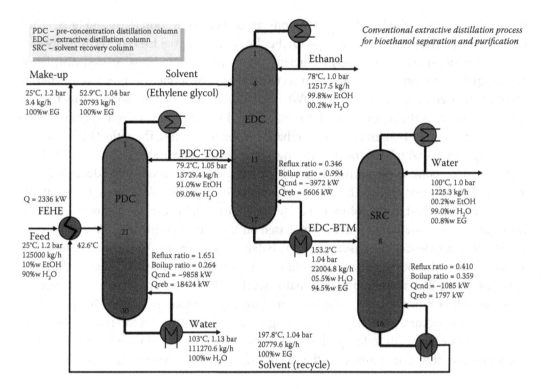

FIGURE 10.1 Flowsheet including the mass balance and key parameters of the classic bioethanol purification sequence based on extractive distillation (the numbers on columns indicate the top, bottom and feed stage).

(Kiss and Ignat, 2013; Luo et al., 2015). The process consists of three distillation units: PDC, EDC and SRC. The first column (PDC) separates water as a bottom stream and a near-azeotropic composition mixture as distillate (91%wt ethanol being the optimal value as reported by Kiss and Ignat, 2013). The second column (EDC) makes use of EG (a high-boiling solvent), which is added at a solvent-to-ethanol ratio of 1.25 mol/mol, on a stage above the feed stage of the near-azeotropic ethanol-water mixture. The presence of the solvent changes the relative volatility of ethanol-water such that their separation becomes possible. High-purity ethanol is collected as the top distillate product of the EDC, while the bottom stream contains solvent and water. The third column (SRC) separates the remaining water as distillate and completely recovers the solvent as the bottom product. Note that a key difference as compared to previous work (Kiss and Ignat, 2013) is the use of the bottom product of SRC to preheat the feed in the feed-effluent heat exchanger (FEHE) unit and then recycle the cooled solvent stream to the EDC. Remarkably, this minor heat integration reduces the specific energy requirements from 2.11 to 2.07 kWh/kg bioethanol.

10.2.2 Vapor Recompression–Assisted Extractive DWC

As all distillation columns of the classic separation sequence (Figure 10.1) operate at atmospheric pressure, the use of a DWC was explored as an attractive process alternative, for example, to combine EDC and SRC into an E-DWC standard configuration or combine all three columns into a single E-DWC with nonstandard configuration (Kiss and Ignat, 2012; Kiss and Suszwalak, 2012). Here, we go further by combining the E-DWC technology with VRC to further increase the energy efficiency. Note that heat pump–assisted distillation can be applied not only to classic columns (Kiss et al., 2012) but also to a lesser extent to DWCs (Chew et al., 2014). In case of DWC systems, the temperature span across the column is larger, thus the temperature lift required by the heat pump is also larger and hence less efficient (Kiss, 2013a). In the case of ethanol-water separation, the ratio $Q/\dot{W} = 16$ ($T_r = 100°C$, $T_c = 78°C$), which means that adding a heat pump is favorable (Plesu et al., 2014).

The starting point is the optimized single-step separation in an E-DWC described elsewhere (Kiss and Ignat, 2012). The VRC part is added on top of E-DWC based on sensitivity analysis as described in the following material. Figure 10.2 presents the flowsheet of the novel process for bioethanol purification based on VRC-assisted ED in a DWC (Luo et al., 2015). For convenience, the mass and energy balance, as well as the key process parameters, are also provided. Note that the liquid split ratio r_L is defined as the ratio between the liquid flow rate going down to the prefractionator section (PDC section, Figure 10.2) over the total liquid flow rate available at the bottom of the top common section (EDC) just before the liquid split occurs. Similarly, the vapor split ratio r_V is defined as the ratio between the vapor flow rate going up to the prefractionator section (PDC section in Figure 10.2) over the total vapor flow rate available at the top of the bottom common section (SRC).

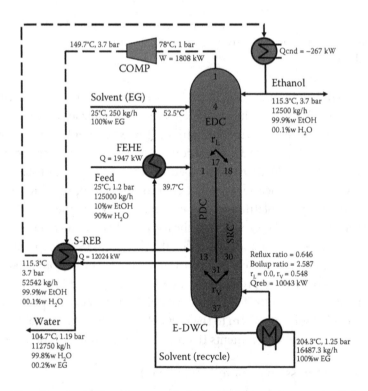

FIGURE 10.2 Flowsheet including the mass balance and key parameters of the novel process for bioethanol purification based on VRC-assisted extractive distillation in a DWC. S-REB, side reboiler.

As shown in Figure 10.2, the feed side (prefractionator) has the role of the PDC unit of the classic sequence. Water is collected as a liquid side stream, but an extra side reboiler is needed to return to the column the required amount of water vapor. The diluted ethanol feed is fed as liquid stream on top of the prefractionator side and therefore serves as liquid reflux to the PDC section. The vapor flow leaving the feed side (PDC) of the E-DWC is enriched in ethanol. Heavy solvent is fed at the top of the E-DWC unit, with this common top section playing the role of the EDC unit of the classic sequence. The solvent-to-ethanol ratio considered is 1.0 mol/mol due to the different ethanol concentration at the top of the PDC. Ethanol is separated in the top section (EDC) as high-purity vapor distillate that is compressed to a higher pressure and temperature level and then used to drive the side reboiler of the column and eventually is condensed and collected as the main product. The liquid flowing down the top section (EDC) is collected and distributed only to the (SRC) side located opposite the feed side. This complete redistribution of the liquid flow (liquid split ratio $r_L = 0$) is required to avoid the presence of solvent on the feed side (PDC), which would lead to a loss of solvent in the water side stream. In the bottom common section (SRC), the solvent is removed as a bottom product, then cooled in an FEHE and recycled to the E-DWC unit. Due to the large difference of volatilities between water and EG, the separation is easy; therefore, the recovery of solvent is practically complete; hence, all the recovered solvent is recycled. The vapor coming from the bottom part of the E-DWC to the lower part of the dividing wall consists mainly of water, but this amount is not sufficient for the PDC section, hence the need for an extra side reboiler, which can be effectively driven by a heat pump (VRC in this case). Note that in spite of the high degree of integration of DWC technology and extractive DWC, the controllability of such systems is satisfactory (Kiss and Bildea, 2011; Tututi-Avila et al., 2014).

Figure 10.3 plots the temperature and composition profiles in the E-DWC (Luo et al., 2015). The changes in the composition along the column are clearly illustrated by these profiles, being in line with the functional task of each column section: PDC on the diluted feed side, EDC in the common top part and SRC on the bottom common part of the column. Notably, the temperature difference between the two sides (PDC vs. SRC) of the wall is low (less than 20 K); hence, no practical issues are expected. Also, high purity and

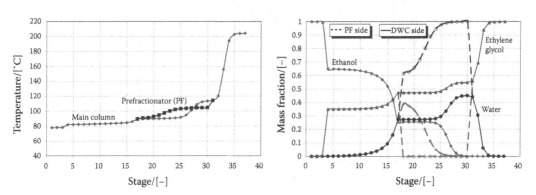

FIGURE 10.3 Temperature (left) and composition (right) profiles along the extractive dividing-wall column.

recovery are possible for all three products: ethanol as top distillate (99.9%wt), water as side stream product (99.8%wt) and EG solvent (>99.9%wt) as recovered bottom product. Note that the column profiles are slightly different as compared to previous work (Kiss and Ignat, 2012). The difference comes mainly from using fewer stages, but a higher vapor split ratio ($r_V = 0.548$ instead of 0.4), which means that more vapor is distributed to the PDC side. This was needed to obtain a feasible match between the duty of the side reboiler and the heat available for recovery from the top vapor stream (VRC loop in Figure 10.2). Although we have started from an optimized system (Kiss and Ignat, 2012), it is worth noting that optimizing a chemical process is typically a mixed-integer nonlinear problem that is nonconvex and likely to have multiple locally optimal solutions. Such problems are intrinsically difficult to solve, and the solution time increases rapidly with the number of variables and constraints. A theoretical guarantee of convergence to the globally optimal solution is not possible for nonconvex problems.

10.2.3 Sensitivity Analysis of VRC System

Integrating a VRC heat pump with an extractive DWC requires the setting of the appropriate discharge pressure from the compressor. The actual figure can be obtained by performing sensitivity analysis. Figure 10.4 shows the dependence of the compressor duty and the inverse of the log-mean temperature difference (LMTD) on the discharge pressure of the compressor of the VRC system (Luo et al., 2015). In terms of reducing the VRC costs, the compressor duty should be as low as possible (smaller and cheaper compressor), while the LMTD should be as high as possible (larger driving force and smaller heat exchanger).

However, practical limits are imposed, and these reduce the available range for the discharge pressure to 3.1–3.7 bar. For the heat exchanger (side reboiler of E-DWC) that is part of the VRC loop, the LMTD must exceed 5 K to obtain a reasonable size and inexpensive side reboiler. Note that the area of the heat exchanger A depends proportionally with the inverse of LMTD (e.g. $A = \dot{Q}/U \cdot 1/\text{LMTD}$), considering a constant heat duty $\dot{Q}$ and

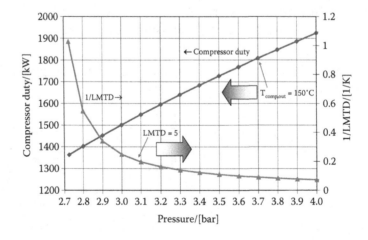

FIGURE 10.4 Dependence of compressor duty and the log-mean temperature difference (LMTD) in the side reboiler on the discharge pressure of the compressor (VRC system).

overall heat transfer coefficient U. Similarly, the compressor is also limited not only by the compression ratio (typically up to 2.5–4.0) but also mainly by the discharge temperature (DCT), which should not exceed 150°C for safety reasons; at higher temperatures, the system may fail from worn rings, acid formation and oil breakdown.

10.2.4 Economic Evaluation and Process Comparison

The total investment costs (TIC), total operating costs (TOC) and total annual costs (TAC) have been calculated according to the procedure described in our previous work (Kiss and Ignat, 2013). Note that the economic estimation considers a 20% extra cost factor for DWC internals that are somewhat more complex as compared to conventional equipment. Concerning the sizing of DWC, each section can be sized for the particular liquid and vapor loads, using available software. In this way, an equivalent cylindrical cross-sectional area needed to accommodate the required vapor and liquid loads can be rather easily calculated. The latitudinal position of the wall is then set in such way that each column section has an equivalent cross-sectional area to that calculated by individual sizing (Kiss, 2013a). The equipment costs are estimated using correlations, based on the Marshall and Swift (M&S) equipment cost index (M&S = 1468.6 in 2012). For columns and heat exchangers (e.g. condensers and reboilers) made of carbon steel, the estimated cost in US dollars is given by

$$C_{shell} = f_p (M\&S/280) d_c^{1.066} h_c^{0.802} \tag{10.1}$$

$$C_{hex} = (M\&S/280)\, c_x\, A^{0.65} \tag{10.2}$$

$$C_{comp} = (M\&S/280)\, (664.1\, P^{0.82} F_c) \tag{10.3}$$

where f_p is the cost factor (2981.68 in this case), $c_x = 1609.13$ (condensers) or 1775.26 (kettle reboilers), A is the heat transfer area (m²), P is the compressor power (kW), h_c is the column height (tangent to tangent) and d_c is the column diameter, calculated using the internals-sizing procedure from Aspen Plus and in case of DWC as equivalent diameter based on cross-sectional area of the two column sections. For compressors, the correction factor F_c varies with design type as follows: 1.00 for centrifugal (motor), 1.07 for reciprocating (steam), 1.15 for centrifugal (turbine) and 1.29 and 1.82 for reciprocating (motor and gas engine, respectively). A price of US$600/m² was used for the sieve tray cost calculations. For the TAC calculations, a plant lifetime of 10 years was used. Furthermore, the following costs were considered for different types of utilities: $0.03/t cooling water and multiple levels of steam rated at $15, $17 and $20/t steam of low, medium and high pressure, respectively (Wu et al., 2013). Note that the CO_2 emissions were calculated according to the method described in earlier studies (Kiss and Ignat, 2013):

$$[CO_2]_{emissions} = (Q_{fuel}/NHV) \times (C\%/100)\, \alpha \tag{10.4}$$

where $\alpha = 3.67$ is the ratio of molar masses of CO_2 and C, NHV is the net heating value and C% is the carbon content, dependent on the fuel. For natural gas, NHV is 48,900 kJ/kg and the carbon content is 0.41 kg/kg. Therefore, the total amount of fuel used can be calculated as follows:

$$Q_{fuel} = Q_{proc}/\lambda_{proc} \times (h_{proc} - 419) \times (T_{FTB} - T_0)/(T_{FTB} - T_{stack}) \qquad (10.5)$$

where λ_{proc} (kJ/kg) and h_{proc} (kJ/kg) are the latent heat and enthalpy of the steam, respectively, and T_{FTB} (K) and T_{stack} (K) are the flame and stack temperature, respectively.

Table 10.1 provides more details about the key performance indicators, including the total investment and operating and annual costs (Luo et al., 2015). Due to the use of a compressor and a larger side reboiler required by the VRC system, the TIC of this VRC E-DWC process is about 29% higher than for the classical process, but this is largely compensated by the significant energy savings, which exceed 60% at a direct comparison.

However, one should note that such a high percentage value in energy savings is based on the direct comparison of thermal duty (heat) versus combined thermal and compressor duty (kilowatts heat and power) of the VRC alternative. When taking into account the inefficiencies in power generation (e.g. by considering that the ratio of heat to electrical kilowatts is ~3 [conservative figure], then the equivalent energy requirements are 0.805 + 3 × 0.145 = 1.24 kWh/kg ethanol), the real savings in primary energy become 40%, which is still a remarkable figure. However, when a ratio of heat to electrical kilowatts equal to 2 is considered (more realistically currently), then the equivalent energy requirements are 0.805 + 2 × 0.145 = 1.095 kWh/kg ethanol, meaning that the real savings in primary energy are 47%.

TABLE 10.1   Process Comparison in Terms of Key Performance Indicators

Key Performance Indicator	Classic Process	E-DWC Process	VRC E-DWC Process	Difference vs. Classic (%)
Equipment cost breakdown (k$)				
– column shells (including internals)	1103	1111	912	+29.3
– condensers (heat exchangers)	1335	1073	71	
– reboilers (heat exchangers)	885	1442	356	
– process-process heat exchangers	137	–	1503	
– compressor (VRC)	–	–	1632	
Total investment costs, TIC (k$)	3462	3626	4477	
Total operating costs, TOC (k$/yr)	5784	5355	4221	−27.0
Total annual costs, TAC (k$/yr)	6130	5718	4668	−23.8
CO_2 emissions (kg CO_2/ton product)[a]	288.94	288.31	173.04 (112.35)	−40.1 (−61.1)
Thermal energy use (kWh/kg product)	2.07	2.07	0.80	−61.1
Electrical energy use (kWh/kg product)	n/a	n/a	0.14	n/a
Equivalent energy requirements (kWh/kg)	2.07	2.07	1.24	−40.1

[a] Values given in parentheses are for the case when electricity is generated from renewable sources.

Considering both the investment and operating costs, the TAC is reduced by about 24%, while the CO_2 emissions, which are closely linked to the primary energy requirements, are reduced by 40% (or even 60% when the electricity for the compressor comes from renewable sources). Note that in case of the VRC-assisted E-DWC process, the investment cost may include also some additional equipment required for the startup procedure. An example is a trim reboiler – of a smaller size, working at a larger temperature approach due to the use of steam – which could be used next to the side reboiler.

The novel heat pump–assisted ED process illustrated in this case study is based on an efficient combination of a VRC heat pump and DWC technology. In this new configuration, the ethanol top vapor stream of the extractive DWC is recompressed from atmospheric pressure to 3.1–3.7 bar (thus to a higher temperature) and used to drive the side reboiler responsible for water vaporization. The results show that the specific energy requirements drop from 2.07 kWh/kg (classic sequence) to only 1.24 kWh/kg ethanol (VRC-assisted extractive DWC); thus, energy savings of over 40% are possible. Considering the requirements for a compressor and use of electricity in case of the heat pump–assisted alternative, about a 24% decrease of the TAC is possible – in spite of the 29% increase of the capital expenditures – for the novel process, as compared to the classic ED process using three columns.

10.3 VAPOR COMPRESSION HEAT PUMP FOR HEAT RECOVERY

Waste heat from a refinery is to be used to heat district heating (DH) water for a local DH network. This case has been investigated by Ravi (2012). The details of the existing DH network may be found in Table 10.2.

The purpose is to investigate the possibility of extracting waste heat from the refinery and constructing a DH substation. The potential of using heat pumps to recover heat from low-temperature process water streams for DH is to be analyzed. The heat pump will have to extract heat from a source at 37°C and add it to the DH water to increase its temperature as much as possible with reasonable values of the coefficient of performance (COP). The process water is presently cooled using cooling towers. Extracting heat from the process water serves two purposes: waste heat recovery for DH and reduction in the load sent to the cooling towers. The available volumetric flow rate of process water is 500 m³/h.

Vapor compression heat pumps are the most widely used type of heat pumps. Current maximum temperatures are limited to 120°C, and because this is the technology closest to market applications, it is the most interesting technology for this refinery case.

Current industrial heat pumps refer to 90°C as the level for high-temperature applications. However, in the case of the considered heating network, the temperatures of interest

TABLE 10.2 Operating Conditions of District Heating System

Parameters		Units
Maximum network load	100	MW
Supply temperature (maximum)	130	°C
Return water temperature	75	°C

are higher than 100°C. To study the potential of using high-temperature heat pumps, first a market study should be conducted to investigate if machines operating within these temperature ranges are available. Thereafter, a study is conducted to analyze the use of different refrigerants for the heat pump.

Commercially available high-temperature heat pumps usually provide water at 90°C. Few manufacturers have the experience to supply large-scale units (greater than 1-MW thermal capacity) in the 90°C range. Friotherm (a Swiss firm) has extensive experience in waste heat recovery and has installed many heat pumps for DH. Also, Thermea and Combitherm (both from Germany) have experience with heat pumps that can be used to produce hot water at 90°C for DH. At the moment, the preferred working fluid is generally R134a (1,1,1,2-tetrafluoroethane). High-temperature heat pumps are custom made, and all three manufacturers had no previous experience in operation of a heat pump with a sink temperature at 120°C. For achieving the high temperature of 120°C on the sink side, Friotherm and Combitherm have proposed to use R245fa (pentafluoropropane) as the refrigerant, whereas Thermea has proposed the use of CO_2 refrigerant.

A market survey has shown that there are no commercially available absorption heat pumps (AHPs) that can deliver water at 120°C. Similarly, mechanical vapor recompression (MVR) heat pumps cannot be used in this case as the source is a liquid.

The DH water that has to be heated from 75°C is considered to be part of the refinery heat exchanger network. The pinch point for the refinery has then been calculated at 75°C for the cold stream and 85°C for the hot stream. Thus, using a heat pump for transferring heat from the stream below the pinch point (cooling water at 37°C) to one above the pinch point (DH water at 75°C) would be an efficient solution. DH water is to be heated from 75°C to 130°C. Some possible temperature levels for the implementation of heat pumps need to be considered. A solution that is proposed is to heat DH water from 75°C to at least 120°C using a heat pump and then to do the additional heating with medium-pressure (MP) steam or waste heat recovered from flue gas. This way, the requirement for MP steam would decrease, and the heat pump will work with reasonable efficiency limits. The aim of the process is twofold: cooling process water and heating water for DH. The process water can be cooled to 30°C. The operating conditions of the heat pump are schematically represented in Figure 10.5.

The method introduced in Chapter 6 can be used to determine the performance of vapor compression heat pumps operating with diverse working fluids. The input variables used are as follows:

- Source (process water) temperature at inlet and outlet ($T_{source,in}$ = 37°C, $T_{source,out}$ = 30°C)

- Source (process water) flow rate ($\dot{m}_{source} = \dot{V}_{source} \times \rho_{source}$ = 137.9 kg/s)

- Sink (DH water) temperature at inlet and outlet ($T_{sink,in}$ = 75°C, $T_{sink,out}$ = 120°C)

- Isentropic efficiency of compressor (η_i = 0.70)

- Pinch temperature in condenser (ΔT_{min} = 3 K)

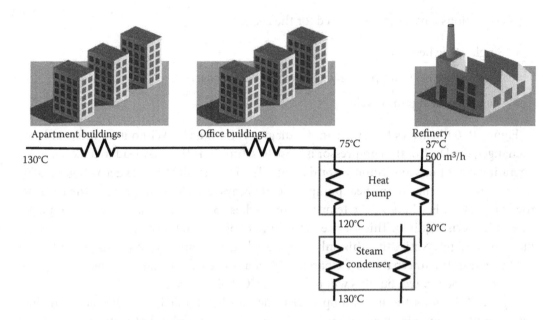

FIGURE 10.5　Waste heat stream of refinery to deliver heat to district heating system.

The method proposed in Chapter 6, making use of REFPROP (Lemmon et al., 2013), gives the following output variables:

- Pressure, temperature, enthalpy and entropy at all relevant states of the heat pump cycle

- COP

- Heating capacity $\dot{Q}_{DH}$

- Compressor power $\dot{W}_{HP}$

The calculation procedure aims to determine the best solution for each working fluid among feasible solutions. To eliminate incorrect solutions, during the course of calculations a few aspects should be verified. If the evaluated conditions do not fulfill one of the requirements, the values of the particular cycle should be discarded.

The following requirements should be met by the evaluated cycle:

- The compressor outlet temperature must be higher than the sum of the sink outlet temperature and the minimum pinch temperature.

- The quality of vapor at the outlet of the compressor must be greater than 1.

- The quality of the two-phase vapor entering the evaporator must not exceed 0.9.

- The operating conditions of the condenser should prevent a temperature cross. A $\dot{Q} - T$ diagram for the condenser is used to verify this.

Three options have been considered for the cycle:

- Single-stage heat pump cycle
- Single-stage heat pump cycle with internal heat recovery
- Two-stage heat pump cycle

Figure 10.6 illustrates the cycles in *T-s* diagrams for butane. Without an internal heat exchanger, the DCT of the compressor is close to saturation. This may lead to unacceptable operating conditions and compressor damage. The internal heat exchanger will generally improve the efficiency of the heat pump cycle. It always leads to an increase of the DCT of the compressor. For butane, the temperature reaches 165.3°C when a hot water temperature of 120°C is required. This is close to the limit of operation for the current compressor/lubricant designs. Notice that this only happens when an approach temperature of 3 K can be realized in the internal heat exchanger. With a less-efficient heat exchanger, the temperature can be limited, but this will reduce the COP of the cycle.

Figure 10.7 shows the heat pump COPs that can be attained with the most suitable refrigerants. Ammonia shows a lower performance when a high hot water temperature is required. It also shows a decrease in performance when an internal heat exchanger is

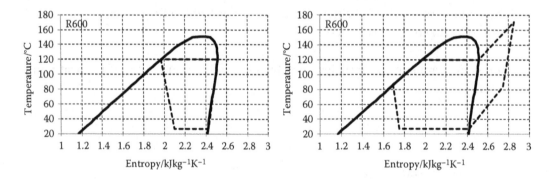

FIGURE 10.6 Single-stage heat pump cycle of butane (R600). Left, without internal heat exchanger; right, with internal heat exchanger.

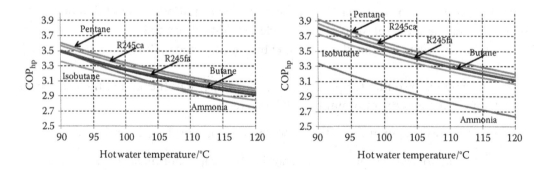

FIGURE 10.7 The COP of a single-stage heat pump cycle of diverse refrigerants: without internal (left) and with internal heat exchanger (right).

TABLE 10.3   Comparison of the Working Fluids in Terms of Performance Indicators

Working Fluids	GWP (-)	$p_{suction}$ (bar)	$p_{ratio}{}^{a}$ (-)	$\dot{V}_{compressor}{}^{a}$ no hex/ with hex (m³/h)	$T_{discharge}{}^{a}$ no hex (°C)	$T_{discharge}{}^{a}$ with hex (°C)
Ammonia	0	10.7	9.0	322/308	270	349
Butane	3	2.6	9.0	1595/1418	124	165
Isobutane	4	3.7	8.0	1248/1079	126	163
R245fa	1030	1.6	12.8	2183/1949	124	167
Pentane	4	0.74	13.1	4530/4054	123	158

ᵃ These values apply when the hot water outlet is 120°C; the volume flow applies per megawatt heating capacity.

added to the ammonia system. What is not visible from Figure 10.7 is that the compressor DCT becomes extremely high when ammonia would be selected. The performance of pentane, R245ca, R245fa and butane is similar. Because pentane and butane are fluids that most probably are already available in the plant, these fluids are much cheaper than the hydrofluorocarbons (HFCs) and the HFCs have high global warming potential (GWP) factors, pentane and butane should be preferred as working fluids.

Table 10.3 provides more details about the performance indicators of the different working fluids. The table makes clear that the operating conditions of butane are more favorable than those of pentane (operation always above atmospheric pressure and significantly lower pressure ratio). From the table, it is also clear that from the selected working fluids, only ammonia shows extremely high DCTs.

The volume flow at the compressor inlet gives an indication of the compressor size required. The smaller the volume flow is, the smaller the compressor and its initial costs will be.

As indicated in Table 10.2, the water needs to be heated to 130°C so that it can be delivered to the DH network. It is always possible that part of the heating is done with the heat pump and that a fuel is further used to reach the required temperature. In this case, it has been assumed that the thermal efficiency of the electrical grid is 0.42 and of a boiler is 0.86. It has further been assumed that electricity costs €65/MWh and gas costs €32/MWh. Heating the water flow from 75°C to 130°C with a boiler would cost €10.1 million/year. Depending on the share of the

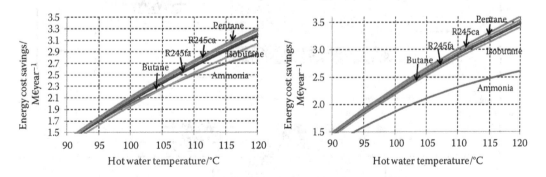

FIGURE 10.8   Yearly energy cost savings when the heat pump is applied to heat the waste heat stream from 75°C to the indicated temperature. The remaining temperature increase up to 130°C is obtained making use of a fuel-fired boiler. Left, without internal heat exchanger; right, with internal heat exchanger.

heat pump and on its COP, the heat pump will give the yearly savings shown in Figure 10.8. Obviously, using the heat pump as much as possible is always advantageous even if the COP decreases to values around 3.0 (see also Figure 10.7). The working fluid with the highest COP gives the largest yearly energy cost savings.

The large pressure ratios in Table 10.3 suggest that two-stage compression will possibly lead to more suitable operating conditions. For this reason, the performance of two-stage systems with open flash tanks has also been investigated. This design considers two internal heat exchangers between the liquid flow and the vapor flow before entering each of the compressors. The schematic of the process is illustrated in Figure 10.9.

An energy balance for the separator, considering it externally adiabatic, gives the ratio of the mass flows through both compressors so that the COP of the cycle can be calculated:

$$COP_{2-stage} = \frac{h_5 - h_6}{(h_2 - h_1) \times \dfrac{h_3 - h_7}{h_2 - h_9} + (h_5 - h_4)} \qquad (10.6)$$

The indexes of the enthalpy correspond to the states indicated in Figure 10.9. The ratio $[(h_3 - h_7)/(h_2 - h_9)]$ is based on the energy balance around the separator considering it externally

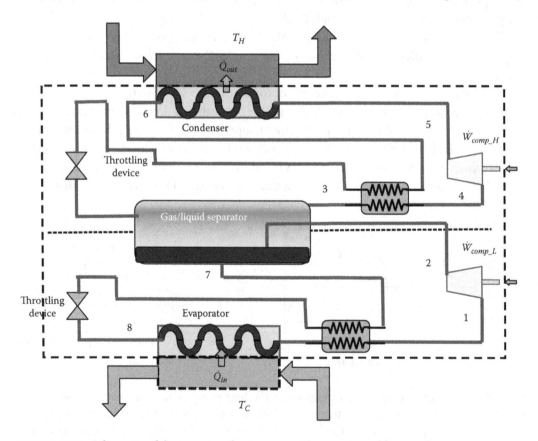

FIGURE 10.9   Schematic of the two-stage heat pump with two internal heat exchangers.

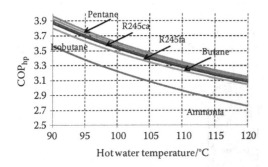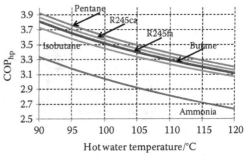

FIGURE 10.10 The COP of two-stage heat pump cycles of diverse refrigerants (left) and of a single-stage cycle with internal heat exchanger (right).

adiabatic and gives the ratio between the mass flow through the low-pressure compressor and the mass flow through the high-pressure compressor. Figure 10.10 shows on the left side the predicted COP for the two-stage heat pump system when diverse refrigerants are applied. To facilitate a comparison with the single-stage performance (when an internal heat exchanger is applied), the results for the single stage are shown on the right-hand side.

From the figure, it is evident that the performance of the ammonia system significantly improves, while the improvement of the other fluids is small (lower temperatures) to negligible (higher temperatures). The calculations have assumed the isentropic efficiency of the compressor to be maintained with the pressure ratio. In reality, at least for screw compressors, the efficiency will decrease with the pressure ratio so that the two-stage solution offers better results.

10.4 COMPRESSION-RESORPTION HEAT PUMP

A distillation column operating under the conditions illustrated in Figure 10.11 has to be integrated with a compression-resorption heat pump (CRHP). The integration of a CRHP with distillation columns, including implementation of parts of the heat pump in the column trays, has been discussed by Taboada and Infante Ferreira (2008a,b). The case discussed here has been investigated by Kuang (2008).

The temperature glides of the process flow in the boiler and in the condenser are extremely large (100°C to 183°C and 155°C to 98°C, respectively), making the application of a CRHP attractive. The CRHP is installed between the condenser and boiler of the distillation column. Depending on the duty values of the specific distillation column case, it is possible that the condenser and reboiler of the column can be completely replaced by the desorption and resorption units of the CRHP. However, energy balance calculations for the considered column indicate that an additional small boiler needs to be added to the system. Electricity-driven and motor-driven (combined heat and power [CHP] cycle) compressor options can be considered (see Figure 10.12). Both electricity and CHP utilization have their own advantages and disadvantages and need to be numerically analyzed to see clearly which one is better for the studied case.

The design includes obtaining the optimum NH_3/H_2O composition for the system specifications and obtaining optimum ΔT values for the resorber and desorber. Optimum

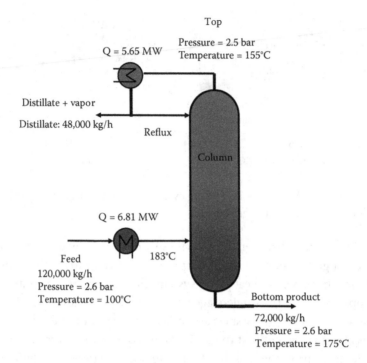

FIGURE 10.11 Operating conditions of a distillation column proposed for integration with a compression-resorption heat pump. A reflux rate of 8270 kg/h is required to attain the required purity at top and bottom.

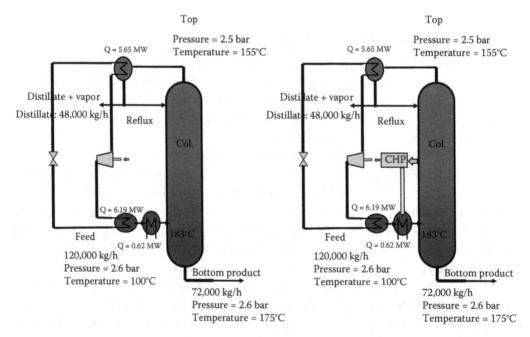

FIGURE 10.12 Possible methods of integrating the CRHP with the distillation column. Left, with additional boiler; right, using a CHP system to produce both electricity to drive the compressor and heat to do the additional heating required by the process.

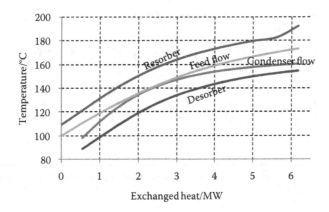

FIGURE 10.13 $\dot{Q} - T$ diagram of resorber and desorber of the considered CRHP.

composition and temperature characteristics are based on the results of a CRHP parameter optimization task. The CRHP conditions giving the lowest payback period determine the optimum. The optimization study has indicated that an ammonia-water mass concentration of 40% and an average temperature difference of 9 K in both desorber and resorber lead to the smallest payback time for this application. Figure 10.13 shows a $\dot{Q} - T$ diagram of the resorber and desorber of the proposed CRHP and illustrates how the process fluids and the ammonia-water temperature change along the length of the heat exchangers. It makes clear that the selected ammonia-water concentration leads to a temperature profile that closely approaches the process fluids profiles. This match of temperature profiles makes it possible that the COP of the heat pump is high while the temperature lift is quite high.

Figure 10.14 shows the CRHP cycle in a temperature-enthalpy diagram of an ammonia-water mixture with 40% ammonia, making clear how the temperature of the mixture

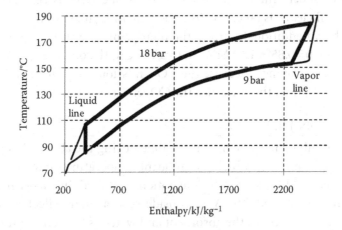

FIGURE 10.14 Temperature-enthalpy diagram of ammonia-water with 40% ammonia showing the operating conditions of the considered CRHP cycle. The diagram makes clear how the isobars behave in the liquid-vapor region of the mixture.

TABLE 10.4   Operating Conditions and Performance of the Optimized CHRP Integrated with the Distillation Column of Figure 10.11

Component	T_{in}^a (°C)	T_{out}^a (°C)	p_{in}^a (bar)	p_{out}^a (bar)	$\dot{Q}^a$ (MW)	$\dot{m}$ (kg/s)	COP (−)
Desorber	89.0	154.6	9.5	9.0	5.65	2.98	
Compressor	154.6	184.4	9.0	18.0	0.54	2.98	
Resorber	184.4	109.0	18.0	17.5	6.19	2.98	
Throttling device	109.0	89.0	17.5	9.5	−	2.98	
Additional boiler	−	−	−	−	0.62	−	
COP$_{heat\ pump}$	−	−	−	−	−	−	11.45

[a] The value indicated for the compressor is the shaft power required by the compressor under the assumption of an isentropic efficiency of 70%.

changes as it advances through the cycle. Resorption takes place from saturated vapor to saturated liquid at approximately 18 bar, while the desorption process starts with low vapor quality and ends at the inlet conditions of the compressor at a pressure of approximately 9 bar. The compression process ends at saturated vapor conditions. The operating conditions and system performance are listed in Table 10.4.

In the boiler-operated column case, the boiler in the feed line requires 6.81 MW to operate the column. Taking a burner efficiency of 85% into account, 8.01 MW primary energy is needed to operate the conventional distillation column. The proposed electricity solution (left-hand side of Figure 10.12), from an economic point of view the best solution, has a primary energy consumption of (6.19/11.45)/0.42 + 0.62/0.85 = 2.02 MW, so that the application of the CRHP leads to 75% primary energy savings. Note that the value 0.42 stands for the average efficiency of all the Dutch power plants. Notice that application of smaller approach temperatures increases the primary energy savings but significantly increases the investment costs, leading to longer payback times.

The proposed CHP-driven option (right-hand side of Figure 10.12) has a primary energy consumption of (6.19/11.45)/0.38 = 1.42 MW, while it also delivers (1.42 × 0.40) = 0.57 MW for the additional boiler. The electrical efficiency of the CHP is considered to be 38%, while its thermal efficiency is taken as 40%. The remaining (0.62 − 0.57) = 0.05 MW needs to be delivered by an additional source. The total primary energy requirement becomes (6.19/11.45)/0.38 + 0.05/0.85 = 1.48 MW, so that the CRHP combined with a CHP leads to 81.5% primary energy savings. Because the CO_2 emissions are linearly related to the primary energy consumption, these are also the CO_2 emission savings.

10.5 ABSORPTION HEAT PUMP

Absorption heat pumps show a temperature glide during the absorption process, allowing for heating applications in which a significant temperature glide of the process fluid needs to be attained. In this section, the integration of an AHP into a benzene-toluene distillation column is illustrated. The AHP considered is a single-effect system that works with lithium bromide–water as the absorbent and water as the refrigerant. This case has been studied by Vasilescu (2008) and discussed in more detail in the work of Vasilescu et al. (2009).

In the benzene-toluene distillation process, the bottom product is considered quite pure so that a limited glide applies in the reboiler. For this reason, the absorber of the heat pump is integrated in the bottom trays of the column. The evaporator of the AHP will be combined with the condenser of the column, while the condenser will be integrated with the reboiler of the column and the absorber will be integrated in the stripping section, obtaining a partially diabatic column. Instead of just one heat source (reboiler) and one heat sink (condenser), a diabatic column includes heat exchangers at some of the trays of the column. In this way, the cycle of the heat pump can be designed to show a temperature glide that corresponds to the temperature glide of the stripping section of the column. Figure 10.15 shows the temperature profile of the column.

Figure 10.16 illustrates the proposed integration between distillation column and components of the absorption cycle. The evaporator of the absorption cycle is the (partial) condenser of the column; the condenser of the absorption cycle is the reboiler of the column; the absorber of the absorption cycle is integrated in the bottom trays of the column; the desorber of the absorption cycle is heated, making use of a burner or using a hot flue gas stream.

A continuous binary distillation column with the following requirements and characteristics is considered:

- A benzene-toluene mixture is separated in the column with the aim of obtaining distilled benzene at the top with purity of 97%wt, at 80.9°C, and toluene at the bottom with purity of 98%wt, at 109.8°C.

- The feed flow is 10,000 kg/h and contains 40%wt of benzene and 60%wt of toluene.

- Feed, distillate and bottom product are liquids at their boiling point (saturated liquid).

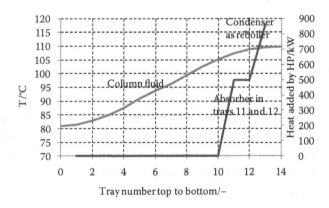

FIGURE 10.15 Temperature profile along the different trays of the benzene-toluene column considered. The figure also illustrates where heat is added: at the two bottom trays of the column, which are served by the absorber of the heat pump, and at the reboiler, which is the condenser of the heat pump.

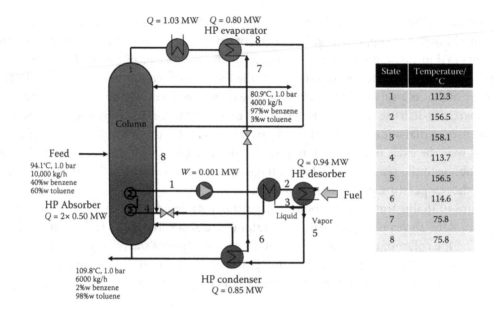

FIGURE 10.16 Absorption heat pump integrated with distillation column in which the absorber is installed in the bottom trays of the column. The table on the right-hand side shows the operating conditions of the heat pump cycle for the solution with largest energy savings.

- The column operates at atmospheric pressure.

- An adiabatic column will have 13 trays with feed at the sixth tray.

- The calculation procedure is as follows:

    1. An adiabatic column is considered as the starting point to identify the operating conditions of the diabatic column model.

    2. The temperature profiles and the heat requirements of the adiabatic column are obtained.

    3. The heat pump cycle has to be determined using the temperatures from the adiabatic column and the heat requirements.

    4. The next step is to integrate the AHP into the column. To make this possible, it is necessary to consider that the absorber is covering some of the trays of the column. Therefore, using the previously calculated values of the heat pump cycle, it is possible to know the values of the heat that will be exchanged within each tray of the column.

    5. A diabatic column is obtained. It is necessary to solve the column again because the properties throughout the length of the column have changed. This issue causes the column to both have a new temperature profile and need a different input of heat to achieve the same quality of the products of the original adiabatic column.

6. Therefore, the heat pump cycle has to be determined again.

7. The values obtained from this cycle lead to a 'new' diabatic column.

8. The final step is repeating steps 5 to 7 as many times as needed until the distillation requirements of the original adiabatic column are obtained.

The main problem of calculating the heat pump integrated with the distillation column is to fit the temperature profiles of the internal and external media. This issue leads to an assumption to simplify the calculations. It will be considered that the heat exchanged between the AHP heat exchangers and each tray of the column is

$$\dot{Q}_n = \frac{\dot{Q}_{abs}}{Number\_of\_trays\_covered\_by\_absorber} \tag{10.7}$$

The tray energy balance for the trays in which part of the absorber is installed will include the additional term $\dot{Q}_n$ given by Equation (10.7). Several combinations for which the heat exchangers of the AHP are covering different numbers of trays of the distillation column have been studied. Table 10.5 gives the calculated results for the conditions studied. The number of trays covered by the absorber is listed in the second column. The temperature difference between the column fluid and the working fluid of the cycle has been varied from 5 to 15 K (column 1). The third column indicates the heat requirement of the distillation column (condenser plus absorber of the heat pump), showing that making the column diabatic implies an increase in heat requirement. As can be seen, the heat

TABLE 10.5 Performance of the Absorption Heat Pump Integrated with the Toluene-Benzene Distillation Column

Temperature Approach of Heat Exchangers (K)	Number of Trays with Part of Absorber (−)	$\dot{Q}_{reboiler}$ (MW)	$\dot{Q}_{cond}$ (MW)	$\dot{Q}_{des}$ (MW)	Reflux Rate (−)	COP (−)
5	1	1.84	1.81	1.01	3.11	1.81
	2	1.85	1.83	0.94	3.14	1.95
	3	1.87	1.85	1.01	3.19	1.84
	5	1.93	1.90	1.07	3.32	1.80
	7	2.01	1.99	1.11	3.52	1.81
10	1	1.84	1.81	1.03	3.11	1.78
	2	1.85	1.83	0.96	3.14	1.92
	3	1.87	1.85	1.04	3.19	1.80
	5	1.93	1.91	1.09	3.32	1.76
	7	2.02	2.00	1.15	3.53	1.76
15	1	1.84	1.81	1.04	3.11	1.76
	2	1.85	1.83	0.97	3.15	1.90
	3	1.87	1.85	1.04	3.19	1.80
	5	1.93	1.91	1.11	3.33	1.73
	7	2.02	2.00	1.18	3.54	1.72

requirement in this column increases with the number of trays of the column that include parts of the absorber. The reflux rate required to maintain the same purity of the top and bottom products increases with the number of trays in which the absorber adds heat. The fourth column indicates the amount of heat that needs to be removed by the heat pump in the partial condenser at the top of the column. The rest of the column condensation heat needs to be removed in a separated condenser. The heating requirement of the heat pump is given in the sixth column.

In the adiabatic column case, the reboiler of the column requires 1.84 MW to operate the column. Taking a burner efficiency of 85% into account, 2.16 MW primary energy is needed to operate the conventional distillation column. The required reflux ratio is 3.11. The best solution, from an energy usage point of view (smallest $\dot{Q}_{des}$ in the fifth column), is obtained for the smallest temperature approach and the absorber installed in the two last (bottom) trays. The primary energy consumption is then 0.94/0.85 = 1.11 MW, so that the application of the AHP leads to 48.5% primary energy savings. Notice that application of larger approach temperatures reduces the investment costs and only slightly reduces the attained savings (for $\Delta T = 15$ K, the energy savings are still 47%).

Figure 10.17 illustrates the operating conditions of the AHP in a pressure-temperature diagram for the water–lithium bromide mixture. As indicated, the cycle operates far from the crystallization limits.

10.6 EJECTOR HEAT PUMP

The application of an EHP in combination with a propylene-propane distillation column (C₃ splitter) is investigated. Because the top product of the column is propylene, this fluid

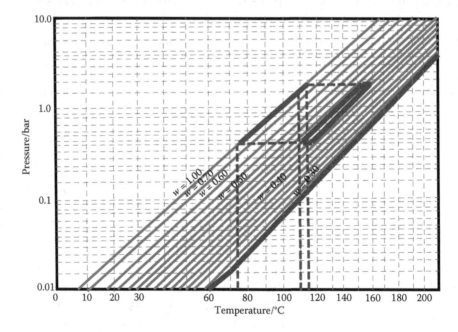

FIGURE 10.17 Pressure-temperature diagram for the mixture water–lithium bromide showing the operating conditions of the case studied. The bold line indicates the crystallization line.

is applied as motive fluid. Propylene vapor is produced in a vaporizer at an elevated pressure. The vaporizer is fuel driven. The energy consumption of the cycle is the heat needed to produce the motive flow plus the pump used to bring the condensate to the vaporizer pressure. Figure 10.18 shows the column before the ejector has been added (left) and with the EHP integrated with the column (right).

The pressure in the vaporizer is taken as 70 bar (supercritical conditions), the vapor temperature as 130°C, with vapor enthalpy of 657.28 kJ/kg at the inlet of the motive nozzle. The procedure introduced in Chapter 7 is used here. The conditions around the ejector are summarized in Figure 10.19.

The figure makes clear that only part of the vapor leaving the top of the column can be upgraded to higher pressure with the ejector. The rest of the vapor needs to be condensed in an extra condenser. The suction flow taken by the ejector is limited by the requirement that the outlet flow of the ejector is fully condensed. Part of this condensate is pumped to the supercritical gas heater to maintain the conditions at the inlet of the motive nozzle. The relevant geometrical data have been selected to match the required operating conditions. Table 10.6 summarizes these data.

Figure 10.20 shows the pressure profiles along the ejector (bottom) and the ejector geometry (top). The pressure at the outlet of the motive nozzle is decreased by accelerating and expanding the high-pressure propylene (70 bar) to supersonic conditions from inlet to outlet of the motive nozzle. Here, the propylene vapor from the top of the column, at 17.9 bar, is drawn and entrained through the suction nozzle. The two flows become mixed in the mixing section, making use of the momentum of the motive fluid flow. In the mixing section, which has a constant area, a shock wave reduces the flow from supersonic

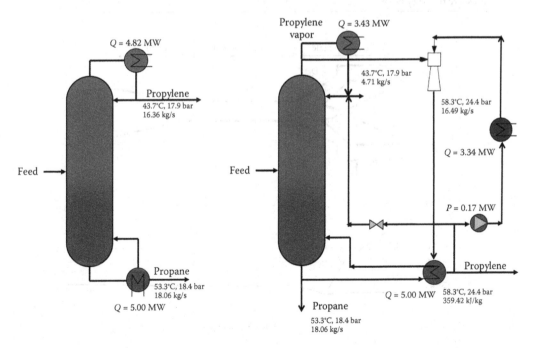

FIGURE 10.18 Integration of ejector heat pump with a propylene-propane distillation column.

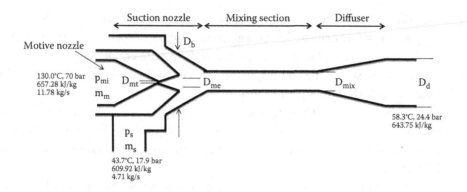

FIGURE 10.19 Operating conditions of the ejector applied to a propylene-propane distillation column.

TABLE 10.6 Geometrical Data and Assumed Efficiencies of the Ejector

Input	Symbol (–)	Area (m²)	η (–)	Ratios (–)
Area throat motive nozzle	A_{mt}	0.0006		
Area outlet motive nozzle	A_{me}	0.00096		
Area outlet suction nozzle	A_b	0.00215		
Area mixing section	A_{mix}	0.004		
Area diffuser outlet	A_d	0.04		
Efficiency motive nozzle	η_m		0.90	
Efficiency suction nozzle	η_s		0.80	
Efficiency mixing section	η_{mix}		0.90	
Entrainment ratio	φ			0.40
Pressure recovery ratio	p_R			1.36

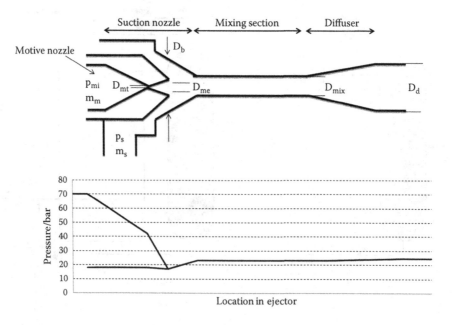

FIGURE 10.20 Pressure profiles along the ejector for the operating conditions under consideration.

velocity to subsonic velocity, and an elevated pressure results (23.0 bar). In the diffusor, the velocity is further reduced and converted into static pressure with the final pressure of 24.4 bar. The motive pressure has been selected relatively high so that a sufficiently high pressure recovery ratio could be attained. This ratio is imposed by the operating conditions of the reboiler.

In the conventional column case, the reboiler of the column requires 5.00 MW to operate the column. Taking a burner efficiency of 85% into account, 5.88 MW primary energy is needed to operate the conventional distillation column. From an energy consumption point of view, the proposed EHP requires 0.17 MW electrical energy to drive the pump and 3.34 MW primary fuel to bring the condensate to motive flow conditions. The primary energy consumption is then 0.17/0.42 + 3.34/0.85 = 4.33 MW, so that the application of the EHP leads to 26.3% primary energy savings. Again, the value of 0.42 stands for the average efficiency of all the Dutch power plants. The COP of this cycle is given by Equation (10.8):

$$COP = \frac{\dot{Q}_{reboiler}}{\dot{Q}_{motive\_flow\_heater} + \dot{W}_{pump}} = \frac{5.00}{3.34 + 0.17} = 1.42 \tag{10.8}$$

The yearly economic savings of the EHP resulting from the energy saving is given in Equation (10.9). For electricity generation, a price of €65/MWh is assumed, whereas for heat a price of €31/MWh is assumed (van de Bor et al., 2015).

$$Cost\_Saving = 5.88 \times 8300 \times 31 - \frac{3.34}{0.85} \times 8300 \times 31 - \frac{0.17}{0.42} \times 8300 \times 65 = 0.283 \text{ M€} \tag{10.9}$$

Because, in comparison to a conventional design, the EHP requires an additional investment in ejector and motive flow heater, this application does not seem viable for this type of heat pump.

10.7 CONCLUDING REMARKS

The case studies presented here illustrated the importance of the main heat pumps and significant impact in achieving energy efficiency in various processes from the chemical process industry. For example, in the case of the bioethanol dehydration by ED, combining VRC with DWC technology led to energy savings of more than 40% with TAC savings of 24% in spite of the additional cost of the compressor.

Table 10.7 gives an overview of the technologies considered in the presented cases. All cases led to large primary energy savings and to large reduction of the associated CO_2 emissions. The COP of the AHP and EHP is significantly lower than for the other technologies, but it should be remembered that these are heat-driven heat pumps, while the others are electricity-driven heat pumps. The COP of the considered VC heat pump is low, but the adopted temperature lift is extremely high. These cases illustrate that heat pumps can operate at 'high' temperatures and high temperature lifts and still deliver significant energy and CO_2 emission advantages.

TABLE 10.7 Overview of Case Studies

Input	Technology (-)	Supply Temperature (°C)	Primary Energy Saving (%)	Temperature Lift (°C)	COP (-)
Extractive distillation in DWC	VRC	105	47	26.7	6.7
Heat recovery from refinery	VC	120	36	90.0	3.2
CRHP in distillation column	CRHP	165	75	65.0	11.5
AHP in distillation column	AHP	110	48	28.9	2.0
Ejector in distillation column	EHP	58	26	14.6	1.4

LIST OF SYMBOLS

A	Area	m^2
C	Cost	US$
d_c	Diameter of column	m
F_c	Correction factor	–
f_p	Cost factor	–
h	Specific enthalpy	$kJkg^{-1}$
h_c	Height of column	m
h_{proc}	Enthalpy of steam	$kJkg^{-1}$
$\dot{m}$	Mass flow	kgs^{-1}
p	Pressure	bar
P	Power	W
p_R	Pressure recovery ratio	–
Q	Heat	J
$\dot{Q}$	Heat flow	W
r_L	Liquid split ratio	–
r_V	Vapor split ratio	–
T	Temperature	K
U	Overall heat transfer coefficient	$Wm^{-2}K^{-1}$
$\dot{V}$	Volume flow	m^3s^{-1}
W	Work	$Jmol^{-1}$
$\dot{W}$	Power	W

Greek Symbols

Δ	Difference	–
η	Efficiency	–
λ_{proc}	Latent heat of steam	$kJkg^{-1}$
φ	Entrainment ratio	–
ρ	Density	kgm^{-3}

Superscripts

0	Reference/standard condition

Subscripts

0	Reference state
abs	Absorber
b	Suction section
c	Cold/condenser
comp	Compressor
cond	Condenser
d	Diffuser
des	Desorber
DH	District heating
e	Exit
FTB	Flame temperature of the boiler flue gases
h	Hot
hex	Heat exchanger
HP	Heat pump
i	Isentropic
in	Inlet
m	Motive section
min	Minimum approach
mix	Mixing section
n	*n*th tray
out	Outlet
r	Reboiler
s	Suction nozzle
t	At throat

Abbreviations

AD	Azeotropic distillation
AHP	Absorption heat pump
CHP	Combined heat and power
COP	Coefficient of performance
CRHP	Compression-resorption heat pump
DCT	Discharge temperature
DH	District heating
DWC	Dividing-wall column
ED	Extractive distillation
EDC	Extractive distillation column
EG	Ethylene glycol
EHP	Ejector heat pump
FEHE	Feed-effluent heat exchanger
GWP	Global warming potential
HEX	Heat exchanger

HFC	Hydrofluorocarbon
HP	Heat pump
ktpy	Kiloton per year
LMTD	Log-mean temperature difference
MP	Medium pressure
MSA	Mass separating agent
MVR	Mechanical vapor recompression
NHV	Net heating value
NRTL	Nonrandom two liquid
PDC	Preconcentration distillation column
SRC	Solvent recovery column
TAC	Total annual cost
TIC	Total investment cost
TOC	Total operating cost
VC	Vapor compression
VRC	Vapor recompression

REFERENCES

Balat M., Balat H., Oz C., Progress in bioethanol processing, *Progress in Energy and Combustion Science*, 34 (2008), 551–573.

Boot H., Nies J., Verschoor M. J. E., de Wit, J. B., *Handboek industriële warmtepompen* (in Dutch), Kluwer Bedrijfsinformatie, Deventer, the Netherlands, 1998.

Chew J. M., Reddy C. C. S., Rangaiah G. P., Improving energy efficiency of dividing-wall columns using heat pumps, organic Rankine cycle and Kalina cycle, *Chemical Engineering and Processing*, 76 (2014), 45–59.

Frolkova A. K., Raeva V. M., Bioethanol dehydration: State of the art, *Theoretical Foundations of Chemical Engineering*, 44 (2010), 545–556.

Hernandez S., Analysis of energy-efficient complex distillation options to purify bioethanol, *Chemical Engineering and Technology*, 31 (2008), 597–603.

International Energy Agency (IEA), *Application of industrial heat pumps. Final report of Annexes 13 and 35 of the International Energy Agency on industrial energy-related systems and technologies and on the heat pump programme. Task 4: Case studies*, International Energy Agency, Paris, 2014.

Kiss A. A., *Advanced distillation technologies – Design, control and applications*, Wiley, London, 2013a.

Kiss A. A., Novel applications of dividing-wall column technology to biofuel production processes, *Journal of Chemical Technology and Biotechnology*, 88 (2013b), 1387–1404.

Kiss A. A., Bildea C. S., A control perspective on process intensification in dividing-wall columns, *Chemical Engineering and Processing: Process Intensification*, 50 (2011), 281–292.

Kiss A. A., Flores Landaeta S. J., Infante Ferreira C. A., Towards energy efficient distillation technologies – Making the right choice, *Energy*, 47 (2012), 531–542.

Kiss A. A., Ignat R. M., Innovative single step bioethanol dehydration in an extractive dividing-wall column, *Separation & Purification Technology*, 98 (2012), 290–297.

Kiss A. A., Ignat R. M., Optimal economic design of a bioethanol dehydration process by extractive distillation, *Energy Technology*, 1 (2013), 166–170.

Kiss A. A., Suszwalak D. J.-P. C., Enhanced bioethanol dehydration by extractive and azeotropic distillation in dividing-wall columns, *Separation & Purification Technology*, 86 (2012), 70–78.

Kuang Z., Compression resorption heat pump assisted distillation. Desorber investigation, MSc thesis, Delft University of Technology, the Netherlands, 2008.

Lemmon E., Huber M., McLinden M., *NIST standard reference database 23: Reference fluid thermodynamic and transport properties – REFPROP*, Version 9.1, National Institute of Standards and Technology, Gaithersburg, MD, 2013.

Luo H., Bildea C. S., Kiss A. A., Novel heat-pump-assisted extractive distillation for bioethanol purification, *Industrial & Engineering Chemistry Research*, 54 (2015), 2208–2213.

Plesu V., Bonet Ruiz A. E., Bonet J., Llorens J., Simple equation for suitability of heat pump use in distillation, *Computer Aided Chemical Engineering*, 33 (2014), 1327–1332.

Ravi R., Refinery case study – Heat pump for district heating, internal report, Delft University of Technology, the Netherlands, 2012.

Taboada R., Infante Ferreira C. A., Compression resorption cycles in distillation columns, presented at the International Refrigeration and Air Conditioning Conference at Purdue, West Lafayette, IN, July 14–17, 2008a, Paper R316.

Taboada R., Infante Ferreira C. A., Heat recovery in distillation columns using compression-resorption cycles, presented at the Eighth IIR Gustav Lorentzen Conference on Natural Working Fluids, Copenhagen, Denmark, September 7–10, 2008b.

Tututi-Avila S., Jimenez-Gutierrez A., Hahn J., Control analysis of an extractive dividing-wall column used for ethanol dehydration, *Chemical Engineering and Processing*, 82 (2014), 88–100.

van de Bor D. M., Infante Ferreira C.A., Kiss A. A., Low grade waste heat recovery using heat pumps and power cycles, *Energy*, 89 (2015), 864–873.

Vane L. M., Separation technologies for the recovery and dehydration of alcohols from fermentation broths, *Biofuels, Bioproducts and Biorefining*, 2 (2008), 553–588.

Vasilescu C., Integration of LiBr-H_2O absorption heat pumps in distillation columns, MSc thesis, Delft University of Technology, the Netherlands, 2008.

Vasilescu C., Infante Ferreira C. A., Tarlea G. M., Integration of LiBr-H_2O absorption heat pumps in distillation columns, in *Proceedings of the 3rd IIR Conference on Thermophysical Properties and Transfer Processes of Refrigeration*, Boulder, CO, June 23–26, 2009, Paper No. 110.

Wu Y. C., Hsu P.H.-C., Chien I.-L., Critical assessment of the energy-saving potential of an extractive dividing-wall column, *Industrial and Engineering Chemistry Research*, 52 (2013), 5384–5399.

Index

Page numbers followed by f and t indicate figures and tables, respectively.

Printed in the United States
by Baker & Taylor Publisher Services